Reflections on Observational Astronomy
in the Medieval Islamic Period

This volume presents comprehensive investigations into various facets of observational astronomy during the medieval Islamic period, spanning from the ninth to the seventeenth centuries. The chapters compiled here, originally published between 2012 and 2018, have undergone significant revisions to enhance their accuracy and explore a broad spectrum of topics organized into five main sections.

Reflections on Observational Astronomy in the Medieval Islamic Period begins with solar astronomy, providing a detailed evaluation of Islamic astronomers' determinations of fundamental solar parameters. In the realm of lunar astronomy, it examines the gradual endorsement and rationalization of annular solar eclipses, along with an exclusive historical account of predicting and observing such an event in 1283 CE. The section on planetary astronomy scrutinizes empirical discoveries that distinguish between the precession of equinoxes and the motion of apogees, as well as significant enhancements to Ptolemy's parameters for planetary latitudes. Stellar astronomy is explored through a non-Ptolemaic star table that encompasses observations from ninth-century Baghdad to thirteenth-century Marāgha. The final section examines observational instruments, focusing on those constructed during the second period of activities at the Marāgha observatory. A critical analysis of astronomical observations conducted at the Marāgha and Istanbul observatories is a key focus of this work.

This book will be invaluable to those interested in the historical progression of exact sciences; the scope, distinctive aspects, and caliber of experimental activities in medieval times; and the interplay between theory and observation throughout history. It is intended for historians, scientists (including astronomers and physicists), and particularly, historians of astronomy.

S. Mohammad Mozaffari is an Iranian historian of medieval astronomy currently serving at the Research Institute for Astronomy and Astrophysics of Marāgha (RIAAM), University of Maragheh, in Iran. He is also a research associate in the project of the Ptolemaeus Arabus et Latinus (Bavarian Academy of Sciences and Humanities, Munich). His primary research focus lies in the growth and development of observational astronomy, particularly its interplay with theoretical astronomy, during the medieval Islamic period. He is an active member of the International Astronomical Union and holds editorial roles as an advisory editor for the *Journal for the History of Astronomy*, an associate editor for *SCIAMVS* (*Sources and Commentaries in the Exact Sciences*), and an associate editor for the *Journal of Astronomical History and Heritage*.

Reflections on Observational Astronomy in the Medieval Islamic Period

S. Mohammad Mozaffari

VARIORUM COLLECTED STUDIES

Routledge
Taylor & Francis Group

LONDON AND NEW YORK

First published 2025
by Routledge
4 Park Square, Milton Park, Abingdon, Oxon OX14 4RN

and by Routledge
605 Third Avenue, New York, NY 10158

Routledge is an imprint of the Taylor & Francis Group, an informa business

© 2025 S. Mohammad Mozaffari

British Library Cataloguing-in-Publication Data
A catalogue record for this book is available from the British Library

ISBN: 978-1-032-77234-9 (hbk)
ISBN: 978-1-032-77236-3 (pbk)
ISBN: 978-1-003-48196-6 (ebk)

DOI: 10.4324/9781003481966

Typeset in Times New Roman
by Apex CoVantage, LLC

VARIORUM COLLECTED STUDIES SERIES CS1130

CONTENTS

ACKNOWLEDGMENTS

I wish to express my deepest appreciation to Benno van Dalen, Julio Samsó Moya, and the late Noel M. Swerdlow (1941–2021). Their unwavering support, positive encouragement, constructive criticisms, insightful suggestions, and valuable comments on earlier versions of the chapters presented collectively in this volume have been instrumental in shaping my academic career. I cannot thank them enough. Additionally, I extend my deepest gratitude to David A. King, the late Michael Hoskin (1930–2021), the late Paul Kunitzsch (1930–2020), Alexander Jones, James C. Evans, John M. Steele (co-author of Chapter 5), George Saliba, Bernard R. Goldstein, Glen R. Van Brummelen, Peri Bearman, F. Richard Stephenson, Dennis Duke, the late José Chabás Bergón (1948–2024), Richard L. Kremer, and Matthieu Husson. Their unparalleled knowledge has been pivotal in my learning journey. Special thanks go to Shi Yunli and Wayne Orchiston, whose impressive enthusiasm has fueled my work. I also gratefully acknowledge the assistance of Georg Zotti (co-author of Chapter 11).

INTRODUCTION

Early in the ninth century, astronomy in the Middle East entered a new stage of its historical development through the integration by astronomers (of various ethnic origins and religions) of borrowings from both the occidental and oriental traditions that had come down to them mainly through the medium of texts, but also orally and on the basis of material culture. Despite the paucity of the surviving empirical data (especially, in the classical period, before 1050 CE), the definite role played by experimental activities both in shaping the conceptual framework of early Islamic astronomy and in the development of astronomical ideas and theories in the late Islamic period is undeniable.

The chapters collated for this volume aim to substantiate and clarify the traceable relations between *observation* and *theory* and the entire process, starting with obtaining fresh quantitative data from observations, through the determination of the fundamental parameters in Ptolemaic astronomy (which provided the standard models for practical use in the period in question), to building new theories in Islamic astronomy. This creative and productive enterprise led to the assessment of existing theories based on observations, the discovery of new anomalies, the emergence of new concepts, the formulation of new hypotheses regarding celestial motions, and the definition and justification of previously unknown phenomena. Furthermore, the extensive observational work compelled medieval astronomers to innovate novel methods, devise unprecedented strategies, and create new instruments to surmount the diverse challenges they encountered during their experiments.

The central theme running through all these topics is *observational astronomy*. Each of these topics finds representation within the chapters. Pivotal to all contributions in the book is a comprehensive textual study of the Islamic astronomical corpus. Equally crucial to our approach is the *technical* evaluation of the preserved materials, which frequently involves a nuanced analysis considering historical, astronomical, and conceptual aspects. This assessment adapts to the unique features of each topic and every celestial object under scrutiny.

DOI: 10.4324/9781003481966-1

Here, I will provide an overview of how I assemble the pieces of the puzzle into a coherent and cohesive whole, discuss the technical approach I have employed, and highlight my accomplishments.[1]

The key principle every historian should bear in mind is that, unlike the myriad potential futures, we are left with only one immutable past. However, this historical backdrop can be interpreted from diverse perspectives, and its disparate fragments can be juxtaposed in numerous different ways. Many of these arrangements will only yield distorted or unclear images that can greatly mislead us, leading to conclusions far from the truth. Historians of science tackle their historical jigsaw puzzles using their distinctive approaches, depending on their interests and abilities. Regardless of how they choose or are able to adhere to general historiographical rules, it is important to remember that in reconstructing the mechanisms involved in the advancement of an exact science in ancient and medieval times, while many primary aspects are necessary, from codicology to chronology, no study can be limited to them. Narrating the entire story solely through the editing and translation of texts, or merely reflecting social and historical changes, with all sorts of complexities, does not absolve a historian from directly engaging with what an astronomer actually did. Therefore, a technical approach focusing on specific details is not only necessary – especially since minute details can have significant impacts on multiple levels in the advancement of science – but it is also the main focus that a historian of an exact science must maintain. A technical approach can challenge our intellectual assumptions at almost every turn.

The paucity of reliable data from the medieval Islamic period poses a significant challenge. Islamic astronomers, as evidenced by their works, are a highly heterogeneous group of individuals. The major developments and progress highlighted in this volume were made by a small number of astronomers, who were known as observational astronomers (*rāṣid* or *muhandis al-raṣadī*). Our knowledge about the life and intellectual careers of these renowned Islamic observational astronomers is limited.[2] This is not to mention the completely unknown[3] or enigmatic, shadowy figures.[4] Furthermore, we cannot always rely on an astronomer's judgment of his predecessors' careers or achievements.[5]

---

1 Hereafter, I refer to this book's chapters and my other papers listed in the bibliography in bold characters.

2 For instance, we only know that Ibn al-Aʿlam (d. 985) and Ibn Yūnus (d. 1009) were perceived as eccentric.

3 For example, a certain al-Ḥasan observing in Baghdad about 830 (see following) and an anonymous "trustworthy" colleague of Ibn Yūnus, who provided him with empirical data on the observation of the conjunction between Saturn and Mars on June 11, 995.

4 For example, Muʿīn al-Dīn al-Kāshī, a member of the Samarqand observatory, who is referred to as "the observer" in a marginal gloss on a copy of Ghiyāth al-Dīn Jamshīd al-Kāshī's (d. 1429) *Khāqānī zīj*.

5 For example, Bīrūnī (d. 1048) describes al-Nayrīzī (d. 922) as an incompetent astronomer, while through Ibn Yūnus, we know that he was one of the first astronomers to detect a critical issue with lunar eclipse timings (see following).

Unfortunately, the information available to us about their empirical findings and practical activities is also quite limited. This is not only due to the vicissitudes of time but also because they intentionally chose not to document them, for various reasons, such as avoiding verbosity, as explicitly stated by al-Battānī (d. 918) and Ibn Yūnus. Bīrūnī criticizes his predecessors for this, and for instance, he refrains from using Ibn al-Aʿlam's new values for the solar, lunar, and planetary data, under the principle that "the explanation of how they were derived is a prerequisite for their adoption" (شرط القبول له ايضاح العمل).[6]

Drawing an accurate and comprehensive picture of the practical, technical, and conceptual developments in astral sciences from the limited information available is no easy task. So how can we gather more data or indirect evidence to bolster our resources when faced with mysterious problems or situations? One approach is to increase our raw data by thoroughly analyzing numerical and quantitative information passed down through preserved textbooks, handbooks, astronomical tables, and treatises of various sizes. The available sources provide us with *clues* that require a *deep technical approach, rigorous analysis,* and *extensive experience* – rather than mere vague intuitions – to lead us to pieces of *indirect evidence* that offer *standard inductive conjectures.*

Based on this evidence, we can theorize about the historical mechanisms and influential factors involved in the genesis of observational astronomy during the Islamic period in a secure and safe manner. Otto E. Neugebauer (1899–1990) emphasized that "only the most intimate knowledge of details reveals some traces of the overwhelming richness of the processes of intellectual life."[7] Great importance lies in the technical details, enabling us to establish precise relations in the field of science. This precision is akin to a "proof" in the sense of mathematical rigor.[8] Strict adherence to this methodology ensures that we remain highly immune to the undesirable consequences of adopting arbitrary attitudes and approaches. Otherwise, we might perceive "the elephant in darkness" as a python, a pillar, a throne, or something entirely different from its true shape.

For instance, when analyzing Muḥyī al-Dīn al-Maghribī's (d. June 1283) trio of lunar eclipses (Chapter 9), I discovered systematic time errors of about −5 minutes. In Kāshī's triple lunar eclipses, it was puzzling that a deviation of −8 minutes in the time of his second eclipse – of the same order as the error made by Muḥyī al-Dīn – was inconsistent with the discrepancies in his first and third eclipses, which were about one hour (**2020–2021**). Some calculations showed that, in the latter two eclipses, the eclipsed Moon set before the eclipse ended, so he could not determine the time of the maximum phase as the midpoint between the first and last contacts. Creating a profile of the Kashan horizon and comparing the Moon's nocturnal path revealed that a geographical barrier obstructed

---

6 al-Bīrūnī 1948, 3: p. 30.
7 Neugebauer [1941a] 1983, p. 5.
8 Neugebauer [1941b] 1983, p. 23.

the Moon's view. So how did he measure the times? The derivation of the times from the Marāgha *zījes* indicated that his times are likely theoretical output, *not* fresh observational data. This is just one example that underscores the importance of a technical approach, granting us an unparalleled understanding of what lies beneath the surface.

By strictly adhering to this methodology, I began my journey into understanding the practical aspects of astronomical knowledge in medieval Islam. As noted by the late N. M. Swerdlow, this is "hard history of astronomy" (private communication in 2017), but it carries the advantage of "the deeper, the more beautiful" (after reading an early draft of Chapter 9 in 2014). Over the past decade, my work has gradually evolved into a lifelong project titled *Observational Astronomy in the Medieval Islamic Period* (*OAMIP*). This volume presents some parts of its initial phase (prior to 2018).

A General Outline of the History of Islamic Astronomy

The medieval Islamic period witnessed a multitude of transformations in astronomy, both *developmental* and *generative*. This means that changes were not only in terms of data accumulation but also in the introduction of *new approaches, perspectives, methods*, and even *a new system of thought*. This period also saw the recurrence of some fundamental topics from the ancient era:

- Physical astronomy *vs.* geometrical astronomy, and its close connection to the relation between mechanics and astronomy
- Dynamicity *vs.* stacity of the fundamental parameters (an issue our modern cosmology is also confronted with)
- Hipparchus's *similarity vs.* Ptolemy's *simplicity* in hypotheses for the celestial motions

In terms of methodology, for instance, a conceptual shift occurred in the simple term *sýnkrisis* (comparison) when Ḥajjāj b. Yūsuf b. Maṭar translated it into Arabic as *i'tibār* (experiment) in *Almagest* VII (827–828). This term encapsulates four aspects of an empirical activity – observation, comparison, consideration, and measurement – in a single word. Identified as a primary criterion and forming the core of meaningful astronomical practice, it became a key term frequently used (along with *istidrāk* = "correction") in influential works in the Islamic astral corpus. Early in the twelfth century, al-Khāzinī, in his *Experimental astronomy*, defines *ṭarīq i'tibār* (experimental method) in comparison with its building block, that is, *observation* (*raṣad*):

> The difference between it and [astronomical] observation is that in observation, all of the computation[al procedure]s are invented anew, and nothing is taken for granted (or as a fact, *musallam*), [whereas] in the

experiment, observed facts (*musallamāt marṣūda*) are taken, upon which the desired outcomes (*maṭlūbāt*) are based.[9]

In the *traditional narrative* of the development of astronomy in the classical period (prior to around 1050), we learn that the *Mumtaḥan* astronomers, based on their observations starting from 829, found that Ptolemaic astronomy held a definitive and absolute advantage over Indian traditions. Consequently, Ptolemaic astronomy established the broad, constitutional framework of Islamic astronomy. This belief seems justified and indisputable in itself, as it is entirely based on a primary source (a statement by Ḥabash, who died after 869, in the prolegomenon of his *zīj*). However, this view makes it extremely difficult to explain, for instance, how Muḥammad b. Isḥāq al-Sarakhsī (active around 876) conducted his observations in the latter part of the same century to correct Saturn's period relations in Indian traditions, and why he combined the values for the orbital elements from both Greek and Indian sources. It also raises questions about why the Khurāsānian astronomers, as reported by Bīrūnī, were still using the *Sindhind* tradition to calculate eclipse timings at the turn of the eleventh century; why notable figures like Abū Naṣr Manṣūr b. ʿAlī b. ʿIrāq (died before 1036) and Kūshyār b. Labbān (*fl.* tenth–eleventh century) were involved in compiling astronomical tables based on Hindu parameters around that time; why the Indian hypothesis of the identicality of the orbital elements of the Sun and Venus was dominant in the classical period; and so on.

Most importantly, it would be challenging to explain whether and why the skilled *Mumtaḥan* astronomers (judged by the accuracy of their solar observations, which constitute the only subset of empirical data passed down from them to us) did not discover that the Ptolemaic numerical theories, as adapted to the meridian of Baghdad, suffer from significant errors in longitude at positions where Ptolemy's models were intended to provide reasonably accurate results.[10] The problem becomes more puzzling since, in *On Solar Motions*, written by Ibrāhīm b. Sinān (d. 946; the grandson of Thābit b. Qurra, d. 901), we are explicitly told that the Mumtaḥan observations uncovered errors of "about 4°" in the solar and planetary longitudes as derived from the *Almagest*.

The preference for Ptolemaic astronomy is understandable, as the *Almagest* explains the models, their characteristic features, and how to quantify them, while Indian textbooks primarily contain instructions, often without explaining the rationale behind them and their underlying hypotheses and models. Therefore, a logical choice for the *Mumtaḥan* observers could be to re-determine the solar, lunar, and planetary parameters based on Ptolemy's models. Of course, it could

---

9 MS Vatican, Biblioteca Apostolica Vaticana, Arabo 761, f. 4r.

10 Larger than −3° for the Sun; at oppositions: −2.5° for Jupiter, and −3° for Saturn and Mars; and at the inferior planets' greatest elongations (>40° for Venus and >20° for Mercury), the errors, although they may occasionally disappear, often reach several degrees, about −6° for Venus and −9° for Mercury.

remain somewhat of a mystery what the exact source of "errors" was in Ptolemy's numerical theories: his parameter values or the problem lies with the models themselves? Regardless, treating them as the "*standard system of astronomy*," the *Mumtaḥan* group re-quantified the Ptolemaic models, according to the Ptolemaic "*standard methods*," using Ptolemy's observations as the initial data for deriving mean motions (as demonstrated in a pictorial form in **2023a**).

However, it did not take long (until the first half of the tenth century) for Islamic astronomers to *discover*, based on their observations, *nine critical empirical problems* (*decisive but non-refuting anomalies*) in the *applicative domain* of Ptolemaic astronomy.[11] That is, the empirical problems which raised reasonable doubts about the empirical adequacy of its hypotheses and theories in situations where they were thought to provide satisfactory results. These can be categorized with respect to the Sun, Moon (L), and planets.

> **S1.** Length of the seasonal solar year
> **S2.** Rate of precession
> **S3.** Decrease in the obliquity of the ecliptic
> **S4.** Progressive motion of the solar apogee
> **S5.** Decrease of the solar eccentricity
> **L1.** Times of lunar eclipses
> **L2.** The orbital inclination of the Moon
> **L3.** Annular solar eclipses
> **P1.** The variation of longitudinal errors

The first critical problems arose in solar astronomy, which forms the cornerstone of any astronomical system. This occurred when the authors of *On the solar year* (attributed to both the Banū Mūsā and Thābit b. Qurra) derived two distinctly different sets of values for **S1** and **S2**. They did this by comparing the solar observations made by an elusive figure (known only by the name al-Ḥasan, through the medium of Bīrūnī) in Baghdad around 830–832 (simultaneously with Khālid b. ʿAbd Al-Malik al-Marwarūdhī's observations in Damascus) with Hipparchus's and Ptolemy's solar data. So what was the source of this *difference*? Who was in error: Hipparchus or Ptolemy? They believed that Ptolemy was at fault. This work sparked much controversy for at least a century.

This exposed a complex issue in the scientific thought for Islamic astronomers: *the problem of error*. They did not have any readily available methods to cope with errors.

Battānī believed that this problem could only be solved over an extended period, during which continuous observations would lead to progressive refinements in the parameter values derived from Ptolemy's observational data. He felt that astronomy had a long journey ahead before it could unravel this mysterious

---

11 For these concepts, see Lauden 1977, pp. 26–30.

situation. This gave rise to the *"istidrāk approach"* added to *standard astronomy*, which views astronomy as a successive accumulation of data and a lengthy iterative process of applying repeated corrections to earlier theories. This perspective was widely upheld in the Eastern branch of Islamic astronomy in the following centuries.

Almost concurrently, another approach emerged from a fundamentally different perspective, seemingly first by Abū Jaʿfar al-Khāzin (d. after 970), who exchanged theories and ideas with the aforementioned Ibn Sinān. The main question was: What were the errors committed by either Hipparchus or Ptolemy? As clearly described by al-Khāzin and Ibn Sinān, astronomy in that era (in the first part of the tenth century) was indeed in a severe state of critical controversy,[12] primarily because no real progress could be made in any other branch of astronomy until the problems in solar astronomy were resolved. In the single surviving manuscript of al-Khāzin's *Ṣafāʾiḥ zīj*, which is currently under study by the present author, our author appears to put forth the first *theory of error* in the medieval period and constructs a new physical, spherical model (*hayʾa*) to account for **S1–S4**. This approach paved the way for the view, as expressed over two centuries later by Ibn Rushd (Averröes, d. 1198), that astronomy is the science of trust in the past. An astronomical theory must make use of all data gathered over time, across all space, and be constructed in such a way as to explain all of them. This gave birth to *"long-term astronomy,"* as opposed to *"standard astronomy."* Two other models were developed within its framework in the Middle East (see following text).

This *new astronomy* was established with two distinct *approaches*:

1. *New* models could fundamentally be composed of *new* basic hypotheses. The second model within the context of *long-term astronomy* was the trepidation model proposed in the anonymous *On the motion of the eighth sphere* (attributed to Thābit b. Qurra) to explain **S1–S4**. This model, extensively studied by numerous scholars in the latter part of the past century, like B. R. Goldstein, J. Samsó, R. Mercier, and others, is spherical in structure, like Khāzin's–Ibrāhīm's model, and has a definite physical identity.

2. Another approach was that the building blocks of new models could be borrowed from available hypotheses, but they would be *newly* defined and *differently* structured, thus finding new applications. The third model, created within the framework of *long-term astronomy* and unknown until recently, is a solar model constructed based on Ptolemy's model for the latitudinal motion of the superior planets to account for **S3** and **S5**. It has been fully explained in Quṭb al-Dīn al-Shīrāzī's (d. 1311) cosmographical works and was studied in depth in **2016**. As can be inferred from a statement in Ibn Sīnā's (d. 1037) prologue to the astronomical part of his *Shifāʾ*, the model

---

12 This may be favored by those interested in the Kuhnian narrative of scientific progress. The exact terms used by Khāzin are "a great apprehension" (*shaghl ʿaẓīm*) and "a prolonged controversy" (*baḥth ṭawīl*).

should have been established in the latter part of the tenth century. Its inventor may be Abū Maḥmūd al-Khujandī (d. 1000), considering that in a brief treatise surviving from him, he draws attention to the oscillatory deferents used by Ptolemy in his model of the latitudinal motion of the inferior planets as an auxiliary hypothesis to explain the continuous decrease in the obliquity of the ecliptic. This model, whether described as *economy of thought* or a *conservative attitude*, demonstrates how a theoretically *anomalous* situation in a part of a coherent set of theories constituting a scientific tradition can be resolved by using a specific relevant element firmly established and clearly defined in another part of the same tradition. In **S4** and **L3**, demanding astronomers to rethink current hypotheses used in a single tradition, Islamic astronomers extended this approach to cover the use of elements borrowed from competitor traditions (e.g., Indian) – see our discussion on **S4** in Section 8.5.1, and Chapters 2 and 3 for **L3**.

Adherents of *standard astronomy* were divided into two groups. Bīrūnī drew attention to two important factors. First, *the limitations of the methods*, primarily due to their high sensitivity to input data, and, second, the *inadequacy of observational instruments* (due to size, calibration, etc.) for deriving the solar parameters beyond a certain level of accuracy (Chapter 1). The second group consisted of astronomers who eventually identified and/or maintained that Hipparchus's solar data were reasonably accurate and reliable, while Ptolemy's observations were erroneous and unreliable. This group included, for example, Ibn al-Aʿlam, Ibn Yūnus, Ibn al-Shāṭir (1306–1375/6), and the astronomers at the Marāgha and Samarqand observatories (we will revisit this later in the review of my studies on solar astronomy). Thus, **S1**, **S2**, and **S4** were resolved, as these three parameters were considered fixed and constant; only **S3** and **S5** were left for the inventor of the third model mentioned earlier to account for.

I briefly address **L1–L3** and **P1** in the review of my studies that follows. Medieval astronomers did not manage to resolve all these problems; **L2** and **P1** remained unsolved. The materials presented earlier, in general, are sufficient to demonstrate that the classical period was an era of significant controversies and major discoveries.

Long-term astronomy reached its peak in Ibn al-Zarqālluh's tradition (d. 1100). He developed a variety of models to account for the long-term periodic variations and the secular proper motions in the fundamental solar parameters. He also substantially corrected Ptolemy's lunar model, in relation to **L1**, with the discovery of the annual equation (**2024b**). Through the transmission of his works to the Latin world, this *new astronomy* became *the greatest contribution of medieval Islamic astronomy to the entire science of astronomy*. Of course, it was discredited in confrontation with Middle Eastern standard astronomy when Ibn al-Shāṭir found notorious errors in the quantitative output of Ibn al-Zarqālluh's solar theory. Needless to say, the problem did not lie with the approach but with the insufficient primary empirical data available to the Andalusian and, particularly, with the serious errors in Ptolemy's solar observations.

Beyond Muḥyī al-Dīn al-Maghribī's detailed account of his observations and measurements at Marāgha in the latter part of the thirteenth century, little is known about how Middle Eastern astronomers addressed errors in the theories of their predecessors within Ptolemaic standard astronomy, either according to the *istidrāk approach* or by Ptolemy's *standard methods*. This leads to considerable uncertainty about the origin of non-Ptolemaic values for both the structural and motional parameters. Moreover, without a thorough technical analysis, it would be unknown whether the evaluation of theories was conducted *systematically* or *randomly* based on a minor subset of data, whether preference was given to *timing astronomy* or *positional astronomy*, and so on. Khāzinī's *Experimental astronomy* is the most comprehensive manual on the *istidrāk* approach and theory of error within Ptolemaic standard astronomy. This treatise offers a comprehensive guide on how to identify errors in the fundamental structural or motional parameters of a theory for the Sun, Moon, and planets, developed within the context of Ptolemy's models. It does so by examining patterns of longitudinal errors and strategically planning observations to isolate or amplify an error stemming from a discrepancy in a particular basic parameter. This is achieved by either neutralizing or significantly reducing probable errors due to other parameters. Also, the concept of *inevitable error*, occurring due to the unavoidability of human error or due to unquantified, but necessary, adjustments (e.g., atmospheric refraction), is introduced in it.

OAMIP

I divide *OAMIP* into the following five subfields.

I. Solar Astronomy

Chapter 1 delves into the precision of solar orbital elements spanning from 800 to 1500. An alternative method for deriving solar and planetary orbital elements, proposed by Ibrāhīm b. Sinān, is the focus of Chapter 6.

In **2018**, I embarked on a study of 11 solar theories originating from the medieval Middle and Near East. This led to the emergence of two distinct categories: **Type I**, grounded in Ptolemy's data, and **Type II**, built upon the observations of Hipparchus and/or early Islamic astronomers. By the close of the tenth century, it became evident that issues **S1, S2**, and **S4** were deeply rooted in Ptolemy's equinox observations, casting doubts on the accuracy of Ptolemy's solar theory. Concrete, direct evidence was unearthed in al-Kāshī's lunar measurements (**2020–2021**:107). In his calculation of the Moon's mean double elongation at the time of Ptolemy's second lunar eclipse, he employed a figure of 207;41° for the Sun's mean longitude, a departure from Ptolemy's 206;42°. A recalculation using a modified version of Ibn Yūnus's solar theory unveiled an error of approximately −1° in Ptolemy's solar theory. The discovery of this error in Ptolemy's solar theory and its link to the Marāgha astronomers' application of two different precession rates to update the stellar longitudes of Ptolemy and their Islamic predecessors are discussed in Chapter 10.

In addition to the solar model examined in **2016**, a mechanical solar model is briefly explained in Khāzinī's *Experimental Astronomy* (**2022b**). This paper also put to rest debates surrounding Khāzinī's ethnic origin and the completion date of his work *Mu'tabar zīj*.

II. Lunar Astronomy and Theory of Eclipses

Three empirical challenges were:

L1. Early Middle Eastern astronomers al-Māhānī (d. ca. 880), al-Nayrīzī (d. 922), and the Banū Amājūr (*fl. ca.* 920) identified discrepancies in eclipse timings. Ibn Yūnus proposed a solution by adjusting the lunar epicycle's radius. Ibn al-Zarqālluh recognized that they correlated with the seasons and were contingent upon the Moon's distance from the solar apsidal line. This phenomenon corresponds to a specialized case of the Moon's annual equation (or, more precisely, the cumulative impact of lunar inequalities influenced by lunisolar anomalies). This discovery was documented in **2024b**.

L2. During their periodic lunar observations from June to September 918, the Banū Amājūr discovered that the Moon's orbital inclination is "highly divergent and irregular," deviating from a fixed value (**2024b**: 297–300).

L3. According to the *Almagest*'s range of variations in the apparent angular sizes of the Sun and Moon, the annular solar eclipse remains undefined. Islamic astronomers defined and justified the phenomenon by integrating Indian hypotheses into Ptolemy's models (Chapter 2 **and 2015a**). In this context, Shams al-Dīn Muḥammad al-Wābkanawī predicted[13] the annular solar eclipse of January 30, 1283, based on the solar and lunar parameter values obtained by Muḥyī al-Dīn al-Maghribī (Chapter 3).

Bīrūnī's evaluation of Ptolemy's lunar model revealed that the center of the epicycle describes a path shaped like an ellipse (Chapter 4).

Four determinations of the parameters of Ptolemy's lunar model have survived from the eleventh to the sixteenth centuries: Bīrūnī (a trio of lunar eclipses in 1003–1004), Muḥyī al-Dīn al-Maghribī (1262–1274; **2014** and Chapter 9), al-Kāshī (1406–1407; **2020–2021**), Taqī al-Dīn Muḥammad b. Maʿrūf (1576–1577; Chapter 5).

---

13 It does not matter whether this prediction was *proactive* or *retroactive* (see Chapter 3, note 3).

III. Planetary Astronomy

Concerning **P1**, from the tenth century, it was known (seemingly discovered by the Banū Amājūr) that errors in a planet's longitude, as calculated from a numerical theory within the Ptolemaic context, are variable and oscillatory.

The relationship between theoretical considerations and observational practices in medieval planetary astronomy is a complex matter. It carries significant implications for our comprehensive understanding of Islamic astronomy as a whole. Three sets of observations have endured: about 50 observations from early Islamic astronomers by Ibn Yūnus, extensive systematic observations by Muḥyī al-Dīn al-Maghribī at Marāgha in the 1260s and 1270s, and al-Kawāshī's observations in Egypt in the 1270s and 1280s.

Ibn Yūnus's wealth of planetary observational data from the classical period is currently under investigation. Muḥyī al-Dīn's astronomical work has been the focus of my studies over the past decade. A comprehensive evaluation of his observations is presented in Chapter 9 and his measurements of Mars in **2018–2019**.

In Chapter 6, I investigate an alternative method for measuring the orbital elements of the Sun and planets, proposed by Ibrāhīm b. Sinān.

Chapter 7 explores how the two observatories at Marāgha and Samarqand provided practical astronomers with the means to test and measure new values for the orbital inclinations of the planets.

Chapter 8 discusses a significant breakthrough in the late Islamic period, highlighting the distinct difference between the apogeal and processional motions as established by Jamāl al-Dīn al-Zaydī (thirteenth century) and Ibn al-Shāṭir.

In **2019a**, I analyze the accuracy of the values determined for the orbital elements of Venus.

In **2019b** and **2023a**, I examine Ibn al-Fahhād's astronomy, which offers valuable insights into the relationship between theory and observation in Islamic astronomy.

IV. Stellar Astronomy

Ptolemy's *Almagest* VII and VIII served as the primary reference for determining the positions of fixed stars in Islamic *zījes*. ʿAbd al-Raḥmān al-Ṣūfī (903–986) measured stellar magnitudes and recorded his findings in his work *Kitāb Ṣuwar al-kawākib (al-thābita) al-thamāniya wa 'l-arbaʿīn* (*The Book of images of the 48 (fixed) stellar constellations*).[14]

Non-Ptolemaic star tables/catalogues appeared in the *Mumtaḥan zīj* (studied in **2016–2017**), Ibn Yūnus's *Ḥākimī zīj* (currently under study), the *Īlkhānī zīj* (Chapter 10), the star list in Muḥyī al-Dīn al-Maghribī's *Talkhīṣ al-majisṭī* (Chapter 9), the star catalogue produced by Jamāl al-Dīn al-Zaydī, Ibn al-Shāṭir's *Jadīd zīj* (analyzed in **2022a**), and Ulugh Beg's *Sulṭānī zīj*.

---

14 In the ninth-century Arabic translations of the *Almagest*, the term *kawkaba* (pl. *kawākib*; cf. *kawkab* = star/planet) is to render the Greek *asterismos* = constellation.

For a long time, it was believed that Ṣufī's *magnum opus* was the sole attempt to reorganize stellar brightness within the framework of the six-class magnitude system. However, a new set of values emerges from Jamāl al-Dīn's catalogue and, as I demonstrated in **2022a**, also from Ibn al-Shāṭir's star table. Interestingly, these two sources significantly improved the stellar magnitudes after Ṣūfī. For instance, they assign unprecedented values to certain bright stars – for instance, $m = 1$ to α Sco (modern visual magnitude $V = 0.96$) and $m = 2$ to both β Cas ($V = 2.3$) and α Oph ($V = 2.1$) – contrasting with Ptolemy and Ṣūfī, who, respectively, have 2 and 3.

In **2021**, I investigated (in collaboration with Jeremy J. Drake) a historical dimming of Algol in the latter part of the thirteenth century. Our findings underscore the importance of uncovering such alleged events, as demonstrated by the recent "Great Dimming Event" observed in Betelgeuse (α Ori). I have presented and evaluated Abu 'l-Wafā''s (d. 997) dated observations and quantitative data for five clock stars in **2023b**. In **2024a**, I identified the north pole star in Euclid's *Phaenomena* as the star HR 4646, and the one mentioned in the Pahlavi *Bundahišn* and al-Ṣūfī's *Ṣuwar al-kawākib* as HR 4893.

V. Observational Instrumentation

I have authored four articles in this subfield, with assistance from Georg Zotti (Vienna).

Three of these articles are connected to the observatories of Marāgha and Samarqand. The first one explores a treatise on 11 observational instruments, presumably invented by Ghāzān Khān (1271–1304), the seventh ruler of the Īlkhānid dynasty of Iran (r. 1295–1304). These instruments utilized long straight rulers and catgut or copper wires to collect observational data. This study is documented in two complementary versions (Chapter 11 and **2013**).

The central instrument at the Samarqand observatory has been a topic of frequent discussion. Our study, which sheds light on this intriguing instrument, was published in **2020**.

2015b focuses on Bīrūnī's large sighting tube, shaped like a telescope. The instrument was equipped with a graduated quadrant, allowing alignment at any arbitrary altitude.

The remnants of the Samarqand and Marāgha observatories proved useful in my studies for throwing new lights on the real construction of an instrument in the second period of the Marāgha observatory (**2013**:77–79) and the configuration of the central instrument of the Samarqand observatory.

Last Words

The chapters included in this volume have undergone significant revisions due to subsequent reflections, the necessity for a new organization, and the availability of new sources that were not accessible to me at that time, among other things. We still have a considerable journey ahead to elucidate all facets of observational astronomy, the thematic issues, and the nine critical problems introduced in this introduction. Furthermore, there is still a wealth of material awaiting exploration.

Bibliography

al-Bīrūnī, A. R., 1948, *Rasā'il al-Bīrūnī*, Hyderabad: Osmania Oriental Publication Bureau.

Laudan, L., 1977, *Progress and Its Problems: Towards a Theory of Scientific Growth*, Berkeley [etc.]: University of California Press.

Mozaffari, S. M., 2014, "Muḥyī al-Dīn al-Maghribī's lunar measurements at the Maragha observatory", *Archive for History of Exact Sciences*, **68**, pp. 67–120.

Mozaffari, S. M., 2015a, "Annular eclipses and the considerations about the solar and lunar angular diameters in the medieval astronomy", in: Orchiston, W., Green, D. A., and Strom, R. (eds.), *New Insights from Recent Studies in Historical Astronomy: Following in the Footsteps of F. Richard Stephenson*, New York: Springer, pp. 119–142.

Mozaffari, S. M., 2016, "A forgotten solar model", *Archive for History of Exact Sciences*, **70**, pp. 267–291.

Mozaffari, S. M., 2016–2017, "A revision of the star tables in the *Mumtaḥan zīj*", *Suhayl*, **15**, pp. 67–100.

Mozaffari, S. M., 2018, "An analysis of medieval solar theories", *Archive for History of Exact Sciences*, **72**, pp. 191–243.

Mozaffari, S. M., 2018–2019, "Muḥyī al-Dīn al-Maghribī's measurements of Mars at the Maragha observatory", *Suhayl*, **16**, pp. 149–249.

Mozaffari, S. M., 2019a, "The orbital elements of Venus in medieval Islamic astronomy: Interaction between traditions and the accuracy of observations", *Journal for the History of Astronomy*, **50**, pp. 46–81.

Mozaffari, S. M., 2019b, "Ibn al-Fahhād and the Great Conjunction of 1166 AD", *Archive for History of Exact Sciences*, **73**, pp. 517–549.

Mozaffari, S. M., 2020–2021, "Kāshī's lunar measurements", *Suhayl*, **18**, pp. 69–127.

Mozaffari, S. M., 2022a, "An analysis of Ibn al-Shāṭir's star table", *Journal for the History of Astronomy*, **53**, pp. 163–196.

Mozaffari, S. M., 2022b, "A mechanical concentric solar model in Khāzinī's *Mu'tabar zīj*", *Archive for History of Exact Sciences*, **76**, pp. 513–529.

Mozaffari, S. M., 2023a, "Sources of the planetary theories in Fahhād's *'Alā'ī zīj*: Solving a medieval case of intellectual fraud", *Suhayl*, **20**, pp. 143–221.

Mozaffari, S. M., 2023b, "A survey of Abu 'l-Wafā''s solar and stellar observations", *Journal of Astronomical History and Heritage*, **26**, pp. 469–488.

Mozaffari, S. M., 2024a, "Ancient and medieval north pole stars: A review and new identification", *Journal of the American Oriental Society*, **144**, pp. 23–40.

Mozaffari, S. M., 2024b, "Ibn al-Zarqālluh's discovery of the annual equation of the Moon", *Archive for History of Exact Sciences*, **78**, pp. 271–304.

Mozaffari, S. M. and Drake, J. J., 2021, "Algol anomaly or careful observations of its brightness? The values recorded for the magnitude of Algol in the medieval astronomical corpus", *Journal for the History of Astronomy*, **52**, pp. 77–103.

Mozaffari, S. M. and Zotti, G., 2013, "The observational instruments at the Maragha observatory after AD 1300", *Suhayl*, **12**, pp. 45–179.

Mozaffari, S. M. and Zotti, G., 2015b, "Bīrūnī's telescopic-shape instrument for observing the lunar crescent", *Suhayl*, **14**, pp. 167–188.

Mozaffari, S. M. and Zotti, G., 2020, "New light on the main instrument of the Samarqand observatory", *Journal for the History of Astronomy*, **51**, pp. 255–271.

Neugebauer, O. E., 1941a, "Some fundamental concepts in ancient astronomy", [1941b], "Exact science in antiquity", both reprinted in Idem, 1983, *Astronomy and History; Selected Essays*, New York: Springer, pp. 5–31.

Part I

SOLAR ASTRONOMY

LIMITATIONS OF METHODS

The Accuracy of the Values Measured for the Earth's/Sun's Orbital Elements in the Middle East, 800–1500 CE[1]

This chapter addresses the methods proposed and the values achieved for the eccentricity and the longitude of apogee of the (apparent) orbit of the Sun in the Ptolemaic context in the Middle East during the medieval period. The main goals of this research are as follows: first, to determine the accuracy of the historical values in relation to the theoretical accuracy and/or the intrinsic limitations of the methods used; second, to investigate whether medieval astronomers were aware of the limitations and, if so, which alternative methods (assumed to have a higher accuracy) were then proposed; and finally, to determine what was the basis of the substitution in terms of improving the accuracy of the values achieved.

In Section 1.1, the Ptolemaic eccentric orbit of the Sun and its parameters are introduced. Its relation to the Keplerian elliptical orbit of the Earth, which is used as a criterion for comparing historical values, is briefly explained. In Section 1.2, three standard methods for measuring the solar orbital elements during the medieval period found in primary sources are reviewed. In Section 1.3, more than 20 values for the solar eccentricity and the longitude of apogee from the medieval period are classified and provided with historical comments. The discussion and conclusions are presented in Section 1.4.

1.1 Introduction

Ptolemy, in *Almagest* III, presented a solar model consisting of an eccentric circle whose center is displaced from the center of the Earth by an amount of eccentricity $e = TO$ expressed in terms of an arbitrary length $R = 60$ for the radius of the eccentric (Figure 1.1). The Sun revolves on the eccentric around the Earth – but its motion is uniform with respect to the center of eccentric – and completes one

---

1 Original publication: S. M. Mozaffari, "Limitations of methods: The accuracy of the values measured for the Earth's/Sun's orbital elements in the Middle East, A.D. 800–1500," *Journal for the History of Astronomy*, **44** (2013), Part 1: pp. 313–336, Part 2: pp. 389–411. © 2013 Sage, and republished by permission.

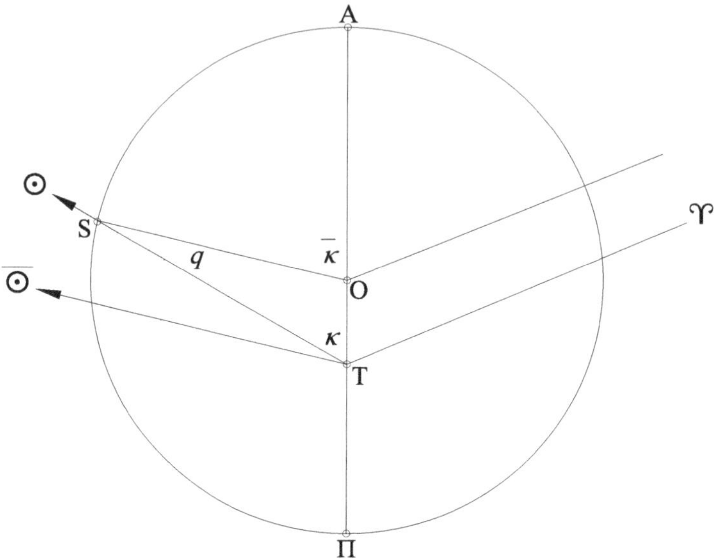

Figure 1.1 The eccentric solar model with constant eccentricity.

revolution in a tropical year *TY* (counted in days), that is, with a mean motion of $\omega = 360°/TY$. Due to the eccentricity of its orbit, the true longitude of the Sun (i.e., its longitude as seen from the Earth *T*) does not match its mean longitude (its longitude as it appears from the center *O* of uniform motion) except at apogee (*A*) and perigee (*Π*). The difference between the two, called the "equation of center" *q*, is defined as a one-variable function of the solar mean anomaly *c*, which is the difference between its mean longitude and the longitude λ_{ap} of the apogee:

$$q = \tan^{-1}\left(\frac{e\sin\bar{\kappa}}{R+e\cos\bar{\kappa}}\right), \quad q_{max} = \sin^{-1}\left(\frac{e}{R}\right) \tag{1.1}$$

Therefore, the solar eccentric model has three parameters (*TY* or ω, *e*, and λ_{ap}) but only one anomaly due to its eccentricity.

Ptolemy assumes that all three parameters are constant.[2] Since, at least, the latter part of the ninth century, the Middle Eastern astronomers found (see Tables 1.1 and 1.2) that the longitude of the solar apogee increases with the passage of time, and they (seemingly, first of all, the Banū Mūsā or Thābit b. Qurra and Ḥabash) assumed its rate of change to be the same as the rate of precession, as is the case

---

2 For the solar eccentric model as presented in *Almagest* III (Toomer 1984, pp. 133–172), see Neugebauer 1975, vol. 1, pp. 54–61, Pedersen 1974, chap. 3. For its evolution through pre-Ptolemaic data, see Jones 1991, and Duke 2008.

with the motion of the apogee of the planets in the Ptolemaic context (the relevant discussion will be presented in Section 1.4). Also, different values for e and TY were observed during the medieval period. The variety in the values obtained for the two was so large that, for example, Copernicus, in the prologue of his *De revolutionibus*, complains that "[the astronomers] are uncertain of the motion of the Sun and Moon to such an extent that they cannot demonstrate or observe a constant magnitude for the tropical year,"[3] while Bīrūnī was forced to resolve his readers' worries about the huge differences in the values measured for the solar eccentricity over a short period (see Section 1.4).

In fact, both e and TY are changing with the passage of time: e decreases slowly by the average amount of 4.2×10^{-5} per century. TY is also subjected to a slight time-dependent change whose rate and sign depend on which of the four cardinal points (solstices and equinoxes) is adopted as the zero point for measuring the length of TY.[4] It is worth noting that the solar apogee also progresses with a rate of $61.9''$ per year or, approximately, $1°$ in 58 years, that is, faster than the rate of precession, which is $1°$ in 71.6 years.[5]

The Earth, in reality, travels on an elliptical orbit with eccentricity e' (expressed in terms of the length of the semi-major axis $a = 1$) around the Sun, where the Sun is located at one of the foci of the orbit. By exchanging the (positions of the) Sun and the Earth, the system is simply transferred to the geocentric mode, and therefore the Sun apparently orbits around the Earth in an elliptical orbit. The difference in the solar geocentric longitude between the eccentric and elliptical models and the relation between the two eccentricities, e and e', are as follows:[6]

$$\Delta\lambda \approx \left(e - 2e'\right)\sin M + \left(\frac{1}{2}e^2 - \frac{5}{4}e'^2\right)\sin 2M$$

$$e \approx 2e' - \frac{3}{2}e'^2 \cos M$$

(1.2)

---

3 Copernicus 1543, English translation: Dobrzycki and Rosen 1978, p. 4.

4 For example, if TY is measured from one vernal equinox to the next, it *increases* with the average rate of 1 second per century (s/cent) from 0 to 2000 CE. If it is counted from one autumnal equinox to the next, it *decreases* with the average rate of 2 s/cent in this period. The lesser changes in TY will result if it is counted with reference to solstices: from a summer solstice to the next: -0.4 s/cent; from a winter solstice to the next: -0.6 s/cent (see Meeus and Savoie 1992, pp. 40–42, Meeus 2002, pp. 357–366). Ancient and medieval astronomers measured the solar year with the autumnal equinox as the zero point over long periods; thus, the values achieved by them should be considered "mean/average." The mean values for each of the four types of the solar year from 0 to 2000 CE are as follows:

Two vernal equinoxes: 365;14,32
Two summer solstices: 65;14,30
Two autumnal equinoxes: 365;14,32
Two winter solstices: 365;14,34 days.

5 Meeus 2002, p. 361.

6 Neugebauer 1975, Vol. 3, pp. 1097–1102; Maeyama 1998, p. 4.

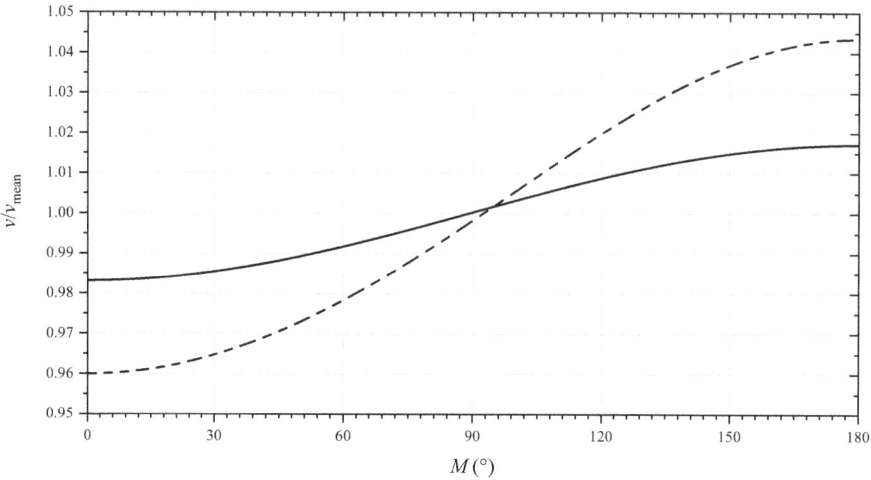

Figure 1.2 The instantaneous velocity of the Sun according to the eccentric model based on Ptolemaic eccentricity $e = 2;30$ (dashed curve), and to the Keplerian elliptical orbit based on the modern eccentricity $e' = 0.017$ ($a = 1$) (continuous curve).

where M is the mean anomaly ($M = \bar{\kappa}$). The formula for $\Delta\lambda$ in (1.2) is accurate up to arc seconds, which is more than sufficient for medieval astronomical studies. If the eccentricity of the eccentric model is taken as equal to the distance between the foci of the elliptical orbit, that is, $e = 2e'$ (for either $R = a = 1$ or $R = a = 60$),[7] then the difference in the solar longitude given by the two models will be maximum when the Sun reaches its apparent orbital octant, that is, for $M = 45°$. In practice, the difference will be of the order of 1 second. This implies that the elliptical orbit does not match completely with the eccentric orbit, and under the condition $e = 2e'$, a deviation of about 1 arc second occurs. In the case of historical values for the eccentricity of the eccentric model, we expect in an ideal manner that the values measured are twice as long as the elliptical eccentricity. However, because of the small incongruence of the two models, this relation $e = 2e'$ may not be expected to hold exactly. We shall later note and take into account the effect of the mismatch of the two models on the accuracy of the historical values for e obtained from different methods.

The eccentricity is responsible for the alternative deviation of the solar angular velocity from its mean value as seen from the Earth. For instance, the dotted curves in Figure 1.2 depict the rate of change in the solar velocity v with respect to its mean value ω as a variable of the mean anomaly (counted from the apogee) for

---

7 The intention to bisect the Ptolemaic solar eccentricity can be traced out in medieval astronomy (Goldstein and Sawyer 1977) and is found, for example, in Ibn al-Shāṭir's solar model (Saliba 1987, p. 42).

the Ptolemaic eccentricity 2;30 and Muhyī al-Dīn al-Maghribī's (d. 1283) 2;6. The continuous curve shows the true rate of change in the elliptical orbit with eccentricity $e' = 0.017$. The correction for this anomaly, that is, the equation of center q (formula (1.1)), and the true longitude of the Sun depend on its eccentricity.

In general, any inaccuracy in one of the solar parameters can cause a systematic error in the true longitude of the Sun. In all systems for modeling the planetary motions (geocentric or heliocentric, based on either circular orbits or elliptical ones), the solar model occupies a central position. In the Ptolemaic models, the longitude of the Moon, planets, and stars depends on the longitude of the Sun; for example, the mean motion of the inferior planets is equal to that of the Sun. Furthermore, the superior planets' motion in anomaly directly depends on the solar mean motion (the planet's *mean anomaly* is the difference between the mean longitudes of the Sun and of the planet). Moreover, in order to determine the parameters of the Ptolemaic orbit of the Moon (eccentricity and radius of epicycle), the longitude of the Moon is directly derived from the longitude of the Sun at the instant of the synodic phenomenon involved (i.e., the maximum phase of a lunar eclipse for determining the radius of the lunar epicycle and half-moon for determining the eccentricity of the lunar deferent). Therefore, a simple error in any solar parameter may *ipso facto* produce a chain of systematic errors in determining the longitude of other celestial objects. A historical example is Ptolemy's error in determining the moment of the vernal equinox, by $+1$ day, which caused the well-known systematic error of around $1°$ in the longitude of the Sun; this error affected the longitude of the stars, and as a consequence, these were listed in *Almagest* VII and VIII with a systematic error of about $-1°$.[8]

1.2 Historical Methods for Measuring the Solar Orbital Elements

The three main historical methods for measuring the solar orbital elements are the seasons method, the mid-seasons (or mid-signs) method, and the three-point method, as will be described later. All three methods either are explained in the *Almagest* (III.4) or can be extracted from the mathematical methods by which Ptolemy determined the parameters of his own planetary models (IV.6, IV.11, X, and XI).[9] Nevertheless, Bīrūnī attributes the three-points method to his master, Abū Naṣr Manṣūr b. 'Irāq, and also mentions that in his (now lost) treatise titled *al-Istishhād bi-ikhtilāf al-arṣād* (*Producing witness for the difference in the observations*), he had proved the superiority of this method over the other two.[10] Muhyī al-Dīn al-Maghribī's account of his measurement of the solar eccentricity in the

---

8 For the analysis of the error, see Newton 1973, pp. 369–370. Seven centuries later, this caused a remarkable error in measuring the rate of precession in early Islamic astronomy; see Grasshoff 1990, pp. 19–20.

9 Toomer 1984, pp. 153–157, 190–203, and 469f; Saliba 1985a, p. 114; Duke 2008, pp. 284–285, 291.

10 al-Bīrūnī 1879, p. 167.

Marāgha observatory is one of the excellent surviving accounts of the utilization of this method in the medieval period (see following text). His older colleague at the observatory, Mu'ayyad al-Dīn al-'Urḍī (d. 1266), wrote a preliminary treatise to sketch the main outline of the method.[11] The mid-sign method seems to have been somewhat innovative to medieval astronomers. Although Yaḥyā b. Abī Manṣūr knew and made practical use of this method in the early ninth century, it is not known if he proposed it. In what follows, the methods are explained, and some historical and technical considerations concerning each are outlined.

1.2.1 The Seasons Method

In this method, the length of the tropical year (TY), and therewith the mean angular velocity ω ($= 360°/TY$) of the Sun, and the length of two neighboring seasons (e.g., the interval between an equinox and its successive solstice and the time from that solstice to the next equinox) are measured by observation. Then, the magnitude of the mean angular motion of the Sun on its geocentric orbit in these two intervals (i.e., the lengths of the arcs a and b in Figure 1.3) is calculated. (The Sun travels 90° on the ecliptic in each of these intervals.) The eccentricity $e = TO$ and the longitude of apogee λ_{ap} = angle ΥTA in Figures 1.1 and 1.3 are obtained by the following formulas:

$$e = R\sqrt{\sin^2\left(\frac{a-b}{2}\right) + \cos^2\left(\frac{a+b}{2}\right)} \tag{1.3}$$

$$\alpha = \tan^{-1}\left(\sin\left(\frac{a-b}{2}\right) / \cos\left(\frac{a+b}{2}\right)\right)$$

where α is the angle between the solar apsidal line and the line passing through the solstices/equinoxes that lies between the two seasons, from which the longitudes of the solar apogee and perigee can be derived.

In order to determine the moments of equinoxes and solstices, medieval astronomers measured the meridian altitude of the Sun by means of meridian instruments, for example, quadrants, sextants, or the Ptolemaic parallactic instrument. These instruments were installed in the line of meridian and were utilized for measuring the noon altitudes of the Sun, the latitude φ of the place of observation, the inclination ε of the ecliptic from the celestial equator, and so on. We know that such instruments in different sizes were constructed in the early Islamic period.[12]

In the equinoxes, the solar declination $\delta_{eq} = 0$, and in the solstices, $\delta_{sl} = \pm\varepsilon$. Thus, the solar noon altitude h in the days around the equinoxes and solstices is measured in a given place with the latitude φ. If $h = 90° - \varphi$ or $h = 90° - \varphi \pm \varepsilon$, then the instant of the true noon will be simply the time of, respectively, the equinox or solstice. If

---

11 Saliba 1985b.
12 Charette 2006.

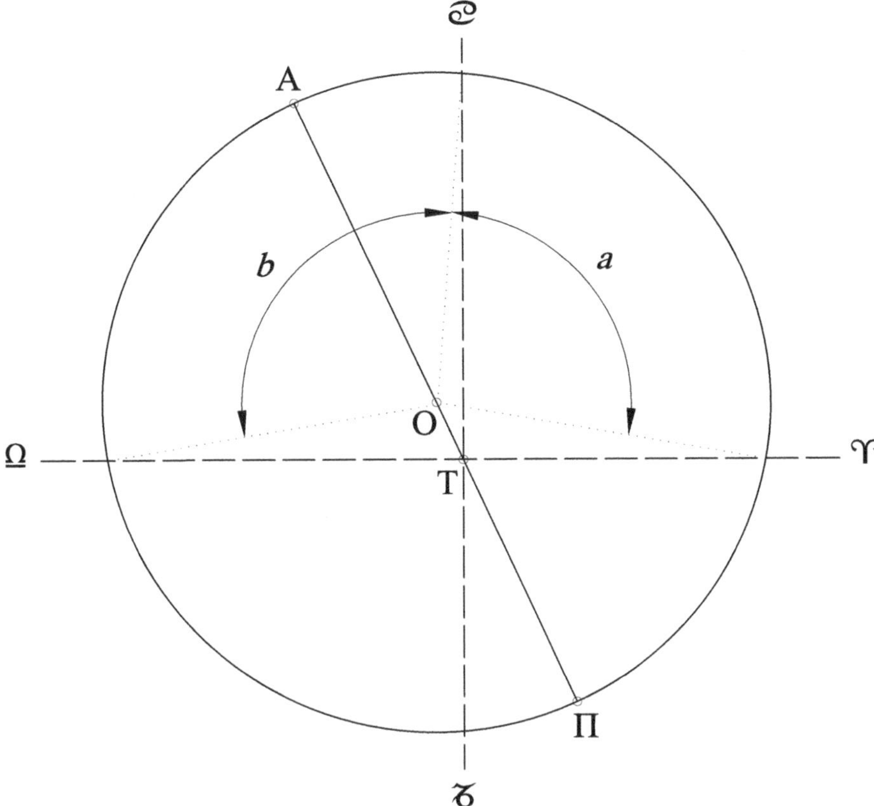

Figure 1.3 The measurement of the parameters of the eccentric solar model based on the seasons method.

not, in order to determine the time of an equinox or of a solstice, the Middle Eastern astronomers utilized some techniques adopting the linear interpolation as follows.

The equinox occurs between the true noon of two successive days when $h_1 < 90° - \varphi$ and $h_2 > 90° - \varphi$ (in the case of vernal equinox) and $h_1 > 90° - \varphi$ and $h_2 < 90° - \varphi$ (in the case of autumnal equinox). Then, since the rate of variation in the solar declination (and so noon altitude), $|\Delta\delta_{eq}| = |\Delta h_{eq}| \approx 0;24°$, is known, the instant of the equinox is determined as $|(h - 90 + \varphi)/\Delta\delta_{eq}|$-th of a day after the true noon of the first day or before that of the second one. Another variant is that after the measurement of h, one derives δ and next λ; then $|\lambda - \lambda_{eq}|/\omega$ (where $\omega = 0;59,8°/d$ is the solar daily mean motion) will be the time elapsed since or remaining before the equinox. Al-Maghribī,[13] for example, employed it in order to determine the instant of the autumnal equinox of 1264 and that of the vernal equinox of 1265;

---

13 al-Maghribī, *Talkhīṣ* III.1: f. 58r.

on September 16, 1264 (JDN 2182993): $h = 52;21°$ (true modern: $52;17°$); with $\varphi = 37;20,30°$ for Marāgha, $\delta = -0;18,30°$; then with $\varepsilon = 23;30°,$[14] $\lambda = 180;46,25°$ (recomputed: . . .,24). Thus, the autumnal equinox of the year occurred 18;50,20 (text: 18;24,46; recomputed: 18;49,55) hours before the true noon of September 16 or on September 15, ~17:02 MLT (true modern: ~15:25). On March 14, 1265 (JDN 2183172): $h = 53;8,30°$ (true modern: $53;5°$); then $\lambda = 1;12,44°$. Thus, the vernal equinox of the year occurred 1 day and 5;31 hours before the true noon of March 14 or on March 13, ~6:37 MLT (true modern: ~7:37). In the case of solstices, since the variation in δ/h is very small, $~0;0,14°$, such a simple method of interpolation is of no practical application. In order to determine the instant of a solstice, Bīrūnī proposed[15] two techniques for the solstice interpolation (although he himself declared he was not satisfied with either).

(1) Observing the solar meridian (noon) altitudes on some days prior and posterior to the solstice. Find, first, the meridian altitude h_1; next, the declination δ_1; and then the longitude λ_1 of the Sun on a day t_1 prior to the occurrence of a solstice by a few days; also find the meridian altitude h_2, δ_2, and λ_2 for a day t_2 after that solstice by a few days. The time interval between t_1 and t_2 is then known. Consider the third point on the ecliptic symmetric to λ_2 with respect to that solstice, through which the Sun passed at an unknown time t_3; it is evident that $\delta_3 = \delta_2$ and $\lambda_3 = 180° - \lambda_2$. Thus, the time interval between t_1 and t_3 is $(180° - \lambda_2 - \lambda_1)/v_\odot$. Therefore, the time of the occurrence of the solstice is $t_1 + (t_2 - t_1)/2 + (180° - \lambda_2 - \lambda_1)/2v_\odot$. Bīrūnī illustrates the method as follows. In Jurjāniyya with $\varphi = 42;17°$ (today, Kunya-Urgench, $42;18°$), on June 7, 1016 (JDN 2092310): $h_1 = 70;58°$ (modern: $70;59°$),[16] so $\delta_1 = 23;13°$ [recomputed: $23;15°$], and since Bīrūnī has $\varepsilon = 23;35°$ (modern value for Bīrūnī's time: $\varepsilon_0 = 23;34°$), $\lambda 1 = 80;11°$ [$80;38°$]. On June 23, 1016 (JDN 2092326): $h_2 = 71;4°$ (modern: $71;5°$),[17] so $\delta_2 = 23;21°$, and then $\lambda_2 = 98;6°$ [$97;50°$]; then $\lambda_3 = 81;54°$ [$82;10°$]. Since in the days sufficiently near the solstice the solar velocity has its minimum value ($v_\odot = 0;57°/d$), the time interval between t_1 and t_3 is $(81;54–80;11)/0;57 = 1;48$ [$(82;10–80;38)/0;57 = 1;37$] days. The time interval between the two observations (i.e., $t_2 - t_1$) is 16 days. Therefore, the moment of solstice is 8;54 [8;49] days after the first observation, that is, 21;36 hours after the true noon on June 15 or June 16, 9:36 [7:36] (true modern: 10:48, that is, an error of $-1;12$ [$-3;12$] hours).

In the end, Bīrūnī remarks about the method that it is theoretically sound but practically unreliable, and that it should be employed in a gradual manner (*'alā*

---

14 Measured by the altitude observations in the three consecutive days near the solstice days of 1264: June 12–14: $h = 76;9,30°$ (modern: $76;8,30°$); and December 7–9: $h = 29;9,30°$ (modern: $29;10,30°$). The modern value for the mean obliquity of the ecliptic at that time: $\varepsilon_0 \approx 23;32°$. Note that these altitude observations are about $\pm 1'$ in error, while his equinox altitudes and other three altitudes given in the following (see Table 1.3, no. 9) are about $+4'$ in error.

15 al-Bīrūnī 1954–1956, vol. 2, pp. 621–623.

16 Said and Stephenson 1995, p. 123.

17 Said and Stephenson 1995, p. 123.

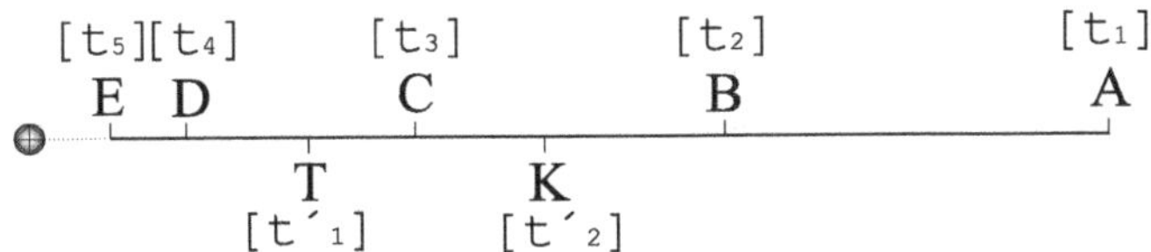

Figure 1.4 The solstice interpolation: the noon shadow technique.

sabīl al-tadarruj; an iterative process with the aid of other pairs of the observational data?) until one arrives at the most exact value of the instant of solstice.

(2) Measuring the shadows of a gnomon erected perpendicularly on the ground or on a wall, respectively, for the winter and summer solstice observations, during some days around solstice. Assume A, B, C, D, and E (Figure 1.4) to be the apex of gnomon shadow in noon on five successive days (t_1 to t_5) before the solstice, and T and K to be the apex of the gnomon noon shadow on two successive days (t'_1 and t'_2) – of course, not essentially and necessarily immediately – after that solstice. If the instant of solstice is denoted by S, the point C corresponds to the time difference $S - t_3$ before the solstice, and the point T corresponds to the time difference $t'_1 - S$ after that solstice. The difference Δt between the two aforementioned time intervals is taken to be $(CT/CD) \cdot (t_4 - t_3)$, where $t_4 - t_3 = 1$ day. Then, $S = t_3 + (t'_1 + \Delta t - t_3)/2$. Or similarly, if $\Delta t'$ is the difference between the time interval $S - t_3$ and $t'_2 - S$, then $\Delta t' = (CK/CB) \cdot (t_3 - t_2)$, where $t_3 - t_2 = 1$ day. Therefore, $S = t_3 + (t'_2 - \Delta t' - t_3)/2$. No example was given for the method.

The method is evidently based on the linear interpolation in the length of gnomon shadows. Nevertheless, the solar declination (and so its meridian altitude), and thus the length of its shadow, does not vary in linear proportions between two successive days around the solstice. Bīrūnī seems to notice the problem[18] and, consequently, remarks that the method is not guaranteed. (Some centuries later, Edmond Halley developed a method of the solstice observation based on the hyperbolic interpolation.[19]) Bīrūnī, however, suggests that in order to obtain a desired value, the gnomon should be erected in the highest places from the ground (i.e., where the length of the shadow is greater) and/or the shadow scale should have a long length to be graduated to the seconds (i.e., the whole length of the scale to be divided into 3,600 equal parts).

In both of the methods, it is assumed that the time intervals between any of the two points are symmetric with respect to a solstice and that solstice are equal while, because the solar apogee does not coincide with the summer solstice, this may not be the case.

A comparison with the altitude observations by other Islamic astronomers for determining the time of solstices[20] shows that only Bīrūnī and Abū al-Ḥasan

18 al-Bīrūnī 1954–1956, vol. 2, p. 621.
19 Halley 1695.
20 Said and Stephenson 1995, pp. 122–123 (table 3).

al-Hirawī have the observed altitudes for the days sufficiently far from the solstices, which might be applied as the input data in these (or other similar interpolation) methods in order to determine the time of solstice, while the others have solely the observed altitudes at the successive days close to solstices.

The formulas (1.3) are highly sensitive to changes in the lengths of the two time intervals: for $a = 93°$ and $b = 92°$, we have $e = 2.67$ and $\lambda_{ap} = 78.7°$; but if a is increased by only $1°$ (as previously mentioned, this is the size of the systematic error in Ptolemy's solar theory, corresponding to the motion in approximately 1 day), we obtain the value $e = 3.31$ (i.e., an error of around 24%) and $\lambda_{ap} = 71.6°$ (an error of around 9%). The sensitivity to a small change in the input data caused drastically different values to be determined for the solar eccentricity and the longitude of the solar apogee from the observations performed even within a short period. This was a main focus of medieval observational astronomers; for example, Bīrūnī noticed the sensitivity of the method as "a major amount of change in the result caused by/due to a minor difference [from the true value] entered into the observation[-al data]."[21] He listed all values of the solar eccentricity obtained from the seasons method by his predecessors as the evidence for his statement (see Table 1.1). Moreover, as seen earlier, there are technical and empirical difficulties in the exact determination of the moment of occurrence of solstices. This may make a sound application of the method problematic and may drastically decrease the accuracy of the value achieved. The other two methods that are found in the primary sources try to remedy this problem.

1.2.2 The Mid-Seasons (or Mid-Signs) Method[22]

This method is a substitute for the seasons method. In order to avoid the probable large error in the determination of the solstices, medieval astronomers remedied the seasons method. The equinoxes and solstices were replaced by the middles of seasons, where the daily rate of difference in the solar meridian altitude is only around $0.3°$. In the other words, instead of measuring the intervals between the equinoxes and solstices, the two intervals are observed between the middle of three successive seasons. The astronomers thus had to determine, first, the interval between the times when, for example, the Sun reached a longitude of $\lambda = 315°$ (the middle of the sign Aquarius) and a longitude of $\lambda = 45°$ (the middle of the sign Taurus). Then they would determine the time interval during which the Sun travelled from $\lambda = 45°$ to $\lambda = 135°$ (the middle of the sign Leo), and so on. In this way, four quadrants on the ecliptic are marked: Eastern: mid-winter to mid-spring; Northern: mid-spring to mid-summer; Western: mid-summer to mid-autumn; and Southern: mid-autumn to mid-winter.

---

21 al-Bīrūnī 1954–1956, vol. 2, p. 653.

22 It is frequently and wrongly named the *fuṣūl* (= seasons) method in the secondary literature, while it is evidently and correctly called *anṣāf al-fuṣūl* or *awsāṭ al-burūj* in Bīrūnī's work (al-Bīrūnī 1954–1956, vol. 2, p. 657).

In order to determine the times of mid-seasons, the Islamic medieval astronomers were again using the interpolation techniques like those described earlier. The declination of the Sun at the mid-seasons, when the Sun reaches the longitudes 45°, 135°, 225°, and 315°: $\delta_{mid} = \delta_{45} = \delta_{135} = -\delta_{225} = -\delta_{315} = \sin^{-1}(\sin 45° \cdot \sin \varepsilon)$. The solar noon altitude in a given place is $h_{mid} = 90° - \varphi + \delta_{mid}$ in mid-spring and mid-summer and $h_{mid} = 90° - \varphi - \delta_{mid}$ in mid-autumn and mid-winter. Now, if in a day the solar noon altitude equals h_{mid}, then the moment of the true noon is the time of mid-season. Example by Bīrūnī: from the latitude $\varphi = 42;17°$ for Jurjāniyya and $\varepsilon = 23;35°$, $\delta_{mid} = 16;26°$; for mid-summer (and also mid-spring): $h_{mid} = 64;9°$; Bīrūnī reports that he measured the Sun's noon altitude on August 2, 1016, and obtained $h = 64;9° = h_{mid}$. Thus, the mid-summer of the year of 1016 in Jurjāniyya occurred on the true noon of August 2. Modern values for the noon: $h = 64;7°$ and $\lambda = 135;1°$; mid-season, $\lambda = 135°$, occurred at 11:35, that is, $\sim +29$ minutes before true noon at 12:4 MLT.

If the solar noon altitude does not equal h_{mid}, a simple linear interpolation is applied to determine the time of mid-season. Note that since we know the rate of variation in the Sun's declination (and thus in its noon altitude) between two days around a mid-season, only the measurement of the solar altitude in a day around mid-season is needed in order to compute the time of mid-season. The daily variation in the solar declination (noon altitude) around the mid-seasons is $|\Delta\delta_{mid}| = |\Delta h_{mid}| = 0;18°$, and the measured noon altitude is h, so that $|h - h_{mid}| < 0;18°$; thus, the mid-season occurs $|(h - h_{mid})/\Delta\delta_{mid}|$-th of a day prior or posterior to the true noon of the day of measurement. Example by Bīrūnī: in order to determine the time of mid-autumn of the year 1016 in Jurjāniyya, on November 1, 1016, Bīrūnī found the Sun's noon altitude to be more than $31;18°$, by $0;0,20°$[23] (this gives the impression that the instrument used was of the calibration of 1/3'). $h_{mid} = (90° - 42;17°) - 16;26° = 31;17°$. Thus, the mid-autumn occurred $(31;18,20° - 31;17°)/0;18° = 0;4,27°$ day (Bīrūnī rounded it as $0;4,30$), or around 1;48 hours after the true noon. Modern values for the noon: $h \approx 31;19,40°$ and $\lambda = 224;55°$; the mid-autumn occurs at 13:47 MLT, or 2;2 hours after the true noon at 11:45 MLT, that is, an error of ~ -14 minutes.

The mid-signs method passed into the Western branch of the medieval Islamic astronomy (Spain and north of Africa) and then into the European literature. A possible way of the transmission appears to have been Ibn al-Zarqālluh's works: in a Latin text of the late thirteenth century, Bernardus de Virduno's *Tractatus super totam astrologiam*, we are told that "he made four observations while the Sun was

---

23 The printed text is not clear here (al-Bīrūnī 1954–1956, vol. 2, p. 619, lines 5–6) and may read "$31;18° + 1^1/_3'$," as Said and Stephenson (1995, p. 126, n. 6) did. Nevertheless, it should be considered that the altitude $31;19,40°$ results in the time of mid-autumn being $0;8,53$ hours after the true noon of November 1, 1016, which by no means agrees with the value $0;4,30$ hours explicitly mentioned in the text. As a result, I prefer to assume that, in spite of the confusing phrase in line 6 (more likely due to the clear fact that we do not have a trustworthy edition of al-Bīrūnī's work at our disposal), the value al-Bīrūnī wanted to mention is $31;18° + 1/_3'$.

in the middle of the quadrants bounded by the solstitial and equinoctial points." In it, the values $e = 1;58$, $\lambda_{ap} = 85;49°$, and the date of observations as 193 years after Battānī (see following, Table 1.1, no. 3), that is, 1075 CE, are associated with him.[24] With respect to difficulties that his medieval predecessors confronted in order to determine the time of the solstices, Copernicus used a mixed version of the two methods of seasons and mid-signs, as already described by Regiomontanus in his *Epitome of the Almagest* III.14[25]: in 1515, he measured the interval from the autumnal equinox to mid-autumn (45;16 days), and then to the consecutive vernal equinox (autumn + winter = 178;53,30 days), and finally determined $e = 1;56$ and $\lambda_{ap} = 96;40°$.[26] Later, around 1584–1588, Tycho Brahe applied the same method to intervals from the vernal equinox to mid-spring and to mid-summer and obtained $q_{max} = 2;3,15°$, which corresponds to $e = 2;9$, and $\lambda_{ap} = 95;30°$.[27]

1.2.3 The Three-Point Method

According to the surviving account by al-Maghribī,[28] after determining *TY* (and thus ω), the Sun is observed at three different times other than those near equinoxes and solstices during one tropical year. The meridian (noon) altitude of the Sun, and thus its declination, is measured. Then, the solar longitude in the three moments is computed by

$$\lambda = \sin^{-1}(\sin \delta / \sin \varepsilon)$$

The differences in the Sun's true longitude between each two successive observations are thus known (the angles between the trio positions, I, II, and III, at point *O*). The differences in its mean longitude between every two consecutive observations can also be calculated by multiplying the mean velocity ω with the corresponding time intervals (the angles between the trio positions, I, II, and III, at point *T*). Then, the eccentricity and the orientation of the apsidal line can be simply determined with the aid of plane trigonometry (Figure 1.5).

The seasons and mid-seasons methods are special cases of the three-point method. The basic formula in the first two methods, (1.3), is sensitive to the input data; similarly, the necessary trigonometric steps in the three-point method are also somewhat sensitive to the input data and, thus, are not guaranteed to produce

24 Toomer 1987, p. 516. In other sources, the varied values between $q_{max} = 1;52°$ and $1;53°$, corresponding to $\frac{1}{2}e = 0.01629 - 101643$ ($R = 1$), are associated with Ibn al-Zarqālluh; see Samsó 1987, p. 471. The value $\lambda_{ap} = 85;49°$ is in agreement with the later Islamic sources, for example, Ibn al-Hā'im; see Calvo 1998, p. 55.
25 See Swerdlow 2010, p. 155.
26 Copernicus 1543, English translation: Dobrzycki and Rosen 1978, pp. 157–159.
27 Dreyer 1890, p. 333; Thoren 1990, p. 224.
28 See Saliba 1985a.

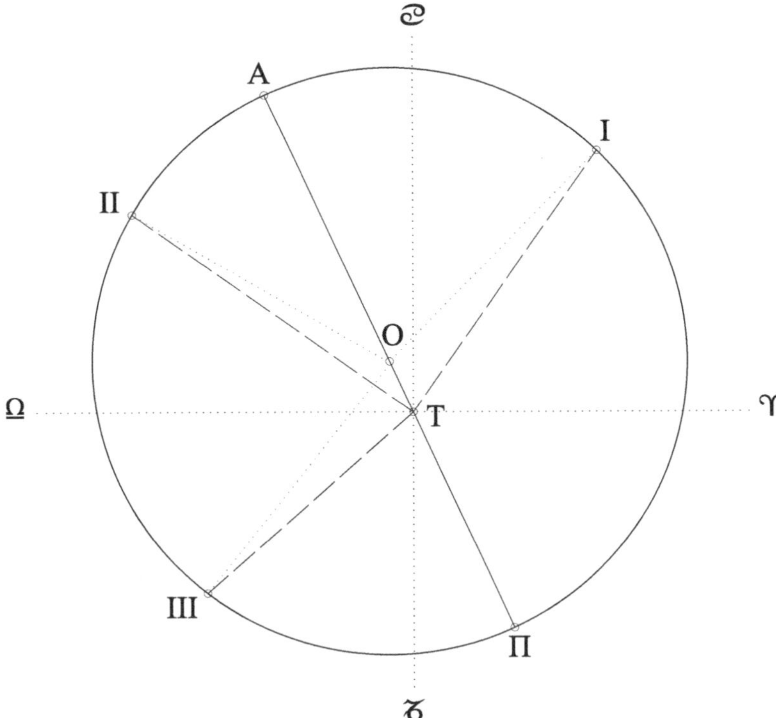

Figure 1.5 The measurement of the parameters of the eccentric solar model based on the three-point method.

results more exact than the other two methods.[29] Nevertheless, considering the observational essentials of each method, it becomes evident that in order to determine the length of seasons, we need, at least, to determine the times of two equinoxes as well as to cope with the measurement of the instant of a solstice; for this, one must observe the solar meridian altitude/shadow during a good number of days and do some interpolations. In order to determine two pairs of the time intervals between two successive mid-seasons, we need to measure, at least, three noon altitudes and carry out, at most, three interpolations. In the general three-point method, we have to measure only three noon altitudes without needing to do any interpolation. The most important feature of the method is that all the observations are well-timed (from noon to noon), and it is only required to take into account the correction due to the equation of time. No special event needs to be observed, and thus, no difficulty is expected to arise, and also the mean motion of the Sun

29 Ptolemy pointed to the sensitivity of the method for the input data in *Almagest* IV.11, where he applies it to determine the radius of the Moon's epicycle (Toomer 1984, pp. 215–216).

between any two observations can be easily computed. Thus, the input data needed may be obtained more conveniently and accurately. Therefore, it seems that the three-point method has a more secure observational basis which may appreciably increase the accuracy of the results in comparison with the other two methods.[30]

Astronomers of the Islamic period generally did not explain their observations and the input data that led to the specific parameters underlying the tables in their *zījes*.[31] Nonetheless, it is somewhat probable that in those *zījes* of the late Islamic period that were based on direct observations, the three-point method was most likely used for determining the desired values for the structural parameters of the solar orbit. Because the accuracy of the method had been known at least since the latter part of the tenth century, it seemingly had become a "standard" method which was explained in *zījes* and astronomical treatises as the method for determining the longitude of the Sun from its declination in order to rectify and correct its mean velocity and eccentricity.[32]

1.3 Classification of the Historical Values for the Solar Orbital Elements

The most important historical sources used in the present chapter are Bīrūnī's *al-Qānūn al-mas'ūdī*, al-Khāzinī's *Wajīz* (= Compendium of) *al-Zīj al-mu'tabar al-sanjarī*, Anonymous *Sulṭānī zīj*, and Kamālī's *Ashrafī zīj*.[33,34] Bīrūnī's *al-Qānūn*,

30 For another method invented by Ibrāhīm b. Sinān Thābit b. Qurra (909–946 CE), see Chapter 6.

31 al-Bīrūnī (*al-Qānūn* 1954–1956, vol. 3, p. 1193) declares his discontent with his Islamic predecessors' attitude of not explaining their observational data and measurement procedures as Ptolemy mentioned his procedures, and adds that the one who elucidates his computational procedures is the worthiest to be followed.

32 For example, al-Maghribī, *Talkhīs* IV.5: ff. 58v–61v.

33 Khāzinī, *Wajīz*, I: f. 1v, *Sulṭānī zīj*, f. 77r; Kamālī, F: f. 49r; Tables in Kamālī, F: ff. 236r–v, 238r–239r, 240r, 241v, and 243r. The latter two works were written contemporarily about the turn of the fourteenth century, the first in Yazd and the latter in Shīrāz (central Persia). The first is different from both Ulugh Beg's *Sulṭānī zīj* (called also *Gūrkānī zīj*) and Shams al-Dīn Muḥammad al-Wābkanawī's *Zīj al-muḥaqqaq al-sulṭānī* (completed about 1316–1324; see Chapter 3). It is attributed to Quṭb al-Dīn al-Shīrāzī (Kennedy 1956, no. 25); however, in the canons of this *zīj* (e.g., ff. 78r, 141r, 142r), the author mentions that he was based in Yazd, and thus, since al-Shīrāzī is not known to have worked in Yazd, he is less likely to be identified as its author. Based on Kennedy (*ibid.*, no. 32), the *Sulṭānī zīj* may be identical with the *Shāhī zīj* by a thirteenth-century astronomer named Ḥusām al-Dīn al-Sālar (killed by Hülegü, the first patron of the Ilkhanid dynasty of Iran, on 8 Muḥarram 661 H/November 22, 1262; Rashīd al-Dīn, vol. 2, p. 1045). However, there are direct references in the first to the latter, as well as some tables of the latter in the first (e.g., the table for the equation of time on f. 7v). Therefore, the two cannot be identical.

34 The values for the maximum solar equation of center used in the pre- and early Islamic periods are also mentioned in the last three sources; in the prologue of al-Khāzinī, *Wajīz*, I: f. 1v: "Ancient observations": 2;24°; Ptolemy: 2;23°; Persians: 2;14°; and Indians: 2;11°. The value of 2;23° in the *Almagest* is drawn from $e = 2;30$ calculated based on the Hipparchan values for the length of seasons. The solar eccentric model more likely existed before Hipparchus, and it is probable that Apollonius knew this model (Pedersen 1974, pp. 135, 340f). The term "ancient observations" is thus interesting; the value associated with it can be found in the papyrus PSI inv. 515 (Jones 2000).

VI.7,[35] provides invaluable comparative studies on the values measured by medieval astronomers in the period 800–1000 CE for the length of the tropical year, the length of the seasons, the solar eccentricity, and the longitude of its apogee. Therefore, our discussion about the awareness of medieval astronomers as to the limitations of the methods (sensitivity to input data and difficulties in obtaining the required correct observational data) concentrates on the analysis of the contents of this source.

Some values for the solar orbital elements based on the seasons method measured by the Middle Eastern astronomers between 800 and 1200 CE are summarized in Table 1.1. Bīrūnī presents the results of the measurements of the solar orbital elements based on the mid-seasons method performed by his Islamic predecessors, as summarized in Table 1.2. As can be clearly seen, the application of the method may be traced back at the very least to the observational program conducted in the reign of al-Ma'mūn (early ninth century). (The Yazdigird era, June 16, 632, is used with the Egyptian/Persian year consisting of 12 months of 30 days plus 5 epagomenal days which, in the early Islamic period, were put after the eighth month. In the late Islamic period, they were transferred to the end of the year.) The modern values for the lengths of spring and summer in the periods are indicated in bold type.

It is interesting that the values recomputed by Bīrūnī based on the input observational data obtained by the early Islamic astronomers are different from the values reported from, or ascribed to, those astronomers. Nonetheless, in our comparative study that follows (Section 1.4), the formal values are used in order to avoid making any possible anachronistic conclusion.

Note that each difference of 0;0,1 in e when $R = a = 60$ corresponds to a difference of $\sim 2.3 \times 10^{-6}$ in e while $R = a = 1$. Accordingly, the deviations up to 4 seconds in e ($R = a = 60$) may be of much less effect when we wish to express the corresponding value for it by the factor of 10^{-5} when $R = a = 1$ (Section 1.4).

The values adopted or reported for the solar eccentricity in the Islamic *zījes* can be deduced from the tabulated values of q_{max}, according to formula (1.1). See Table 1.3.

Apparatus to Table 1.1:

1. *On the solar year (Fī sanat al-shams/De anno solis*; Carmody 1960, Neugebauer 1962a, Morelon 1987) is attributed both to the Banū Mūsā (the eldest

---

According to Jones (2000, p. 85), the treatise of which this papyrus is a fragment was composed before the publication of the *Almagest*. The value of 2;24° is not significantly different from Ptolemy's, and as Jones argues, it might have been computed from the Hipparchian eccentricity 2;30 with the use of a table of chords tabulated for fewer angles than Ptolemy's. It is, nonetheless, interesting that it was transmitted to medieval astronomy as a value that predates Ptolemy's. The value attributed to the Iranians is that used in the *Shāh zīj* (written in Sāsānīd Persia, 224–651 CE). 2;11° is indeed found in the Indian traditions from the year 400 onward (Pingree 1970, pp. 97–98, 100, 105).

35 al-Bīrūnī 1954–1956, vol. 2, pp. 650–661.

Table 1.1 The solar orbital elements measured on the basis of the seasons method

Astronomers	Year	Spring	Summer	Formal values		Recomputed by Bīrūnī		Recomputed	
				e	λ_{ap}	e	λ_{ap}	e	λ_{ap}
1 Mentioned in *On the solar*	199–201 y	93;40^d	93;2,30^d	——	——	2; 7,40,49	81;38,22,28°	2; 7,53	81;17,46°
year	830–832 CE	**93;32,50**	**93;2,40**			2;10,14,19	81;23,10,10	2; 8, 2	81;18,24
2 Khālid al-Marwarūdhī	212	93;54,35	93;9,20	——	——	2;19,17, 6	82; 9,55	2;19,29	80;21,43
ʿAlī b. ʿĪsā of Ḥarrān	843–844	**93;32,24**	**93;2,47**						
Sanad b. ʿAlī									
3 Battānī	251	93;35	93;1,52	2;4,45	82;14°	2; 4,29,19	82; 7,38,23	2; 4,50	82; 8
	882–883	**93;30,55**	**93;4,18**						
4 Sulaymān b. ʿIşmat of	257	93;27, 3,45	93;2,25,25	——	——	2; 0,28,15	83;51, 1, 1	2; 0,30	83;56,34 ≈ 83;57
Samarqand	888–889	**93;30,34,12**	**93;4,23,31**						
5 Abu ʾl-Wafāʾ of Būzjān	343	93;30, 8	93;7,10	——	——	2; 4,10,49	84;34,45,50	2; 4,26	84;32, 2 ≈ 84;32
	974–975	**93;27,46**	**93;7,33**						
6 Abu al-Rayḥān al-Bīrūnī	385	93;28	93;8	2;3,36,34	85;13, 5,24	——	——	2; 3,37	85;12,38 ≈ 85;13
	1016–1017	**93;26**	**93;9**						
7 Muḥamamd b. Ṭabarī	447	93^d 12^h	93^d 3^h 13^m	2;4,45	85;15	——	——	2; 4,47	84;47
	1078–1079	**93^d 9^h 24^m**	**93^d 4^h 28^m**						
8 Ibn al-Fahhād	539	93^d 9^h 33^m	93^d 5^h 51^m	2;4,35	87;48,54	——	——	2; 4,35	87;48,14
	1170–1171	**93^d 7^h 55^m**	**93^d 5^h 57^m**						

brother, Abū Jaʿfar Muhammad, died in January–February 873; Bīrūnī 1954–1956, vol. 2, p. 654) and to Thābit b. Qurra (836–901 CE; al-Nadīm 1973, p. 331; Ibn Yūnus, pp. 103, 106, 107; see also Goldstein 1994, p. 190, Goldstein and Chabás 2001, p. 345). The season lengths are on the basis of the equinoxes of 831 and the vernal equinox and the summer solstice of 832. The first set of values is calculated from the solar year of 365.25 days, while in the second, the length of the year is taken as equal to 365.25 − 1/100 days.

2. Bīrūnī 1954–1956, vol. 2, p. 653. Bīrūnī quotes the result of this measurement by al-Maʾmūn's astronomers from (the now lost) *Exposition of the* Almagest (*Tafsīr al-majisṭī*) of Abū Jaʿfar al-Khāzin (d. *ca.* 970–980). For this measurement, Bīrūnī says, the Ptolemaic solar year length of 365;14,48d was used.

3. Nallino 1899–1907, vol. 1, pp. 40–49, commentary on pp. 204–223, vol. 2, pp. 19–23, 72–83, vol. 3, pp. 61–75. His value for e is repeatedly referred to in Islamic astronomical corpus (e.g., al-Ṭūsī, *Mu ʿīniyya*, p. 42), and also in the Western literature (e.g., Copernicus's *De revolutionibus*: Dobrzycki and Rosen 1978, pp. 158).

4. Bīrūnī 1954–1956, vol. 2, pp. 654, 659.

5. Bīrūnī 1954–1956, vol. 2, pp. 654–655, Kennedy 1956, nos. 29 and 73, Mozaffari 2023a.

6. In the same year, 1016, Bīrūnī re-determined the solar eccentricity by means of the mid-signs method (cf. Table 1.2, no. 5).

7. Kennedy 1965.

8. Besides the anonymous *Sulṭānī zīj*, MS Iran, Parliament, no. 184 contains (on ff. 243v–244r) part of another anonymous treatise named *On the proof of the solar equation (of center), its (longitude of) apogee, its mean motion, and its eccentricity by the observations*. In it, the author gives the result of his observations performed in 539 y (1170–1171 CE) as follows: a rounded value for *TY* as 365^d 5^h 48^m = 365;14,30^d (which is identical to the value observed by al-Maghribī at the Marāgha observatory around one century later; see following, Table 1.3, no. 9), and another value for it as 365^d 5^h 46m 52^s 27$'''$; the time interval from the vernal equinox to the autumnal equinox as 186^d 15^h 24^m (near to the value given by Ṭabarī; true: 186^d 13^h 52^m); and the length of spring as mentioned in the table (given in the text as 93^d 9^h 13^m in Abjad numerals and 93^d 9^h 39^m in written-out numbers, both of which are scribal errors, as shown by the value given for the corresponding angle a in what follows). From them, he computes the angles $a + b$ = 183;57,50° (+1$''$ in error) and a = 92;3,27°, and computes e = 2;4,35, which is, interestingly, the same value used in the *Mumtaḥan zīj*, and λ_{ap} = 87;48,54° (recomputed: . . .,14; at first sight, it seems to be a simple scribal error; however, this is not the case, because the value given by our author agrees with the magnitude he gives later for the daily motion of the solar apogee). Our author compares his measured value with λ_{ap} = 82;39° adopted in the *Mumtaḥan zīj* and, giving the time interval between the two as 124,511 days, computes the daily motion of the apogee as 0;0,0,8,57,36,. . .° (≈ 1°/66y). Our author appears

to achieve success, as the title of the treatise indicates, in proving that the values obtained during al-Ma'mūn's observational program are correct. As shown elsewhere (see Section 8.5.2.1), it is almost certain that the author is Ibn al-Fahhād, and the piece of writing is a lost part of his *'Alā'ī zīj*. Ibn al-Fahhād, pp. 70–76; Kamālī, F: f. 240r. See Kennedy 1956, no. 84, Pingree 1985, van Dalen 2004a, Mozaffari 2019, 2023b.

Notes to Table 1.2:

1. Bīrūnī 1954–1956, vol. 2, pp. 656–657. The surprising results of the first observational program in al-Ma'mūn's day, simply because λap is less than the ancient determinations, for example, the Hipparchan and Ptolemaic value (65;30°) and the Indian values. The source of error, as Bīrūnī notes, was in the solar observation in the middle of the summer.

2. Neugebauer 1962a, p. 275, Morelon 1987, vol. 3, p. 49, Bīrūnī 1954–1956, vol. 2, p. 658. **Remark:* It is curious that none of the four sets of the solar observations from the early Islamic period (Tables 1.1 and 1.2: nos. 1 and 2) favorably result in the values for the solar orbital elements adopted in the available manuscripts of the *Mumtaḥan zīj* (van Dalen 2004b) or associated with the *Mumtaḥan* observations in Baghdad ($e = 2;4,35$, $\lambda_{ap} = 82;39°$) or in Damascus ($e = 2;5,32$, $\lambda_{ap} \approx 80°$). Yaḥyā, E: ff. 14v–15r, L: ff. 67r–68r; Ḥabash, I: f. 90v, B: f. 30v; Ibn Yūnus, *Zīj*, L: p. 121, 124; Caussin 1804, p. 221; Kamālī, F: f. 236r.

3–4. The values Bīrūnī quotes from Abu 'l-Wafā' and al-Ṣaghānī are mentioned nowhere else.

5. Bīrūnī mentions that he repeated three times the measurement of the time interval of the Northern quadrant and that of the Western quadrant in Jurjāniyya and obtained two sets of values (see the two first lines in row 5). He attributes the great deviation in these values to a defect in the instrument used as well as to his inability to determine a single amount. He then mentions that since it was impossible for him to repeat his measurements in Ghazni, he decided to determine the eccentricity using the results of the mid-signs method found in Kh^wārizm. Bīrūnī only reports in detail the measurement of the solar meridian altitude on April 30, 1016, for determining the moment of mid-spring.[36] He adopted $e = 2;4,39$ (the average of his measured values by the seasons method and mid-seasons method) in order to construct his table of the solar equation of center.[37]

Notes to Table 1.3:

1. From the pre-Islamic Persian astronomy. Neugebauer 1962b, p. 20.
2. Ibn Yūnus, p. 104.

---

36 See Said and Stephenson 1995, p. 129.

37 The table is asymmetric and has max = 3;59,3,21° for mean anomaly 266° (al-Bīrūnī 1954–1956, vol. 2, p. 716) and min = 0;0,56,39° for mean anomaly 90° (al-Bīrūnī 1954–1956, vol. 2, p. 710), and so $q_{max} = 1;59,3,21°$, which strictly corresponds to $e = 2;4,39$.

Table 1.2 The solar orbital elements measured on the basis of the mid-sign method

Astronomers	Year	East Q.	North Q.	West Q.	South Q.	Formal values		Recomputed by Bīrūnī		Recomputed	
						e	λ_{ap}	e	λ_{ap}	e	λ_{ap}
1 *Shammāsiya* Observations	199 Y 830–831 CE			94;48,20^d **90;55,50**	88;35,50^d **88;32, 2**	—	—	2;14,28,21	61;23,22,40°	—	—
2 Mentioned in *On the solar year*	200–201 832	91;45,20 **91;38,10**	94;11,15 **94; 8,23**			2;2, 6	82;45	2; 7,38,40	81; 1,50,32	2;7,28	81;12,37°
3 Abu 'l-Wafā' of Būzjān	345 976–977	91;34,25 **91;30,55**	94; 9, 7,30 **94; 8,52**			—	—	2; 5,11,17	85; 0,15,32	2;4,58	84;42,34
4 Abū Ḥāmid al-Ṣaghānī	355 986–987	91;46,40 **91;30,10**	94;10 **94; 8**			—	—	2; 6,33,17	81; 2,29,45	2;6,44	80;42,32
5 Abū al-Rayḥān al-Bīrūnī	385 1016–1017		93;16,40 93;33 94; 8,30 **94; 8,25**	91; 3,16 91;10,40 91; 4,30 **91; 5,30**		2;4,43,25	84;59,11,9	—	—	2;4,24	85;14,52

Table 1.3 The solar orbital elements in some of the Islamic *zījes*

		q_{max}	e	λ_{ap}	Time
1	Khwārizmī	2;14°	2;20	77;55°	Fixed
2	Abū al-Qāsim Aḥmad b. Shākir	2; 0, 8	2; 5,47	84;33	Beginning of 220 Y = April 23, 851
3	al-Nayrīzī	2;13	2;19	82;39	Fixed
4	Ibn Amājūr	2; 0,20	2; 5,59	84; 5	[880–920?]
5	Ibn al-Aʿlam	2; 0,10	2; 5,49	84;25	365 H/345 Y = 975–976 CE
6	Ibn Yūnus	2; 0,30	2; 6,10	86;10	Beginning of 372 Y = March 16, 1003
7	al-Khāzinī	2;12,23	2;18,36	85;53	January 1, 1120
8	Muḥyī al-Dīn al-Maghribī	2; 0,21	2; 5,59	88;50,21	26–2–634 Y = March 5, 1265
9	*Īlkhānī zīj*	= Ibn Yūnus		88;24,51	Beginning of 601 Y = January 18, 1232
10	Ibn al-Shāṭir	2; 2, 6	2; 7,50	90;10,10	Beginning of 750 H = March 21, 1349
11	al-Kāshī	= Ibn Yūnus		From the *Īlkhānī zīj*	Beginning of 781 Y = December 4, 1411
12	Ulugh Beg	1;55,53,12	2; 1,20	92;25,58	Beginning of 841 H = July 4, 1437

3. Kamālī, F: f. 49r; Khāzinī, *Wajīz*, I: f. 1v. Two *zījes* presumably written by al-Nayrīzī have not been preserved (Kennedy 1956, nos. 46 and 75). Bīrūnī reports some confusion in al-Nayrīzī's works concerning the motion of the solar apogee. His value for e belongs to the pre-Islamic Persian astronomy, and for λ_{ap} is adopted from the *Mumtaḥan zīj*.

4. According to Ibn Yūnus (p. 105), in the *Badī'* *zīj* of Ibn Amajūr (Kennedy 1956, no. 8), q_{max} = 2;0,50° (e = 2;6,31), and the solar mean motion in a Persian year = 359;45,39,45° (corresponding to $TY \approx 365;14,32,47^d$), while Mufliḥ b. Yūsuf (an assistant to Ibn Amājūr) ascribes $q$max = 2;0,20° and λap = 84;5° to Ibn Amajūr. The year of measurement is not indicated. The value of eccentricity is equal to al-Maghribī's one (following, no. 8).

5. Ibn al-A'lam's *Zīj al-'Aḍudī* is now lost. His solar parameter values are quoted by Shīrāzī (*Ikhtiyārāt*, f. 50v; *Tuḥfa*, f. 38v) and Kamālī (F: ff. 232v, 236v, G: ff. 249r, 251v).

6.,9. Ibn Yūnus, p. 174; al-Ṭūsī, *Īlkhānī zīj*, C: pp. 60–65, T: ff. 28r–30v, P: ff. 21v–23r.

7. al-Khāzinī, *Zīj*, V: ff. 129r, 131v, 163v; L: ff. 102v, 113v, 125v; *Wajīz*, S: pp. 53–54, 58–59; Kamālī, F: f. 238r, G: f. 250v.

8. According to his *Talkhīṣ* IV.5: ff. 58v–61v (Saliba 1985a), Muḥyī al-Dīn measured the solar orbital elements on the basis of the direct observations made at the Marāgha observatory in the 1260s. For his observations, see Chapter 9.

10. Ibn al-Shāṭir, K: ff. 52r–v, O: ff. 31r–v, L1: ff. 50v, 52r, L2: ff. 65r–66r. Roberts 1957, Saliba 1987.

11. Kāshī, IO: ff. 127v, 128v, 131r, P: pp. 107, 109, 114.

12. Ulugh Beg, *Sulṭānī Zīj*, P1: ff. 116r, 117v–123v, P2: ff. 133r, 134v–141r.

1.4 Discussions and Results

The relatively accurate observational data obtained by the early Islamic astronomers resulted in the good values for the length of the seasons and for the time intervals between successive mid-seasons that were achieved (Tables 1.1 and 1.2) and, consequently, as we shall see presently, the reasonable values for the solar eccentricity that were obtained in the Middle East during the medieval period. Since the accuracy of the observational data has already been discussed by Said and Stephenson (1995), we turn our main attention to making comparisons between the methods and the corresponding recorded historical values.[38] In order to reach the main goals defined at the beginning of this chapter, it should be mentioned that, concerning the varying historical values for the solar orbital elements, we are confronted with a threefold problem consisting of the following factors:

---

38 In other available studies (e.g., Petersen and Schmidt 1968, Neugebauer 1975, vol. 3, pp. 1097–1102, Maeyama 1998, Hughes 1989), the solar parameter values obtained by Hipparchus, Ptolemy, Copernicus, and Tycho Brahe, together with Battānī, were scrutinized. Moesgaard (1974) also analyzed the solar theory in *On the solar year*.

1. The sensitivity of methods and observational data
2. The methods and theoretical errors due to the mismatch between the two models, eccentric and elliptical
3. The secular decrease in the eccentricity of the Earth

(1) As explained in Section 1.2, the three methods are all sensitive in varying degrees to the input data, but the essential observational data for them are substantially different. This makes, hypothetically, the three-point method more grounded and the seasons method less trustworthy.

It can be seen from Tables 1.1 and 1.2 that the immense observational activities in the Middle East between 800 and 1000 CE brought clearly into focus the sensitivity of the seasons and mid-sign methods to the observational data, by leading to different values for the solar eccentricity in a relatively short period (as remarked by Bīrūnī himself; see later text). The sensitivity in the case of the seasons method led, for instance, to an eleventh-century astronomer's obtaining Battānī's value for e (Table 1.1, no. 7), and to a twelfth-century astronomer's producing the *Mumtaḥan* value (Table 1.1, no. 8).

Basically, the possible sources of the observational errors in measuring the time interval between an equinox and its subsequent solstice or between two successive mid-seasons may be due to (a) the skill of the observers, (b) the probable deficiencies and systematic errors in the instruments applied, and (c) the inevitable difficulties and possible errors in exact determinations of the instants of the equinoxes, mid-seasons, and solstices.

It is difficult to distinguish between the errors caused by factors (a) and (b). Said and Stephenson analyzed the measurements of the solar meridian altitude related to equinoxes, solstices, and mid-seasons by a group of early Islamic astronomers (al-Marwarūdhī, the Banū Mūsā, Ibn ʿIṣmat, al-Ṣūfī, Abū Maḥmūd al-Khujandī, and Bīrūnī in both Ghazni and Jurjāniyya) and confirmed that the results achieved are highly accurate. The smallest error was observed in the case of al-Marwarūdhī ($\pm 0.01°$), while none of the errors exceeded $\pm 0.02°$, which falls within the maximum resolution power of the unaided eye (around 1 arc minute).[39] Besides high observational accuracy, as Maeyama notices,[40] two other factors were effective in achieving these results: the neglect of the Ptolemaic solar parallax (except in the case of Ibn ʿIṣmat),[41] and the fact that the observations were made in localities (in the Middle East) with a low geographical latitude, which caused the meridian solar observations not to be affected too much by refraction.[42]

In view of this remarkable accuracy, the problem of distinguishing factors (a) and (b) appears to be of less significance. However, the recorded observational

---

39 Said and Stephenson 1995, pp. 125, 130.
40 Maeyama 1998, p. 39.
41 Said and Stephenson 1995, p. 131, n. 4. A reference to Ibn al-Shāṭir's correction of the meridian solar altitudes due to the parallax in a Latin source is found in Hartner 1977, p. 5.
42 Said and Stephenson 1995, pp. 120–121.

data may inspire the idea that the instruments applied involved mistakes and deficiencies; for example, Said and Stephenson found that Bīrūnī's observation in Ghazni was exposed to a systematic error of around −0.04°. These errors may be due to a slightly different location of the observational site with respect to the modern coordinates used by Said and Stephenson for Ghazni (a short distance north of its present location), or to a structural deficiency in the instrument used. (This systematic error does not exist in Bīrūnī's observations in Jurjāniyya, for which the mean error is less than ±0.02°.)[43]

In the case of factor (c), we know that the mean error of the early Islamic astronomers in determining the instant of the equinoxes was around 1.2 hours as an immediate result of the aforementioned mean error of 0.02° in measuring the Sun's meridian altitude. At the solstices, the error is *ipso facto* much larger because the exact determination of the time of the solstices by observing the solar meridian altitude is particularly hard to achieve; as already mentioned, the rate of difference in solar meridian altitude between the day of the solstice and a day prior or posterior to it is about 0.004°, while it is 0.4° at the equinoxes. This is therefore the main source of error in determining the time of solstices. Accordingly, a minor error in measuring the Sun's meridian altitude on the day of a solstice usually causes a considerable difference (from half an hour up to a day) in fixing the instant of the solstice and thus in the length of the seasons. Since the seasons method is sensitive to the input data, a considerable deviation from the correct value for the eccentricity may therefore occur. This explains why, in various observations, even if made within a year or in a period of a few years, quite different values may be obtained for the solar eccentricity and the longitude of its apogee. The medieval astronomers paid some attention to factors (a) and (b) by commenting on their predecessors' abilities in making sound observations (e.g., Bīrūnī's opinion about the observations of Ibn 'Iṣmat of Samarqand: Table 1.1, no. 4) and the accuracy of the instruments used (e.g., Bīrūnī's awareness of his instrument's defectiveness: Table 1.2, no. 5). But they correctly understood that factor (c) is usually the main source of error in determining the time between an equinox and a solstice, and therefore in determining the parameters of the Ptolemaic solar model by the seasons method.[44] This was the main reason that the mid-signs and three-point methods were developed.

Nevertheless, it seems that factor (c) also plays a role with the mid-seasons method: Said and Stephenson found the mean error in the mid-seasons time measurements in the early Islamic period to be about 2.1 hours (i.e., about 1 hour more than the error in the time of the equinoxes), which is probably due to the lower hourly rate for the change of the solar altitude at mid-seasons, that is, 45″ (1′ at equinoxes).[45] Table 1.2 shows that even the use of the mid-seasons method did not make it possible to obtain a single reliable value for the solar parameters, as Bīrūnī

---

43 Said and Stephenson 1995, pp. 123 (table) and 125.
44 For example, al-Bīrūnī 1954–1956, vol. 2, p. 657.
45 Said and Stephenson 1995, pp. 125, 127.

and probably other Islamic astronomers will have expected. Before expounding his own measurements, Bīrūnī finds it necessary to explain an important note to his audience:

> A thinker must not become agitated because he does not understand the nature of entities. Instead, he must keep in mind that even if the eccentricity is variable, or the [solar] apogee is subject to trepidation, it is impossible that they vary [so noticeably] within a year.[46]

In order to demonstrate that these changes are due to limitations in the methods, he reports different amounts of the lengths of seasons measured by Ibn 'Iṣmat and the time intervals between the mid-seasons by Ṣaghānī and then proceeds to calculate the solar eccentricity based on each set of data. The solar eccentricity values obtained from the data produced by a single person within a year were so different that they confirmed conspicuously the limitation in both methods.

Since there is no potential source of major errors in the three-point method, the results of observations made by different astronomers are expected to be of a considerable congruence. As explained earlier, the early Islamic astronomers were able to measure the meridian solar altitude on average with an accuracy of about 1′. The observatories in Marāgha and Samarqand in the late Islamic periods had central quadrants with an accuracy (the steps on their scales) of 0.5′ and 5″, respectively. However, we cannot be certain that this accuracy was achieved by the late Islamic astronomers; for example, as we will see in Chapter 9 (Table 9.1 on p. 221), a systematic error of +4′ occurred in the solar observations of Muḥyī al-Dīn. Also, accurate altitude measurements may not lead to similarly accurate computations of the declination, because the latitude of the locality needs to be known precisely (e.g., Muḥyī al-Dīn's value of 37;20,30° for the latitude of the Marāgha observatory is in error by about −3′). Moreover, the value widely adopted for the obliquity of the ecliptic in the late Islamic period (23;30°) had a mean error of around −2 arc minutes compared to the true amounts (23;34–23;30°). Since we do not possess observational data related to the three-point method from the Islamic period and the values for e associated with this method, except for Muḥyī al-Dīn's, we cannot proceed further to make evidential comparisons between it and the other two.

(2) The sensitivity of the methods is a prominent factor in achieving the diverse values. Nevertheless, there is still another effective factor in producing such results, which again is rooted in the methods but is due to the mismatch between the circular eccentric and elliptical Keplerian orbits.

In order to take this factor into account, we first mention the formula for finding the theoretical values expected to result from each method for the solar orbital elements under the assumption of a circular eccentric orbit for the Sun (e and λ_{ap}).

---

46 al-Bīrūnī 1954–1956, vol. 2, p. 659.

Then, we investigate the effect of the mismatch of the eccentric solar model with the elliptical Keplerian orbit on the deviations found in the historical values from the corresponding values of $2e'$ and λ'_{ap}. Maeyama[47] dealt in detail with the problem and obtained the following formulas in order to establish a relation between the orbital elements of the elliptical orbit and those of the eccentric one, which allow us to consider the effect of either of the three methods on the accuracy of the values of e and λ_{ap}:

$$\frac{1}{2}e = e'\left(1 - \frac{3}{4}e'\cos(\upsilon_{o})\right)\cos(2\upsilon_{o} + K))$$

$$\lambda_{ap} = \lambda'_{ap} + \frac{3\cos(2\upsilon_{o} + K)}{4\cos(\upsilon_{o} + K)}e'\left(\sin K - \cos\upsilon_{o}\sin(\upsilon_{o} + K)\right) \tag{1.4}$$

in which υ_{o} is the solar true anomaly in the first of the trio observations (i.e., that of the equinox in the seasons method and that of mid-winter or of mid-spring in the mid-signs method), and K the difference in the solar true longitude between the first two of the trio of observations (in the case of the seasons and mid-signs methods, $K = 90°$). For example, for the time of Muḥyī al-Dīn's first observation used for measuring the solar orbital elements (see Table 10.1, no. 5), the Earth's longitude of perigee was about $270;19,7°$; then, from the true modern value mentioned in Table 10.1, $\upsilon_{o} = 262;39,9°$ and $K = 47;46,9°$. Then, from formula (1.4), we have the theoretical values $\tfrac{1}{2}e = 0.01699$ (compared with $e' = 0.01701$) and $\lambda_{ap} = 89;42,40°$ (compared with $\lambda'_{ap} = 90;19,7°$). In the case of the seasons and mid-signs methods, the difference between the theoretical value for $\tfrac{1}{2}e$ and the true modern value for e' will be maximum when $\upsilon_{o} \approx 35.3°$. The difference between the theoretical value for λ_{ap} and the true modern value for λ'_{ap} will be maximum when $\upsilon_{o} \approx 54.7°$. Then, in the case of the seasons and mid-seasons methods, it is expected that for accurate observations, the historical values are bounded by $\sim e' \pm \tfrac{3}{5}e'^{2}$ and $\lambda'_{ap} \pm \frac{3 \cdot 180}{5\pi}e'$ (in degrees), which constitute our tolerance band.

In Table 1.4, the historical values for the solar geocentric eccentricity $\tfrac{1}{2}e$ based on Ptolemy's eccentric model with the orbit's radius R taken equal to 1 are compared with the theoretical values expected from each method with the eccentric orbit and with the modern eccentricity e' of the elliptical orbit for the given dates. Since for the values mentioned in Table 1.3 no method was specified in historical sources, except for Muḥyī al-Dīn, the theoretical value can be given only in his case. We do not know the exact dates of the solar observations performed by some astronomers of the period 1000–1500 CE (except in the case of Muḥyī al-Dīn), but good approximate dates can be given for their careers. Since the solar eccentricity changes by 10^{-5} in about 24 years, no problem is expected to arise if the preceding

47 Maeyama 1998, pp. 4–8.

Table 1.4 The historical values for half the solar eccentricity ½*e* in comparison with the theoretical values, taking into consideration the method used and with the corresponding true modern values computed based on the Keplerian elliptical orbit, *e'*

Method	Nos.	e (measured)	e (theoretical)	e' (modern)
Seasons	1	0.01777	0.01719	0.01718
Table 1.1	2	0.01937	0.01718	0.01718
	3	0.01733	0.01717	0.01716
	4	0.01674	0.01716	0.01716
	5	0.01728	0.01713	0.01713
	6	0.01717	0.01711	0.01711
	7	0.01733	0.01709	0.01709
	8	0.01730	0.01705	0.01705
Mid-signs	1	——		
Table 1.2	2	0.01696	0.01731	0.01718
	3	0.01736	0.01727	0.01713
	4	0.01760	0.01726	0.01712
	5	0.01732	0.01727	0.01711
Table 1.3	1	——		
	2	0.01747		0.01717
	3	——		
	4	0.01750		0.01716
	5	0.01747		0.01713
	6	0.01752		0.01711
	7	0.01925		0.01707
	8	0.01750	0.01699	0.01701
	9	——		
	10	0.01775		0.01698
	11	——		
	12	0.01685		0.01694

modern values for *e'* are expressed to a precision of $\pm 10^{-5}$, corresponding to half a century, enough to cover a lifetime's scientific career.

In Figure 1.6, the continuous thicker curve at the middle depicts the change in the eccentricity *e'* of the elliptical Keplerian orbit of the Earth as a function of Julian years. As is evident, it decreases with the passage of time until it reaches a minimum of 0.0023 about the year 29500.[48] The gray area in the top of and below the curve of *e'* represents the tolerance band. The dotted curve shows the theoretical values for half the eccentricity of the circular orbit (½*e*) produced from the seasons method; the dashed curve, the theoretical values for ½*e* from the mid-seasons method starting from mid-winter; and the dotted-dashed curve, the theoretical values for ½*e* from the mid-seasons method with mid-spring as the zero point. Drawing a single graph for the theoretical values for ½*e* from the general three-point method is impossible because the trio of observations may be selected

48 Meeus 2002, p. 201.

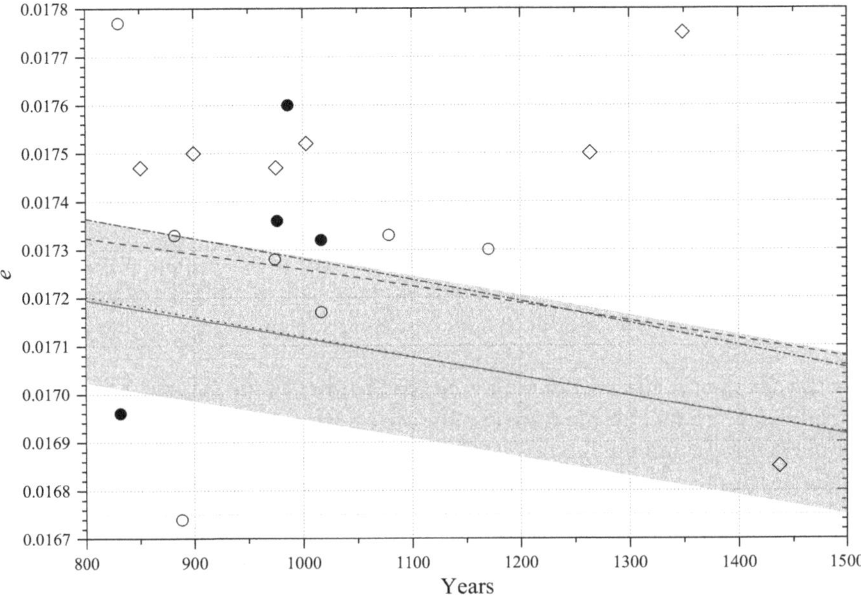

Figure 1.6 Continuous thicker curve: the eccentricity of the Keplerian elliptical orbit of the Earth. *Gray area*: the tolerance band. *Dotted curve*: the eccentricity of the eccentric solar model (e) computed theoretically based on the seasons method. *Dashed curve*: the eccentricity e computed theoretically based on the mid-sign method with the starting point of mid-winter. *Dotted-dashed curve*: the eccentricity e computed theoretically based on the mid-sign method with the zero point of mid-spring. ○: the historical values measured on the basis of the seasons method (excluding the outlier: Table 1.1, no. 2). ●: the historical values measured on the basis of the mid-sign method. ◇: the other historical values in Table 1.3 (except for the three outliers).

arbitrarily. The values shown by the hollow circles (○) are those computed by the seasons method (Table 1.1). The last four values mentioned in Table 1.2, which were obtained by the mid-signs method, are indicated as black circles (●). We have four values measured in the very early Islamic period (Table 1.1, nos. 1 and 2; Table 1.2, nos. 1/2; Table 1.3, no. 2). We give preference to the formal values rather than the computed ones. The hollow diamonds (◇) represent the values listed in Table 1.3. We have excluded the outliers of the *Mumtaḥan* observations of 843–844 (Table 1.1, no. 2), al-Khʷārizmī, al-Nayrīzī, and al-Khāzinī.

Figure 1.6 shows that the theoretical values of ½e computed from the seasons method cause no noticeable difference from the modern values of e′ in the historical period considered in this chapter. In this period, the graphs of the theoretical values for ½e from the mid-seasons method are located in the vicinity of the zones of their maximum distance from the graph of e′, which thus causes considerable differences between these theoretical values of ½e and the corresponding values of e'. (As seen in Figure 1.6, the graph of the theoretical values from the mid-seasons with the zero point of mid-spring began to descend about the beginning of this period, and the graph related to this method with the starting point of mid-winter ascended to reach its maximum difference from e'.) All the graphs of the theoretical values are located at the top of the graph of e'. This, perhaps, is the reason that nearly all the historical values are located above the graph of e'. Therefore, taking the factor (2) into account shows that not only was the sensitivity of the methods a main cause of achieving the varying magnitudes but also that even if the medieval astronomers had strictly accurate observational data, the application of the different methods would cause the different values for the solar orbital elements. They could, by no means, be aware of this second feature of the methods.

The following observations may be made from Table 1.4 and Figure 1.6. It should be noted that besides the possible observational difficulties, the sensitivity of the methods, and the theoretical deviations as the main factors influencing the accuracy of the historical values, there are other contributing factors, such as the observational abilities and instruments of a single astronomer, his computational power, his awareness of the sensitivity of the methods, and so on. The latter factors, which are very difficult (if not impossible) to quantify as well as to trace out, should make us more cautious when interpreting the numeral results gathered here. Accordingly, making comparison between the methods with regard to the accuracy of the historical values may lead to sound conclusions, if (a) a good number of the values strictly and definitely associated with a specific method, and/or (b) values measured by a single astronomer based on different methods, are available. These may possibly neutralize the latter contributing factors but are available solely in the case of the seasons and mid-seasons method.

(2a) The mean absolute error (i.e., without considering the sign of difference) of the values measured by means of the seasons method, excluding the outlier (Table 1.1, no. 2), as compared both with the corresponding theoretical values and with the true modern values for e', is $\sim 2.7 \times 10^{-4}$. The mean absolute error of the mid-seasons method with respect to the values of e' is $\sim 2.9 \times 10^{-4}$, that is, approximately equal to that of the seasons method, but, considering the theoretical values, significantly reduces the mean absolute error to $\sim 2.1 \times 10^{-4}$. It follows that, taking the values presented here, the mid-seasons method is more accurate than the seasons method. Regarding the values of e', the most accurate values are those of Bīrūnī (with an error of $+0.5 \times 10^{-4}$) and Ulugh Beg (-0.9×10^{-4}). A comparison between the couple of values measured by Abu 'l-Wafā' shows that his value obtained from the mid-seasons method (with an error of $+0.9 \times 10^{-4}$) may be considered as a noticeable improvement on that resulting from the seasons method (with an error of $+1.5 \times 10^{-4}$). Bīrūnī's values also indicate a small improvement; of course, both of them are remarkably accurate (error: $\sim +0.6 \times 10^{-4} - +0.5 \times 10^{-4}$). The mid-seasons method was therefore successful in reaching more accurate and congruent (i.e., with a higher average accuracy) values in comparison with the theoretical values expected.

(2b) Interestingly, although Bīrūnī notes that the three-point method is superior to the two others, we do not have any value for e from Bīrūnī himself that was measured by this method. Due to lack of observational reports, it is not known whether astronomers in the period of 1000–1500 (except for Muḥyī al-Dīn) used the three-point method or not, although the potential accuracy of the method had already been known. Despite having rather good observational data (see Table 10.1), the error in Muḥyī al-Dīn's value is $\sim+4.9 \times 10^{-4} - +5.1 \times 10^{-4}$, appreciably larger than the error in the values obtained from the seasons and mid-seasons methods. Since no other values associated with the general three-point method are known to the present author, nothing more can be said about its comparison with the other two.

(2c) The mean absolute error in the values observed in the most influential astronomical tables written between 950 and 1450 (nos. 5–12 listed in Table 1.3), excluding al-Khāzinī's, is about 3.9×10^{-4}. This also shows that the astronomers in the late Islamic period were less successful than their predecessors with regard to the accuracy of the values measured for the solar parameters. *Among these works, the most exact value is that applied to Ulugh Beg's* Sulṭānī zīj.

(2d) Although the mid-signs method as a substitute for the seasons method led to more accurate and consistent results, Bīrūnī's values measured by the two methods, as mentioned earlier, are equally accurate, and interestingly, his *value measured by the seasons method* (compared with the value of e' in his time) *is the most exact of all the values studied in this chapter.* This implies that the sensitivity as a feature of a method does not always cause quantitative results to be less accurate. Generally, in the case of determining the solar orbital parameters in medieval astronomy, the accuracy of observations played a role as important and prominent as the methods used. The theoretically better methods might yield more congruent results but did not guarantee a higher accuracy.

(3) The values measured in the early Islamic period (Tables 1.1 and 1.2) indicated to a good degree the decrease in the eccentricity of the Sun's/Earth's orbit during the centuries. From the aspect of the accuracy of values, one would expect to see the same approximate rate of decrease of the eccentricity in the values computed later and, in particular, that values less than Battānī's (or Bīrūnī's) would be yielded. However, as Tables 1.3 and 1.4 and Figure 1.6 make clear, this is not the case. Only al-Fahhād and the astronomers at the Samarqand observatory observed a value less than those of their outstanding predecessors.

On a theoretical level, the relatively continuous decrease in the solar eccentricity could have the consequence that the solar eccentricity is variable in itself, as the increase in the longitude of the solar apogee led to the acceptance of a moving apsidal line. This could then be considered as a main factor in measuring the varying values for the eccentricity. The passage from Bīrūnī quoted earlier shows that this was initially brought to his attention, but his awareness of the sensitivity of the methods, as a stronger and more evident factor, prevented it from being considered seriously by him. Nevertheless, some values mentioned in Tables 1.1 and 1.2 appear to have been persuasive enough to cause a solar model with a variable

Table 1.5 The historical values for the longitude of the solar apogee in comparison with the theoretical values, taking into consideration the method used and with the corresponding true modern values

Method	Nos.	λ_{ap} (observed)	λ_{ap} (theoretical)	λ_{ap} (modern)
Seasons	1	81.3°	82.7°	82.9°
	2	80.4	82.9	83.1
Table 1.1	3	82.2	83.6	83.8
	4	83.9	83.7	83.9
	5	84.5	85.2	85.4
	6	85.2	86.0	86.1
	7	85.3	87.1	87.1
	8	87.8	88.7	88.7
Mid-signs	1	——		
	2	82.8	83.5	82.9
Table 1.2	3	84.7	85.9	85.4
	4	80.7	86.1	85.6
	5	85.0	85.6	86.1
	1	——		
Table 1.3	2	84.6		83.2
	3	——		
	4	84.1		83.7–84.4
	5	84.4		85.4
	6	86.2		85.8
	7	85.9		87.8
	8	88.8	89.7	90.3
	9	88.4		89.8
	10	90.2		91.8
	11	——		
	12	92.4		93.3

eccentricity to be invented in Spain, where the Western branch of astronomy in medieval Islam was born and developed.

Table 1.5 lists the historical values for λ_{ap}, and Figure 1.7 illustrates them along with the graphs related to the theoretical deviations and to the true modern values of λ'_{ap}.[49] As previously exemplified (Section 1.2) and also noted by Bīrūnī, the erroneous input data may cause e to be in error in the order of minutes, but the corresponding λ_{ap} may be in error by some degrees. The theoretical errors in the values of λ_{ap} measured by the seasons method remain about ±0.2°, but in those obtained by the mid-seasons method, they are, at most, about ±0.6°. For the values measured by the latter method with the starting point of mid-winter (the first three

49 Some of the values studied here have been mentioned in Hartner 1977, p. 8, and have been compared with the corresponding true values of λ'_{ap} (although, with a few errors; for example, assuming 1061 and 1600 CE, respectively, as the years of Ibn al-Zarqālluh's and Tycho's solar observations, or taking the epoch value 77;50° for the longitude of apogee instead of the value Ibn al-Zarqālluh observed).

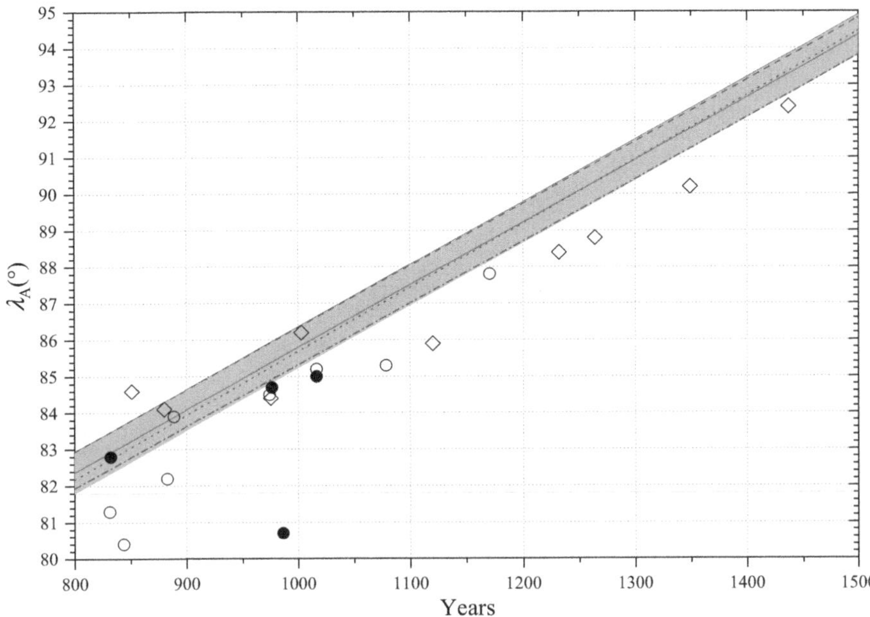

Figure 1.7 Continuous thicker curve: the true modern longitude of apogee (λ'_{ap}). *Gray area*: the tolerance band. *Dotted curve*: the longitude of apogee (λ_{ap}) computed theoretically based on the seasons method. *Dashed curve*: the longitude of apogee (λ_{ap}) computed theoretically based on the mid-sign method with the starting point of mid-winter. *Dotted-dashed curve*: the longitude of apogee (λ_{ap}) computed theoretically based on the mid-sign method with the zero point of mid-spring. ○: the historical values measured on the basis of the seasons method. ●: the historical values measured on the basis of the mid-sign method. ◊: the other historical values in Table 1.3.

values), the theoretical deviation increases the amount of error, and inversely, if mid-spring is adopted as the zero point, it decreases the amount of error. The mean absolute errors in the three sets of the historical values given remain below 2°. It is quite probable that the errors in the values of λ_{ap} have little or nothing to do with the empirical errors and/or with the methods used, as explained in the following. The errors in the lengths of the seasons or in the time intervals may neutralize each other. For example, Bīrūnī's input data (Table 1.1, no. 6), with errors of about $+0.8^{h}$ and -0.4^{h}, are obviously more accurate than Ibn ʿIṣmat's, with errors of about -1.4^{h} and -0.8^{h} (Table 1.1, no. 4); nevertheless, the observational components of the total error, as detectable in Table 1.5, are, respectively, about $-0.8°$ and $+0.2°$. Moreover, the agreements (although remarkable) may be accidental, not due to the observations, but owing to the compensating effect of the theoretical deviation. For example, Ibn ʿIṣmat's observational error of $+0.2°$ is compensated for by the theoretical deviation of $-0.2°$, which results in his value for λ_{ap} being in agreement with the true modern value for λ'_{ap} at that time. The observational error of about $-0.7°$ in the formal value in *On the solar year* (Table 1.2, no. 2) is reduced by the theoretical deviation of $+0.6°$ to a total error of $-0.1°$, while in the case of Bīrūnī's value measured by the mid-signs method, the theoretical error of $-0.5°$ aggregates with the observational error of $-0.6°$ to make a total error of $-1.1°$. It suffices here to say that the values given by Ibn ʿIṣmat, Ibn Amājūr, and Ibn Yūnus are the most accurate of the values considered in this study.

References

Historical Sources

Anonymous, *Sulṭānī zīj*, MS Iran, Parliament Library, no. 184.

al-Bīrūnī, Abū al-Rayḥān, 1879, *The Chronology of Ancient Nations*, C. E. Sachau (En. tr.), London: The Oriental translations fund of Great Britain & Ireland.

al-Bīrūnī, Abū al-Rayḥān, 1954–1956, *al-Qānūn al-masʿūdī*, 3 Vols., Hyderabad: Osmania Bureau.

Ḥabash al-Ḥāsib, *The zīj of Ḥabash al-Ḥāsib*, MSS. B: Berlin, Ahlwardt 5750 (formerly, Wetzstein I 90); I: Istanbul, Süleymaniye, Yeni Cami, no. 784.

Ibn al-Fahhād: Farīd al-Dīn Abu al-Ḥasan ʿAlī b. ʿAbd al-Karīm al-Fahhād al-Shirwānī or al-Bākūʾī, *Zīj al-ʿAlāʾī*, MS. India: Salar Jung Library, no. H17.

Ibn al-Shāṭir, ʿAlāʾ al-Dīn Abu ʾl-Ḥasan ʿAlī b. Ibrāhīm b. Muḥammad al-Muṭaʿʿim al-Anṣārī, *al-Zīj al-jadīd* (*The new zīj*), MSS. K: Istanbul, Kandilli Observatory, no. 238, O: Oxford, Bodleian Library, Seld. A inf 30, L1: Leiden, Universiteitsbibliotheek, Or. 65, L2: Leiden, Universiteitsbibliotheek, Or. 530.

Ibn Yūnus, ʿAlī b. ʿAbd al-Raḥmān b. Aḥmad, *Zīj al-kabīr al-Ḥākimī*, MS. Leiden, Universiteitsbibliotheek, Or. 143.

al-Kamālī, Muḥammad b. Abī ʿAbd-Allāh Sanjar (Sayf-i munajjim), *Ashrafī zīj*, MSS. F: Paris: Bibliothèque nationale, no. 1488, G: Iran–Qum: Gulpāyigānī, no. 64731.

Kāshī, Ghiyāth al-Dīn Jamshīd, *Khāqānī zīj*, MSS. P: Iran, Parliament Library, no. 6198, IO: London, India Office, no. 430.

al-Khāzinī, ʿAbd al-Raḥmān, *al-Zīj al-muʿtabar al-sanjarī*, MSS. V: Vatican, Biblioteca Apostolica Vaticana, Arabo 761 (undated, but certainly after Khāzinī's death, from a reference on f. 81v), L: London, British Library, Or. 6669 (completed on 25 Jumādā II 620 = 26 July 1223; f. 157r); *Wajīz* [Compendium of] *al-Zīj al-muʿtabar al-sanjarī*, MSS. I: Istanbul, Süleymaniye Library, Hamidiye collection, no. 859 (Rabīʿ II 667 H = December 1268–January 1269 CE; f. 38r); S: Tehran: Sipahsālār, no. 682 (Ramaḍān 631 H = May–June 1234 CE; p. 123).

al-Maghribī, Muḥyī al-Dīn, *Talkhīṣ al-majisṭī*, MS. Leiden: Universiteitsbibliotheek, Or. 110.

al-Nadīm, Abu ʾl-Faraj Muḥammad b. Abī Yaʿqūb Isḥāq al-Warrāq al-Nadīm, 1973, *al-Fihrist fī Akhbār al-ʿUlamāʾ al-Muṣannifīn min al-Qudamāʾ wa-ʾl-Muḥaddithīn wa-Asmāʾ Kutubihim* (*The Bio-Bibliography of the Scientists and Authors from the Ancient to the Modern Times*), Tajaddod, R. (ed.), 2nd edn., Tehran: Marwī.

al-Shīrāzī, Quṭb al-Dīn, *Ikhtiyārāt-i muẓaffarī* (*Muẓaffarid Selections*, dedicated to Muẓaffar al-Dīn Bulāq Arsalān, d. 1305), MS. Iran, National Library, no. 3074F, up to f. 176r (copying completed on Monday, 20 Jumādā I 682/16 August 1283).

al-Shīrāzī, Quṭb al-Dīn, *Tuḥfa al-shāhiyya fī 'l-hay'a* (*Gift to the King on Astronomy*), MS. Iran, Parliament Library, no. 6130 (copied in Rabīʿ I 703 H/October 1303 CE).

al-Ṭūsī, Naṣīr al-Dīn, *et al.*, *Īlkhānī zīj*, MSS. C: University of California, Caro Minasian Collection, no. 1462; T: University of Tehran, Ḥikmat Collection, no. 165 + Suppl. P: Iran, Library of Parliament, no. 6517 (Remark: The latter is not actually a separate MS, but contains 31 folios missing from MS. T. The chapters and tables in MS. T are badly out of order, presumably owing to the folios having been bound in disorder); P: Iran, Library of Parliament, no. 181.

al-Ṭūsī, Naṣīr al-Dīn, *Risāla al-muʿīniyya fī ʿilm al-hay'a*, MS. Tehran, Library of Parliament, no. 6347.

Ulugh Beg, *Sulṭānī Zīj*, MSS. P1: Iran, Library of Parliament, no. 72, P2: Iran, Library of Parliament, no. 6027.

Yaḥyā b. Abī Manṣūr, *Zīj al-mumtaḥan*, MS. E: Madrid, Library of Escorial, árabe 927, published in *The verified astronomical tables for the caliph al-Ma'mūn*, Sezgin, F. (ed.) with an introduction by Kennedy E. S., Frankfurt am Main: Institut für Geschichte der Arabisch-Islamischen Wissenschaften, 1986, MS. L: Leipzig, Universitätsbibliothek, Vollers 821.

Secondary Sources

Calvo, E., 1998, "Astronomical theories related to the sun in Ibn al-Hā'im's *al-Zīj al-Kāmil fī 'l-Taʿālīm*", *Zeitschrift für Geschichte der Arabisch-Islamischen Wissenschaften* **12**, pp. 51–111.

Carmody, F. J., 1960, *The Astronomical Works of Thabit B. Qurra*, Berkeley–Los Angeles: University of California Press.

Charette, F., 2006, "The locales of Islamic astronomical instrumentation", *History of Science* **44**, pp. 123–138. 11.

Copernicus, N., 1543, *De Revolutionibus Orbium Coelestium*, English translation: Dobrzycki, J. (ed.), and Rosen, E. (En. tr. & com.), 1978, *Nicolas Copernicus on the Revolutions*, 2 Vols., Poland: Polish Scientific Publishers, Great Britain: Macmillan.

van Dalen, B., 2004a, "The *Zīj-i Naṣirī* by Maḥmūd ibn ʿUmar: the earliest Indian Zij and its relation to the *ʿAlā'ī Zīj*", in: Burnett, C., *et al.* (eds.), *Studies in the History of the Exact Sciences in Honour of David Pingree*, Leiden: Brill, pp. 327–356.

van Dalen, B., 2004b, "A second manuscript of the *Mumtaḥan Zīj*", *Suhayl* **4**, pp. 9–44.

Dreyer, J. L. E., 1890, *Tycho Brahe: A Picture of Scientific Life and Work in the Sixteenth Century*, Edinburgh: Adam and Charles Black.

Duke, D., 2008, "Four lost episodes in ancient solar theory", *Journal for the History of Astronomy* **39**, pp. 283–296.

Goldstein, B. R., 1994, "Historical perspectives on Copernicus's account of precession", *Journal for the History of Astronomy* **25**, pp. 189–197.

Goldstein, B. R. and Chabás, J., 2001, "The maximum solar equation in the Alfonsine tables", *Journal for the History of Astronomy* **32**, pp. 345–348.

Goldstein, B. R. and Sawyer, F. W., 1977, "Remarks on Ptolemy's equant model in Islamic astronomy", in: Maeyam and Salzer 1977, pp. 165–181.

Grasshoff, G., 1990, *The History of Ptolemy's Star Catalogue*, London: Springer.

Halley, E., 1695, "A discourse concerning a method of discovering the true moment of the Sun's ingress into the tropical signs", *Philosophical Transactions of the Royal Society* **19**, pp. 12–18.

Hartner, W., 1977, "The role of observations in ancient and medieval astronomy", *Journal for the History of Astronomy* **8**, pp. 1–11.

Hughes, D. W., 1989, "Hipparchus' spring and summer and the ellipticity of the Earth's orbit", *Journal of the British Astronomical Association* **99**, pp. 90–94.

Jones, A., 1991, "Hipparchus's computations of solar longitudes", *Journal for the History of Astronomy* **22**, pp. 101–125.

Jones, A., 2000, "Studies in the astronomy of the Roman period, IV: Solar tables based on a non-Hipparchian model", *Centaurus* **42**, pp. 77–88.

Kennedy, E. S., 1956, "A survey of Islamic astronomical tables", *Transactions of the American Philosophical Society*, New Series **46**, pp. 123–177.

Kennedy, E. S., 1965, "Applied mathematics in the eleventh-century Iran: Abū Ja'far's determination of the solar parameters", *Mathematics Teacher* **58**, pp. 441–446; reprinted in Kennedy *et al.* 1983, pp. 535–540.

Kennedy, E. S., 1977, "The solar equation in the Zīj of Yaḥya b. Abī Manṣūr", in: Maeyama and Salzer 1977, pp. 183–186; reprinted in Kennedy *et al.* 1983, pp. 136–139.

Kennedy, E. S., colleagues, and former students, 1983, *Studies in the Islamic Exact Sciences*, Beirut: American University of Beirut.

King, D. and Saliba, G. (eds.), 1987, *From Deferent to Equant*, Vol. 500, New York: Annals of New York Academy of Sciences.

Maeyama, Y., 1998, "Determination of the Sun's orbit: Hipparchus, Ptolemy, al-Battânî, Copernicus, Tycho Brahe", *Archive for History of Exact Sciences* **53**, pp. 1–49.

Maeyama, Y. and Salzer, W. G. (eds.), 1977, *Prismata: Festschrift für Willy Hartner*, Wiesbaden: Franz Steiner Verlag.

Meeus, J., 1998, *Astronomical Algorithms*, Richmond: William–Bell.

Meeus, J., 2002, *More Mathematical Astronomy Morsels*, Richmond: William-Bell.

Meeus, J. and Savoie, D., 1992, "The history of the tropical year", *Journal of the British Astronomical Association* **102**, pp. 40–42.

Moesgaard, K. P., 1974, "Thābit ibn Qurra between Ptolemy and Copernicus: An analysis of Thābit's solar theory", *Archive for History of Exact Sciences* **12**, pp. 199–216.

Morelon, R., 1987, *Thābit ibn Qurra: Œuvres d'Astronomie*, Paris: Les Belles Lettres.

Mozaffari, S. M., 2019, "Ibn al-Fahhād and the Great Conjunction of 1166 AD", *Archive for History of Exact Sciences* **73**, pp. 517–549.

Mozaffari, S. M., 2023a, "A survey of Abu 'l-Wafā''s solar and stellar observations", *Journal of Astronomical History and Heritage* **26**, pp. 469–488.

Mozaffari, S. M., 2023b, "Sources of the planetary theories in Fahhād's *'Alā 'ī zīj*: Solving a medieval case of intellectual fraud", *Suhayl* **20**, pp. 143–221.

Nallino, C. A. (ed.), 1899–1907, *Al-Battani sive Albatenii Opus Astronomicum, Publicazioni del Reale osservatorio di Brera in Milano, n. XL, pte. I–III*, Milan: Mediolani Insubrum. The Reprint of Nallino's edition: Minerva, Frankfurt, 1969.

Neugebauer, O., 1962a, "Thābit ben Qurra 'on the solar year' and 'on the motion of the Eighth sphere'", *Proceedings of the American Philosophical Society* **106**, pp. 264–299.

Neugebauer, O., 1962b, *The Astronomical Tables of Al-Khwārizmī*, in Hist. Filos. Skr. Dan. Vid. Selsk., vol. 4, no. 2, Copenhagen.

Neugebauer, O., 1975, *A History of Ancient Mathematical Astronomy*, Berlin-Heidelberg-New York: Springer.

Newton, R. R., 1973, "The authenticity of Ptolemy's parallax data: Part 1", *Quarterly Journal of the Royal Astronomical Society* **14**, pp. 367–388.

Pederson, O., 1974, *A Survey of Almagest*, Odense: Odense University Press.

Petersen, V. M. and Schmidt, O., 1968, "The determination of the longitude of the apogee of the orbit of the sun according to hipparchus and ptolemy", *Centaurus* **12**, pp. 73–96.

Pingree, D., 1970, "On the classification of Indian planetary tables", *Journal for the History of Astronomy* **1**, pp. 95–108.

Pingree, D. (ed.), 1985, *Astronomical Works of Gregory Chioniades*, 3 Vols., Amsterdam: Gieben.

Rashīd al-Dīn Faḍl Allāh al-Hamidānī, 1994, *Jāmi ' al-Tawārīkh* [*The Compendium of Histories*], Rushan, M. and Mūsawī, M. (eds.), 4 Vols., Tehran: Alborz.

Roberts, V., 1957, "The solar and lunar theory of Ibn ash-Shāṭir: A pre-Copernican Copernican model", *Isis* **48**, pp. 428–432.

Said, S. S. and Stephenson, F. R., 1995, "Precision of medieval Islamic measurements of solar altitudes and equinox times", *Journal for the History of Astronomy* **26**, pp. 117–132.

Saliba, G., 1985a, "Solar observations at Maragha observatory", *Journal for the History of Astronomy* **16**, pp. 113–122.

Saliba, G., 1985b, "The determination of the solar eccentricity and apogee according to Mu'ayyad al-Dīn al-'Urḍī", *Zeitschrift für Geschichte der Arabisch-Islamischen Wissenschaften* **2**, pp. 47–67.

Saliba, G., 1987, "Theory and observation in Islamic astronomy: The work of Ibn al-Shāṭir of Damascus", *Journal for the History of Astronomy* **18**, pp. 35–43.

Samsó, J., 1987, "Al-Zarqal, Alfonso X and Peter of Aragon on the solar equation", in: King and Saliba 1987, pp. 467–476.

Swerdlow, Noel M., 2010, "Tycho, longomontanus, and kepler on ptolemy's solar observations and theory, precession of the equinoxes, and obliquity of the ecliptic", in: Jones, A. (ed.), *Ptolemy in Perspective*, *Archimedes*, Vol. 23, Dordrecht-Heidelberg-London-New York: Springer.

Thoren, V. E. and Christianson, J. R., 1990, *The Lord of Uraniborg: A Biography of Tycho Brahe*, Cambridge: Cambridge University Press.

Toomer, G. J., 1984, *Ptolemy's Almagest*, Princeton: Princeton University Press.

Toomer, G. J., 1987, "The solar theory of az-Zarqāl: An epilogue", in: King and Saliba 1987, pp. 513–519.

Part II

LUNAR ASTRONOMY AND THEORY OF ECLIPSES

... و إنّما نصب الله تعالى الكسوفين من أعظم آياته.

– Abū al-Rayḥān al-Bīrūnī
(*al-Qānūn al-Masʿūdī* VIII.3)

. . . eclipses Solis, spectaculum a creatore ordinatum, ut eo
doceretur contemplatrix creatura de ratione cursus siderum.

– Johannes Kepler
(*Epitome Astronomiae Copernicanae* IV.1.4)

2

HOW NATURAL PHENOMENA WERE JUSTIFIED IN MEDIEVAL SCIENCE

The Situation of Annular Eclipses in Medieval Astronomy[1]

2.1 Introduction

In this chapter, we delve into how medieval Islamic astronomers, operating within the Ptolemaic framework – in which the annular solar eclipse is an unjustified and impossible phenomenon – managed to understand, define, justify, and predict annular solar eclipses.[2]

Our context-based study aims to unravel the complexities surrounding annular eclipses in ancient and medieval astronomy. Specifically, we explore the hypotheses used to calculate the apparent angular diameters of the Sun and Moon – a crucial factor in contemplating the feasibility of this phenomenon. Additionally, we delve into the historical mechanisms that eventually led to its justification during the late Islamic period. Notably, this identification resulted from compelling observational evidence and an intriguing interplay of various astronomical traditions available to Islamic scholars.

On a broader scale, we examine the essential conditions that allowed medieval science, rooted in tradition, to entertain and validate phenomena that defied clear definition. These conditions encompass the identification of the phenomenon, empirical evidence, and the justifications provided by underlying traditions.

1 Original publication: S. M. Mozaffari, "A case study of how natural phenomena were justified in medieval science: The situation of annular eclipses in medieval astronomy," *Science in Context* **27** (2013), pp. 33–47. © 2013 Cambridge University Press, and republished by permission.

2 The *annular eclipse* is one type of solar eclipse that occurs when the Moon is sufficiently more distant from the Earth at the time of maximum phase of eclipse so that its angular diameter appears to an observer on the Earth's surface to be less than the Sun's; as a result, the Moon cannot cover the whole solar disk, and thus the rim of the Sun remains unobscured and appears as a bright annulus.

DOI: 10.4324/9781003481966-5

2.2 Annular Eclipses in the Ancient and Medieval Periods

The problem of annular eclipses in medieval astronomy is multifaceted, involving historical mechanisms and intricate technical details both theoretically and observationally. Let us start by revisiting specific undeniable pieces of evidence to grasp the problem's significance within historical and astronomical contexts.

According to Ptolemy (*Almagest* V.14), the solar angular diameter remains permanently constant at 0;31,20°, showing "no noticeable difference due to [variations in] distance," and is equal to the minimum lunar angular diameter. This means, when the Sun and Moon are farthest from Earth, they subtend the same angle.[3] Meanwhile, the lunar apparent diameter fluctuates between 0;31,20° and 0;35,20°. Under these conditions, annular solar eclipses are evidently impossible. Ptolemy's predecessors, notably Hipparchus, believed that the Sun and Moon had equal angular diameters at their mean distances from Earth, allowing for annular eclipses.[4] Intriguingly, a pseudo-Eudoxian papyrus from around 190 BCE[5] asserts that solar eclipses can never be total but must remain, at most, annular.[6] The mystery lies in why ancient astronomers thought the Moon's diameter always appeared smaller than the Sun's.[7] Over a span of about three centuries, Ptolemy's rejection of annular eclipses contrasts sharply with his predecessors' views. However, we must recognize that early Greek reports on annular eclipse possibilities (maybe available to Ptolemy) were vague and unsupported,[8] and he lacked first-hand observations during his career.[9]

3 Toomer 1998, pp. 252, 253.

4 Toomer 1998, p. 252, n. 53.

5 Now known as Papyrus, Paris 1, Col. 19.16–17.

6 For the analysis of it, see Neugebauer 1975, p. 686f.

7 Neugebauer (1975, pp. 668, 688) says nothing about it, but he adds that this could be the result of the observation of an annular eclipse by Polemarchus [*sic*], the younger contemporary of Eudoxus. The observation of an annular eclipse by Polemarchus is mentioned, Neugebauer says, in Simplicius's commentary on *De Caelo*. However, such a thing has never been told therein (see Simplicius 1894, p. 505, lines 20ff); see also translation and critical comments in Bowen 2008, pp. 53 (esp. n. 179), 75, and 107.

8 For instance, Saslaw and Murdin (2004) posed the theory that the double heads of Apollo, the Greek sun-god, engraved on a coin recovered in the ancient city of Isturus (now Costanta, Romania), refers to the observation of the annular eclipses of October 4, 434 BCE and/or August 3, 431 BCE. Thucydides and Xenophon described both the annular eclipses of August 3, 431 BCE and August 14, 394 BCE, respectively, as of "crescent shape." However, in both the cases, it seems that only the partial phase was witnessed (see Stephenson 1997, pp. 346–348, 366–367).

9 The only annular eclipse that was visible near Alexandria during Ptolemy's career took place on November 25, 132 CE, but the Sun set in the longitude of Alexandria before the beginning of the annular phase, and hence, it was not observable. (At the beginning of the partial phase of this eclipse, the Sun was only about 5.5° above the horizon.) Mozaffari (2013, pp. 33–34, 2015) evaluates all the possible annular eclipses about Ptolemy's time.

Although the first account of annular eclipses appeared in Simplicius's commentary on Aristotle's *De Caelo*,[10] what was passed on from the Greek to the medieval astronomers was instead Ptolemy's denial of annular eclipses. Accordingly, the medieval astronomers had to uncover the variety of solar eclipses and their circumstances again for themselves.

Turning our attention to the Orient, Indian hypotheses for determining the solar and lunar angular diameters led to defining them as one-variable functions of their angular velocity.[11] The original formulas are quite complex. However, after substituting constants for the solar and lunar diameters and mean distances from Earth, a simpler formula emerges: $\theta = k \cdot v$. Here, k represents a constant (with different values for the Sun, 0;33, and the Moon, 0;2,26), and v denotes the daily angular velocity in the ecliptic longitude.[12] With a minimum solar velocity of approximately 0;57° per day and a minimum lunar velocity of about 11;30°/d, the Indian formula yields 0;31,20° and 0;28° for the minimum solar and lunar angular diameters, respectively.[13] This suggests that annular eclipses are indeed possible

10 Simplicius 1894, p. 504, line 16–506, line 8. Simplicius used the fact that "sometimes . . . a rim is left appearing during mid-eclipse" as a proof against the idea of the homocentric planetary spheres. That such an account may be due to the real observation of an annular eclipse is not sufficiently clear. Neugebauer (1975, vol. 1, p. 101) has determined the annular eclipse of September 4, 164 CE as the one observed by Sosigenes; however, this does not seem so acceptable (see Bowen 2008, pp. 89–90). Besides considering the possibility of the annular eclipse in Simplicius's treatise, it has also been told in it that some total solar eclipses may show a duration of darkness (Simplicius 1894, p. 505, line 7; Bowen 2008, p. 50), which the majority of the medieval astronomers erroneously assumed to have been neglected by Ptolemy (see following).

11 See *Súrya Siddhánta* IV.1–5: [1860] 1997, p. 41.

12 For the discussion on the original formulas and the procedure for deriving the preceding simple relation from them, see Mozaffari 2015.

13 The angular velocities of the Sun and Moon as computed in the framework of the Ptolemaic models are of main concern here. The values adopted for the parameters of these models in the various medieval astronomical tables were sometimes different from Ptolemy's own, which makes some difference in the extremes of the angular velocities but does not make an appreciable shift in them. For instance, the Ptolemaic values for the solar parameters were mean motion = 0;59,8,17,13, . . .°/d and eccentricity = 2;30^p (in terms of the radius of the solar geocentric orbit, the so-called "eccentric," assumed to be 60 parts). The medieval values for the solar mean motion were different from the Ptolemaic value from the third sexagesimal fractional place on; also, the medieval values for the solar eccentricity were different from Ptolemy's by, at most, −0;30^p. These overall make a difference of about 0;0,30°/d in the minimum solar velocity. In the case of the Moon, Ptolemy has mean motion in longitude = 13;10,34, . . .°/d, the radius of the epicycle $r = 5;15^p$, and the eccentricity of the lunar inclined orb $e = 10;19^p$. The medieval values for the lunar mean motion in longitude differ from Ptolemy's from the second sexagesimal fractional place on. In comparison, for example, with Muḥyī al-Dīn al-Maghribī (Marāgha observatory, d. 1283), whose values for both r and e are appreciably different from Ptolemy's ($r = 5;12^p$ and $e = 9^p$), these differences do not make a shift in the minimum lunar velocity exceeding +0;7°/d. Therefore, the variants in the medieval values for the underlying parameters make no considerable difference in the minimum velocities of the Sun and Moon, and so in their minimum apparent diameters computed according to the Indian formula. We do not deal here with the rules for the calculation of the planetary instantaneous velocity in the Ptolemaic planetary models. The interested reader can consult Pedersen

within the Indian astronomical context. A minor but historically significant consequence is that the Indian hypotheses closely align with Ptolemy's value for the minimum solar diameter.

Four clear records of direct observations of annular eclipses have come down to us from the medieval period. Two of them are outside the Ptolemaic context and do not relate to its intrinsic considerations and limitations: one on October 26, 1147, in Brauweiler, Germany, and another on January 21, 1292, in Beijing, China.[14] The other two are also relevant to the context of our study: the first occurred on July 28, 873 (JDN 2040130), by Abu 'l-ʿAbbās Īrānshahrī in Nīshāpūr, Iran,[15] and the second on January 30, 1283 (JDN 2189703), by Shams al-Dīn Muḥammad al-Wābkanawī al-Bukhārī in Mughān, Iran.

The two later observations differ significantly within their context. The first occurred during the early Islamic period and was preserved as a simple observation, documented only by Bīrūnī (973–1048 CE) without further details. This observation served as evidence for rejecting the concentric solar model proposed by Abū Jaʿfar al-Khāzin (d. *ca.* 961–971 CE).[16] In contrast, the second observation is more than a mere record. Wābkanawī meticulously calculated its parameters – timing, magnitude, and more – based on Muḥyī al-Dīn al-Maghribī's *Adwār al-anwār*, a *zīj* completed in 1275. This *zīj* relied on extensive systematic observations conducted at the Marāgha observatory between 1261 and 1274.[17] Wābkanawī's meticulous work led to the conclusive determination that the eclipse was, indeed, annular (as discussed in Chapter 3). As we transition from the former observation to the latter, it becomes essential to consider historical mechanisms that significantly shaped the annular eclipse as a justified phenomenon within the late Islamic astronomical tradition.

Let us begin with Bīrūnī. He introduces Muḥammad b. Isḥāq al-Sarakhsī's observation of the total solar eclipse on May 27, 876 CE, whose duration of totality was fully perceptible in his hometown. Let us note that since Ptolemy's *Almagest* considers the maximum lunar diameter (0;35,20°) larger than the solar diameter (0;31,20°), the solar eclipse occurring when the Moon is close to the Earth is inevitably expected to show a perceptible duration of totality (Ar. *makth*, "staying"). As a result, the *Almagest*'s table of solar eclipses (VI.8) includes a fifth column (with just one entry) depicting the Moon's motion from complete immersion to maximum eclipse. However, most Arabic manuscripts of the *Almagest* do not have that one-entry column.[18] The passage runs as follows:

[1974] 2010, pp. 223–224, Neugebauer 1975, vol. 1, p. 122, and for the medieval improvements in the calculation of the lunar velocity, Goldstein 1996.

14 See Stephenson 1997, pp. 62, 258–293, 94. Other doubtful or erroneous reports are summarized in Mozaffari 2015.

15 See Said and Stephenson 1997, p. 45, Goldstein 1979.

16 al-Bīrūnī, *al-Qānūn al-masʿūdī*, 1954–1956, vol. 2, p. 632, Samsó 1977, p. 274.

17 Al-Maghribī explained his observations and measurements in a treatise titled the *Talkhīṣ al-majisṭī* (see Chapter 9).

18 Toomer 1998, p. 305, n. 63. Tables on pp. 306–307.

On 12 *Urdībihisht* 245 Yazdigird era (= 27 May 876 CE, JDN 2041164), Muḥammad b. Isḥāq al-Sarakhsī observed a clear duration (*makth*, staying) [of totality] in the solar eclipse that occurred in his native town. It supports what Ptolemy mentioned, and is not taken as criticism of it.[19]

Bīrūnī is evidently correct here; he quite probably knew that the Moon was near its epicyclic perigee (i.e., close to the Earth) at that time,[20] and so a perceptible duration of darkness was in accord with Ptolemy's hypothesis of the solar and lunar angular diameters. Bīrūnī then highlights Īrānshahrī's observation, emphasizing that:

From it (i.e., the Īrānshahrī's observation), it becomes clear that the Sun's apparent diameter may exceed that of the Moon. The principles/hypotheses of the India[n astronomical tradition(s)] (*uṣūl al-Hind*) corroborate it.

Again, he almost certainly knew that the Moon was close to its epicyclic apogee on July 28, 873,[21] but unlike Ptolemy's hypothesis, the lunar disk did not cover the entirety of the Sun's surface. Then, Bīrūnī returned to the problem of the duration of darkness in Sarakhsī's solar eclipse and considered the various possibilities:

The "staying" that we mentioned in the solar eclipse may be because of [1] the decrease [in the apparent diameter] of the Sun from its mean value alone, [2] the increase [in the apparent diameter] of the Moon [from its mean value] by itself, too, or [3] both of them together.

Whatever the case might be, the inescapable conclusion regarding Abū Jaʿfar al-Khāzin's concentric solar model would be:

[The previous considerations] have invalidated what Abū Jaʿfar stated in this regard. [Accordingly,] we do not oblige [to accept Abū Jaʿfar's concentric model of the Sun, on account of Ptolemy's hypothesis that the Sun's apparent diameter remains nearly constant], while it would have been necessary for Ptolemy [to work out such a model].

Early medieval Islamic astronomers were well acquainted with Indian hypotheses and the aforementioned straightforward relation for determining the angular

---

19 al-Bīrūnī, *al-Qānūn*, 1954–1956, vol. 2, p. 632. The duration of this eclipse in Sarakhs (a city located in the northeast of Iran, $\varphi = 36.3°$ N, $L = 60.8°$ E) was about 3 minutes (magnitude = 1.02).

20 For example, both the *Mumtaḥan zīj* and Battānī's *zīj* give a mean lunar anomaly of about 167° at the time of conjunction, about **7:12** MLT for the meridian 16° in the east of Baghdad (in fair agreement with the modern value).

21 At the time of conjunction in Nishāpūr ($\varphi = 36.2°$ N, $L = 58.8°$ E, ~14° in the east of Baghdad), about **7:23** MLT, both *zījes* mentioned in the previous note give the mean lunar anomaly of about 338° (again, in good agreement with the modern value).

diameters of the Sun and the Moon. This familiarity predated their exposure to Ptolemaic astronomy. Interestingly, even after Greek astronomy became the dominant system in the Islamic period, the Indian hypotheses and formulas regarding solar and lunar diameters persisted and even surpassed their Ptolemaic counterparts. They permeated Islamic *zīj* literature, reflecting the mathematical astronomy of that tradition.[22] Bīrūnī's clear reference to the Indian astronomical tradition (the second quotation earlier), coupled with Īrānshahrī's observation of the annular eclipse, reinforces this notion. It thus leaves little doubt that Īrānshahrī's report served as observational evidence that led to the use of the Indian hypotheses and the already-mentioned simple formula as the standards for determining the solar and lunar angular diameters in medieval Islamic astronomy. Remarkably, applying Indian hypotheses appears to be the only viable approach for justifying annular eclipses.

Nevertheless, there is no direct causal link between observational evidence and the range of solar and lunar apparent diameters found in *early* Islamic sources (prior to Bīrūnī). For instance, applying the Indian hypotheses, some early Islamic astronomers, like Khwārizmī, held the Ptolemaic constant value for the solar apparent diameter as its minimum value while accepting a range of 0;31,20° to 0;33,48° for it. He also changed the range of variation in the lunar apparent diameter as 0;29,16° to 0;34,34°. Khwārizmī flourished about half a century before both Īrānshahrī and Sarakhsī made their observations. Another example is al-Battānī (d. 929), who flourished a generation after Īrānshahrī; he held that the solar angular diameter varies between 0;31,20° and 0;33,40°, and the lunar apparent diameter ranges from 0;29,30° through 0;35,20°, but he made no mention of annular eclipses. In general terms, surviving material provides no evidence that astronomers preceding Bīrūnī were aware of Īrānshahrī's observation. There is no evidence to ascertain whether Īrānshahrī wrote in Arabic or Persian. That *only* Bīrūnī referred to Īrānshahrī in his works, and that he spoke of him as a philosopher who followed his own constructed religion rather than any other opinion[23] – neither Islam as the new religion acknowledged by Iranians nor the ancient Iranian religions – gives the strong impression that Īrānshahrī was likely *not* writing in Arabic.

---

22 The examples are so numerous to be counted here; for example, al-Khwārizmī: Neugebauer 1962, pp. 57–58, al-Battānī (d. 929): Nallino 1899–1907, vol. 1, p. 58, Swerdlow 1973, pp. 99–100; ʿAbd al-Raḥmān al-Khāzinī's *Zīj al-muʿtabar al-sanjarī* (completed in 1121, see Mozaffari 2022, p. 517) V.4.2&3: V: ff. 63r–v, L: f. 37v, *Wajīz*, f. 21r; ʿAbd al-Karīm al-Fahhād's (*ca.* 1176) *ʿAlāʾī zīj* (see Kennedy 1956, no. 84, Pingree 1985, King *et al.* 2001, p. 45, Mozaffari 2019, 2023), the tables on pp. 195–199, which are quoted in the anonymous *Sulṭānī zīj*, ff. 172v–174r; the anonymous *Sulṭānī zīj* (written in Yazd, central Iran, *ca.* 1290 CE, not confused with Ulugh Beg's work with the same title), IV.20: ff. 136r–v; Muḥammad b. Abī ʿAbd-Allāh Sanjar al-Kamālī's (known as Sayf al-Munajjim, *ca.* 1300) *Ashrafī zīj* V.11: f. 141r; and Wābkanawī's *Zīj al-muḥaqqaq* III.11.2: T: f. 62v, Y: ff. 112v–113r, P: ff. ff. 94r–v.

23 al-Bīrūnī, *India*, vol. 1, p. 6.

A straightforward examination using NASA's Five Millennium Catalogue of Solar and Lunar Eclipses reveals an intriguing pattern. Between 800 and 950 CE, there were exceptionally few opportunities to witness total or annular solar eclipses in the Middle East's significant cities – places where early Islamic astronomers conducted their work. These cities included Baghdad, Damascus (with the exception of the annular eclipse on August 8, 891, which al-Battānī observed as a partial eclipse in Raqqa), Raqqa, and Antakya (Antioch, one of the two places where al-Battānī performed his observations, except for the annular eclipse on April 26, 906). (This particular eclipse falls outside the period of al-Battānī's documented observations, 883–901 CE, as recorded in his *Ṣābi' zīj*. There is no evidence that he witnessed it.[24]) Given this context, it seems unlikely that an annular solar eclipse or a solar eclipse with a perceptible duration of totality served as direct observational evidence for early Islamic astronomers to explain their adherence to the Indian hypotheses regarding the solar and lunar apparent diameters.[25] However, there is ample evidence supporting the notion that observations by Īrānshahrī and Sarakhsī significantly contributed to firmly establishing the Indian hypotheses and recognizing annular eclipses as a distinct type of solar eclipses during the late Islamic period. Further clarification awaits in the following passages.

Bīrūnī explored three potential ways in which the solar and lunar apparent diameters could change, specifically to explain solar eclipses with perceptible durations of totality (as mentioned in the third quotation earlier). Islamic astronomy had already embraced the third approach two centuries before Bīrūnī. Note that he did not speak of the relation between the apparent diameters and annular eclipses – such consideration was not necessary. To incorporate annular eclipses into the Ptolemaic framework, one simply needed to reject the notion that the Sun's angular diameter remained permanently constant. In Bīrūnī's terms, this meant accepting the Indian hypotheses that the solar apparent diameter varies. Once this adjustment was made, the possibility of annular eclipses became feasible. Put simply, the assumption that the solar and lunar apparent diameters are equal at their farthest points from Earth is not in contradiction with the possibility of annular eclipses. For instance, Muḥyī al-Dīn al-Maghribī gave a comprehensive account of the possible conditions for annular eclipses (as detailed later), acknowledged the equality of solar and lunar apparent diameters at their farthest distances from Earth, but determined them to be 0;31,8°.[26] Although this imposes a seasonal limi-

24 http://eclipse.gsfc.nasa.gov. For the analysis of the solar eclipses of the Islamic period, see Said and Stephenson 1997.
25 The non-dependency of the Indian hypotheses and the observational evidence is not exclusive to early Islamic astronomy but is the case with Indian astronomy itself. Neither the possibility nor the report of the occurrence of annular eclipses is given in the Indian texts (see, for example, *Súrya Siddhánta* [1860] 1997, p. 174).
26 Although 0;31,8° is a non-Ptolemaic value for the minimum solar and lunar diameters, it is the result of a remedy of Ptolemy's computation in *Almagest* V.14. Ptolemy calculated the minimum apparent diameter of the Moon according to the data obtained from the Babylonian observations of the two lunar eclipses that occurred on April 21/22, 621 BCE and July 16/17, 523 BCE (Toomer 1998,

tation on annular eclipses (they can no longer occur during months when the Sun is farthest from Earth),[27] it does not render annular eclipses impossible.

After Bīrūnī's time, we encounter either brief descriptions or detailed accounts of annular eclipses, *kusūf ḥalqat al-nūr* ("bright ring eclipse"), in the Islamic astronomical corpus. An early account, though without mentioning a real example, is found in Khāzinī's *Zīj al-muʿtabar al-sanjarī* VIII.2.6.[28] Numerous references to annular eclipses can be found in the works penned after the founding of the Marāgha observatory. They present a clear account of the annular eclipse and numerate it as a type of solar eclipses. For example, *Īlkhānī zīj* II.9[29]; al-Ṭūsī's *Memoir* II.13,[30] a comment on *Almagest* V.14 in his *Taḥrīr al-majisṭī*[31]; Shīrāzī, *Nihāya*, *Ikhtiyārāt* II.12.3, and *Tuḥfa* II.15[32]; al-Nīshābūrī, *Kashf* and *Sharḥ*[33];

---

pp. 253–254, Pedersen 1974, pp. 208–209). Ptolemy reached the value of 0;31,20° for the angular diameter of the Moon at the instants of the maximum phases of the two eclipses. Al-Maghribī realized that on the basis of the same data presented by Ptolemy, at the times of both the eclipses, the Moon was not located at its maximum distance from the Earth. As a result, 0;31,20° can by no means be the minimum value of the lunar apparent diameter. Then, he proceeds (*Talkhīṣ*, f. 94r) to calculate the minimum apparent diameter of the Moon. Based on Ptolemy's data, in the first and second eclipses, the true epicyclic anomaly of the Moon was, respectively, 20° and 28°. The mean value of 24° then simply results in that the Moon was located at a distance of 64;50ᵖ from the Earth. Thus, the apparent diameter of the Moon at its greatest distance from the Earth should be equal to (64;50 × 0;31,20)/65;15 = 0;31,8°.

27 For more information about this limitation and an optical limitation of the visibility of eclipses as discussed in the medieval period, see Mozaffari 2015, pp. 131–138.

28 al-Khāzinī, *Zīj*, V: f. 85r, L: f. 51r, *Wajīz*, f. 28r. It is worth noting that no annular eclipse was observable in Marw ($\varphi = 37;40°$ N, $L = 62;12°$ E), where Khāzinī was based, between 1070 and 1130 CE.

29 al-Ṭūsī, *Īlkhānī zīj*, C: f. 24r, P: f. 24r. "It is possible for the whole of the Sun's body not to be eclipsed and the ring of light, '*ḥalqah al-nūr*', remains, if the apparent latitude (of the Moon) is not zero." However, the principal condition of the occurrence of annular eclipses (that the Moon's apparent diameter is less than that of the Sun) is not mentioned in the passage.

30 al-Ṭūsī, *Tadhkira* 1993, pp. 238–239.

31 al-Ṭūsī, *Taḥrīr*, P1: p. 167, P2: f. 43v, P3: f. 65v, B: f. 66r, AS: f. 48r (see Saliba 1987, pp. 148–149).

32 al-Shīrāzī, *Nihāya*, f. 134r; *Ikhtiyārāt*, f. 102v; *Tuḥfa*, f. 96r. The rounded values for the solar and lunar apparent diameters (Sun: 0;31–0;33°; Moon: 0;29–0;36°) were very often cited in preliminary treatises of this kind; see, for example, al-Ṭūsī, *Tadhkira* 1993, pp. 236–237; Shīrāzī, *Nihāya*, f. 134r.

33 al-Nīshābūrī, *Kashf*, f. 177v; *Sharḥ*, f. 69v. Nīshābūrī mentions the non-Ptolemaic values for the extremes of the solar apparent diameter: 0;31,3° and 0;33,33°, which are the results obtained from Ptolemy's data in *Almagest* V.14 (Toomer 1998, pp. 253–254) by the application of the Indian hypothesis that the apparent diameter is directly proportional to the angular velocity:

> The [angular] distance of the Sun from its apogee [= solar anomaly] at that time [i.e., the times cited by Ptolemy for the two lunar eclipses of 21/22 April 621 BCE and 16/17 July 523 BCE] was approximately equal to 40.5°. . . [comment: the solar true longitude at the times of the eclipses were, respectively, 27;3° and 108;12°. Since the solar apogee in the Ptolemaic context was fixed at 65;30°, the solar anomaly at the two times were, respectively, about 38.5° and 42.5°. It appears that our author took the mean of the two, i.e., 40.5°]. The hourly velocity of the Sun at that distance is 0;2,23,42°. Thus, if one wishes to determine the [apparent] semidiameter [read: diameter] of the Sun when it is at the apogee, [the procedure is as

al-Maghribī's *Talkhīṣ al-majisṭī* (see later text); Wābkanawī's *Muḥaqqaq Zīj* III.13.22[34]; Kamālī's *Ashrafī Zīj* V.18[35]; and Kāhsī's *Khāqānī zīj* III.14.[36]

In the last three works, a separate chapter is devoted to the annular eclipse and its variants with regard to the apparent (topocentric) latitude of the Moon. Al-Maghribī's text reads as follows:

> I say: a total (solar) eclipse appears as a bright ring when the (apparent) diameter of the Moon is smaller than the (apparent) diameter of the Sun, and the apparent latitude of the Moon is zero at the mid-eclipse (i.e., at the time of the maximum phase of eclipse). In this case, the bright ring encircles the body of the Moon equally (i.e., anywhere around the lunar disk, the thickness of the ring is the same). If the lunar (apparent) latitude is not zero, the thickness of the light (i.e., bright ring) varies. When the (lunar) latitude leans southward (with respect to the ecliptic), the thicker part of the bright ring appears toward the north. [Conversely,] if the (lunar) latitude tilts northward, the thicker portion shifts to the south.[37]

Similar accounts are extant in Wābkanawī's and Kamālī's *zījes*; the former provides additional details about the variations in the annularity, such that his account covers the situation in which the lunar disk is internally tangent to the solar one.[38]

A common characteristic shared by these works is that they were written in the "eastern" Islamic realm, and their authors were well acquainted with Bīrūnī and his reports on the two famous observations by Sarakhsī and Īrānshahrī.[39] This makes a clear difference in the contextual situation of annular eclipses between the late and classic Islamic periods. One may not hesitate to argue that both the observational evidence and the hypothetical grounds were sufficient to accept, justify, and predict annular eclipses in the late period. It was in such a context that

---

follows]. The ratio of 0;31,20 to 0;2,23,42 is equal to the ratio of the unknown [quantity, i.e., the apparent diameter of the Sun at the apogee] to 0;2,22,25 which is the hourly velocity of the Sun at the apogee. Thus, the [apparent] semidiameter [read: diameter] of the Sun at the apogee is 0;31,3°.

By means of a similar procedure, he determined the solar apparent diameter at the perigee (where the hourly velocity of the Sun is 0;2,33,52°) to be 0;33,33° (*Kashf*, ff. 157r–v). Cf. al-Maghribī's other sort of remedy of Ptolemy's calculations in note 26 earlier. Both examples illustrate how the two astronomers of the late Islamic period tried to extract the acceptable ranges of the values for the variation of the apparent diameter of the Sun on the basis of the same data found in *Almagest* V.14, but in two different ways.

34 Wābkanawī, *Zīj*, ff. 126v–127r.

35 Kamālī, *Zīj*, ff. 152v–153r.

36 Kāshī, *Zīj*, f. 89r.

37 al-Maghribī, *Talkhīṣ* VI.17: f. 106r.

38 For the translation of the passage, see Mozaffari 2015, p. 124.

39 For example, Shīrāzī, in his *Ikhtiyārāt* (II.6: f. 49v) and *Tuḥfa* (f. 37r), quoted Bīrūnī's earlier statements with some commentary notes.

Wābkanawī had the opportunity to predict and observe the solar eclipse of 1283 at Mughān as annular (see the next Chapter).

Finally, there are two more remarks left to say about the Indian hypotheses and their interchange with Islamic astronomy. First, in the simple relation $\theta = k \cdot v$, the aforementioned values for k are derived from the values for the diameters of the Sun and Moon and their mean distances from the Earth, as accepted in Indian astronomy. The simple relation appears to be the *standard formula* for determining the angular diameters of the Sun and the Moon in Islamic astronomy; nevertheless, the Islamic astronomers were not aware of the original formulas, and of the fact that the simple relations are derived from them, because they used different values for the planetary magnitudes and distances. The Indian astronomical texts available to them, for example, Brahmagupta's *Khandakhadyaka*,[40] did not supply them with any further information about the procedure but only mentioned the simple relations. Moreover, in his *India*, Bīrūnī only mentioned the simple relation without explaining how it was derived.[41] Second, determining the θ values depends on the instantaneous velocity. Although the medieval formulas for computing the angular diameters originated from Indian astronomy, the velocities of the Sun and the Moon were calculated on the basis of the Ptolemaic solar and lunar models, according to the instructions in *Almagest* VI.4 implemented with a minor refinement made during the medieval period. As a result, it appears that the medieval astronomers had to include the Indian simple relations into the framework of the Ptolemaic models. In other words, here we see a synthesis of Indian astronomy with Greek astronomy or, strictly speaking, a successful setting of a relatively primitive formula with a carefully calculated variable.[42] The adjusted Indian formulas found their way into European astronomy in the late medieval period.[43]

2.3 Discussion and Conclusion

Medieval science was, by and large, tradition-based. Astronomers of that era sought to understand and articulate precisely what had been defined for them. In this context, three essential elements converged: a meticulously "defined" phenomenon, compelling "observational evidence" to bolster it, and a readily available tradition to provide justification. No unidentified or elusive phenomena

---

40 Brahmagupta, pp. 62 and 118–120.

41 al-Bīrūnī, *India* 1910, pp. 79–80.

42 However, there are, in medieval Islamic astronomy, some preceding instances for syntheses like this, although not as successful as our instance; for example, setting a simple sinusoidal function borrowed from the Indian planetary latitude theory with the extremal latitude values applied to Ptolemy's *Handy Tables* for preparing the latitude tables in the *Zīj al-mumtaḥan* (Viladrich 1988, esp. p. 266). For some significant impacts of Indian astronomy on Islamic astronomy, see Mozaffari 2015, pp. 137–138.

43 Swerdlow 1973, pp. 98, 105, n. 9.

could breach this solid framework unless all these conditions – and perhaps a few more – were met.

To delve into the observation of annular eclipses during the pre-telescopic era, we must consider both optical conditions and historical context. First, the optical aspects: during an annular eclipse, the "ring phase" appears less pronounced because a portion of the Sun remains unobscured. Consequently, the loss of daylight is relatively minor.[44] Detecting the eclipse's depth with the naked eye proves nearly impossible. However, the annulus becomes visible when the Sun hovers near the horizon – shortly after sunrise or just before sunset – during the ring phase (as was the case with Īrānshahrī's observation of 873). Additionally, mist or light cloud can reveal the ring due to the atmospheric extinction of sunlight. Ingeniously, pinhole image devices, akin to the one described in Wābkanawī's *zīj*, also capture the ring's clarity and were likely employed during his 1283 observation.[45]

Now, let us explore historical considerations: annular eclipse paths across the globe are typically narrow. To witness such an event, the eclipse trajectory must intersect specific locations and significant cities where astronomers diligently observed. At the precise moment of an annular eclipse, a cadre of knowledgeable astronomers or keen-eyed observers (not mere local astrologers or timekeepers) should have been present to distinguish it from a total eclipse. Yet the rarity of trustworthy historical reports suggests that fulfilling all these requirements simultaneously was unlikely. From Īrānshahrī's 873 observation of an annular eclipse to Bīrūnī's later mention, approximately one and a half centuries elapsed. During this period, extensive eclipse observations occurred in the Middle East (by al-Māhāhnī, the Banū Amājūr, al-Battānī, and others). However, as previously stated, annular eclipses remained elusive in the Middle Eastern locations where these astronomers worked and during their active observation periods. Moreover, it is highly improbable that they had access to Īrānshahrī's report.

Early Islamic astronomers boldly challenged Ptolemy's steadfast hypothesis of an unchanging solar apparent diameter. Instead, they embraced the Indian simple formula, which acknowledged the solar angular diameter's variability. This shift occurred during the nascent stages of Islamic astronomy's flourishing era. Their reasoning was straightforward: Ptolemy's hypothesis seemed both unreasonable and restrictive (as highlighted in the last quotation from Bīrūnī). It risked permitting ill-conceived homocentric solar models (such as Abū Jaʿfar al-Khāzin's). Their line of reasoning was clear: if the solar distance to Earth fluctuated, its angular diameter must follow suit. However, these early astronomers remained unaware of the incidental consequence of denying Ptolemy's hypothesis – the possibility of annular eclipses. Perhaps this oversight arose because the phenomenon itself was not defined for early Islamic astronomers. None of their preceding

44 Stephenson 1997, p. 62.
45 Wābkanawī, *Zīj*, IV.15: T: ff. 92r–v, Y: ff. 159r–v. See Section 11.4.12.

oriental–occidental astronomical traditions had equipped them with knowledge of it, nor was there sufficient observational evidence.

However, everything changed dramatically during the late Islamic period (from approximately 1000 CE). Empirical evidence, confirmed by the remarkable figure Bīrūnī, now played a pivotal role in theoretical astronomy. The Indian hypotheses regarding solar and lunar angular diameters, along with the mathematical aspect – the Indian simple formula for computing angular diameter – were not only known but also well-established. Thus, defining the phenomenon, later coined as the "Bright Ring," as a distinct type of solar eclipse posed no difficulty. It was in such an atmosphere that in nearly all astronomical treatises written in the Middle East during this period (even the most elementary ones), separate chapters were devoted to annular eclipses. Curiously, no account or even a tentative allusion to the phenomenon's possibility can be found in astronomical treatises or *zījes* from the early Islamic period (before ~1,000 CE). This situation also holds true for "Western" Islamic astronomy, where a separate tradition flourished in southern Spain and northwestern Africa.

In summary, early astronomers possessed the foundational elements necessary to "justify" the phenomenon. However, the phenomenon itself remained undefined to them, and they lacked any direct "observational evidence." The annular eclipse only stepped into the spotlight when these three critical factors converged. It is a rare instance where the process – from defining and establishing the phenomenon to justifying and eventually predicting it – unfolded seamlessly within its context. The available traditions interwove like intricate threads, eliminating the need for additional concepts or novel computational techniques.

Notably, beyond medieval astronomy in the Middle East (from the eleventh century onward, when annular eclipses gained widespread recognition), the phenomenon primarily resided within theoretical astronomy during ancient and medieval periods. Its resurgence occurred three pivotal times when the idea of the homocentricity in the planetary configurations arose: in the sixth century, Simplicius challenged the prevailing homocentric models of his era; in the eleventh century, Bīrūnī confronted Abū Jaʿfar al-Khāzin's solar model; and in the fifteenth century, a pseudo-Regiomontani text titled *Refutatio errorum Alpetragii de motibus celestibus* countered al-Biṭrūjī's homocentric models (d. *ca.* 1204).[46] Annular eclipses served as exclusive evidence, exposing serious deficiencies and, ultimately, leading to the rejection of these flawed models.

References

Anonymous, *Sulṭānī zīj*, MS. Iran: Parliament Library, no. 184.
Anonymous, 1860, *Súrya Siddhánta: A Textbook of Hindu Astronomy* (Reprinted in 1997), Gangooly, P. and Burgess, E. (eds. & trs.), Delhi: Motilal Banarsidass.

---

46 The text says that the eclipse of June 17, 1433, was annular and then uses it as evidence against the homocentric models. The eclipse was, of course, total (see Shank 1992, pp. 17–18, 1998, pp. 162, 163, Swerdlow 1999, pp. 5, 22, n. 7).

al-Bīrūnī, Abū al-Rayḥān, 1954–1956, *al-Qānūn al-mas'ūdī* (*Mas'ūdīc canons*), 3 Vols., Hyderabad-Dn: Osmania Oriental Publication Bureau.

al-Bīrūnī, Abū al-Rayḥān, 1910, *Alberuni's India*, Sachau, E. C. (En. tr.), 2 Vols., London: Kegan Paul, Trench, Trübner & Co.

Bowen, A. C., 2008, "Simplicius' commentary on Aristotle, *De Caelo* 2.10–12: An annotated translation (Part 2)", *SCIAMVS: Sources and Commentaries in Exact Sciences* **9**, pp. 25–131.

Brahmagupta, 1970, *The Khandakhadyaka*, Chatterjee, B. (ed. and tr.), New Delhi: Bina Chatterjee.

Goldstein, B. R., 1979, "Medieval observations of solar and lunar eclipses", *Archives Internationales d'Histoire des Sciences* **29**, pp. 101–156.

Goldstein, B. R., 1996, "Lunar velocity in the Middle Ages: A comparative study", in: Casulleras, J. and Samsó, J. (eds.), *From Baghdad to Barcelona: Studies in the Islamic Exact Sciences in Honour of Prof. Juan Vernet*, Barcelona: Instituto "Millás Vallicrosa", pp. 181–194.

Ibn al-Fahhād: Farīd al-Dīn Abu al-Ḥasan 'Alī b. 'Abd al-Karīm al-Fahhād al-Shirwānī or al-Bākū'ī, *Zīj al-'Alā'ī*, MS. India: Salar Jung Library, no. H17.

al-Kamālī, Muḥamamd b. Abī 'Abd-Allāh Sanjar, *Zīj-i ashrafī*, MS. Paris, Biblithèque nationale, suppl. Pers. No. 1488.

al-Kāshī, Ghiyāth al-Dīn Jamshīd, *Khāqānī zīj*, MS. England, India Office, no. Persian 430 (Ethé 2232).

Kennedy, E. S., 1956, *A Survey of Islamic Astronomical Tables*, Philadelphia: American Philosophical Society.

al-Khāzinī, 'Abd al-Raḥmān, *al-Zīj al-mu'tabar al-sanjarī* (*The experimental zīj for* [*sultan*] *Sanjar*, Sanjar b. Malikshāh I b. Alp Arsalān, the last sultan of the Saljuq Empire, b. 1085 CE, reg. 1097–1157 CE), MSS. V: Vatican, Biblioteca Apostolica Vaticana, no. Arabo 761 (undated, but certainly after Khāzinī's death, from a reference on f. 81v), L: London, British Library, Or. 6669 (completed on 25 Jumādā II 620 = 26 July 1223; f. 157r); *Wajīz* [*Compendium of*] *al-Zīj al-mu'tabar al-sanjarī*, MSS. I: Istanbul, Süleymaniye Library, Hamidiye collection, no. 859 (Rabī' II 667 H = December 1268–January 1269 CE; f. 38r); S: Tehran: Sipahsālār, no. 682 (Ramaḍān 631 H = May–June 1234 CE; p. 123).

King, D., Samsó, J., and Goldstein, B., 2001, "Astronomical handbooks and tables from Islamic world", *Suhayl* **2**, pp. 9–105.

al-Maghribī, Muḥyī al-Dīn, *Adwār al-anwār*, MS. Iran, Mashhad, Sacred Shrine Library, no. 332.

al-Maghribī, Muḥyī al-Dīn, *Talkhīṣ al-majisṭī*, MS. Leiden, Or. 110.

Mozaffari, S. M., 2013, "Historical annular solar eclipses", *Journal of the British Astronomical Association* **123**, pp. 33–36.

Mozaffari, S. M., 2015, "Annular eclipses and the considerations about the solar and lunar angular diameters in the medieval astronomy", in: Orchiston, W., Green, D. A., and Strom, R. (eds.), *StephensonFest: Studies in Applied Historical Astronomy and Amateur Astronomy*, New York: Springer, pp. 119–142.

Mozaffari, S. M., 2019, "Ibn al-Fahhād and the Great Conjunction of 1166 AD", *Archive for History of Exact Sciences* **73**, pp. 517–549.

Mozaffari, S. M., 2022, "A mechanical concentric solar model in Khāzinī's *Mu'tabar zīj*", *Archive for History of Exact Sciences* **76**, pp. 513–529.

Mozaffari, S. M., 2023, "Sources of the planetary theories in Fahhād's *'Alā'ī zīj*: Solving a medieval case of intellectual fraud", *Suhayl* **20**, pp. 141–219.

Nallino, C. A. (ed.), 1899–1907, *Al-Battani Sive Albatenii Opus Astronomicum*, 3 Vols., Milan: Milan University Press.

Neugebauer, O., 1962, *The Astronomical Tables of Al-Khwārizmī*, in: Hist. Filos. Skr. Dan. Vid. Selsk. 4, no. 2, Copenhagen.

Neugebauer, O., 1975, *A History of Ancient Mathematical Astronomy*, Berlin-Heidelberg-New York: Springer.

al-Nīshābūrī, Niẓām al-Dīn Aʿraj, *Kashf al-ḥaqāʾiq al-zīj al-Īlkānī* (*Opening of truths of the* Īlkānī zīj, a Commentary on al-Ṭūsī's *Īlkānī zīj*), MS. Iran, Parliament Library, no. 1210 (copied at Jumādā I 825 H/April-May 1422).

al-Nīshābūrī, Niẓām al-Dīn Aʿraj, *Sharḥ-i Tadhkira* (*Explanation of the* Memoir, a commentary on al-Ṭūsī's *al-Tadhkira fī al-hayʾa*), MS. University of Tehran, no. 488.

Pedersen, O., 1974, *A Survey of Almagest*, Odense: Odense University Press, 1974; with annotation and new commentary by A. Jones, New York: Springer, 2010.

Pingree, D. (ed.), 1985, *Astronomical Works of Gregory Chioniades*, Vol. 1: *Zīj al-ʾalāʾī*, Amsterdam: Gieben.

Said, S. S. and Stephenson, F. R., 1997, "Solar and lunar eclipse measurements by medieval Muslim astronomers, II: Observations", *Journal for the History of Astronomy* **28**, 29–48.

Saliba, G., 1987, "The role of the *Almagest* commentaries in medieval Arabic astronomy: A preliminary survey of Ṭūsī's Redaction of Ptolemy's *Almagest*", *Archives Internationales d'Histoire des Sciences* **37**, pp. 3–20; reprinted in Saliba 1994, pp. 143–160.

Saliba, G., 1994, *A History of Arabic Astronomy: Planetary Theories during the Golden Age of Islam*, New York: New York University Press.

Samsó, J., 1977, "A homocentric solar model by Abu Jaʿfar al-Khāzin", *Journal for the History of Arabic Science* **1**, pp. 268–275; reprinted in Samsó 1994, Trace XI.

Samsó, J., 1994, *Islamic Astronomy and Medieval Spain*, Aldershot: Variorum.

Saslaw, W. C. and Murdin, P., 2004, "The double heads of Isturus: The oldest eclipse on a coin?", *Journal for the History of Astronomy* **35**, pp. 21–27.

Shank, M. H., 1992, "The notes on al-Biṭrūjī attributed to Regiomontanus: Second thoughts", *Journal for the History of Astronomy* **23**, 15–30.

Shank, M. H., 1998, "Regiomontanus and homocentric astronomy", *Journal for the History of Astronomy* **29**, 157–166.

al-Shīrāzī, Quṭb al-Dīn, *Ikhtiyārāt-i muẓaffarī* (*Muẓaffarid Selections*, dedicated to Muẓaffar al-Dīn Bulāq Arsalān, d. 1305), MS. Iran, National Library, no. 3074F, up to f. 176r (copying completed on Monday, 20 Jumādā I 682/16 August 1283).

al-Shīrāzī, Quṭb al-Dīn, *Nihāyat al-idrāk fī dirāyat al-aflāk* (*Limit of Comprehension in the Knowledge of Celestial Spheres*), MSS. B: Berlin, no. Ahlwart 5682 = Petermann I 674 (towards the end of Dhi al-qaʿda 726/October 1326), P1: Iran, Parliament Library, no. 6457 (copied at Ulugh Beg's school at Samarqand one month during Dhi al-qaʿda and Dhi al-ḥijja 844/April–May 1441), P2: Iran, Parliament Library, no. 16008 (5 Muḥarram 112 0/27 March 1708).

al-Shīrāzī, Quṭb al-Dīn, *Tuḥfa al-shāhiyya fī ʾl-hayʾa* (*Gift to the King on Astronomy*), MS. Iran, Parliament Library, no. 6130 (copied in Rabīʿ I 703 H/October 1303 CE).

Simplicius, 1894, *Simplicii in Aristotelis De Caelo Commentaria*, Heiberg, I. L. (ed.), Berlin: Reimer.

Stephenson, F. R., 1997, *Historical Eclipses and Earth's Rotation*, Cambridge: Cambridge University Press.

Swerdlow, N. M., 1973, "Al-Battānī's determination of the solar distance", *Centaurus* **17**, pp. 97–105.

Swerdlow, N. M., 1999, "Regiomontanus' concentric-sphere models for the Sun and Moon", *Journal for the History of Astronomy* **30**, pp. 1–23.

Toomer, G. J., 1998, *Ptolemy's Almagest*, Princeton: Princeton University Press.

al-Ṭūsī, Naṣīr al-Dīn Muḥammad, *al-Tadhkira fī [ʿilm] al-hayʾa* (*Memoir on [the Science of] Cosmography*), 1993, *Naṣīr al-Dīn al-Ṭūsī's* Memoir on Astronomy, Ragep, F. J. (ed. & En. tr.), 2 Vols., New York: Springer.

al-Ṭūsī, Naṣīr al-Dīn Muḥammad, *Īlkhānī zīj*, MSS. C: University of California, Caro Minasian Collection, no. 1462, P: Iran, Parliament Library, no. 181.

al-Ṭūsī, Naṣīr al-Dīn Muḥammad, *Taḥrīr al-majisṭī* (*Exposition of the* Almagest), MSS. Iran, Parliament Library, P1: no. 3853 (commented by ʿAbd al-ʿAlī al-Bīrjandī, d. 1525/1526 CE), P2: no. 6357, P3: no. 6395, B: Staatsbibliothek zu Berlin, Sprenger 1838 = Ahlwardt 5655, ff. 1v–152r (copying finished at Marāgha on Wednesday, 22 Ramaḍān 655 ᴴ· = 3 October 1257 CE, for a certain Maḥmūd al-Khāṭīb: f. 152r), AS: Istanbul, Aya Sofia, no. 2583, ff. 1v–112v (copying completed on Thursday, 5 Shawwāl 686 ᴴ· = 13 November 1287 CE: f. 112v).

Viladrich, M., 1988, "The planetary latitude tables in the *Mumtaḥan Zīj*", *Journal for the History of Astronomy* **19**, pp. 257–268.

Wābkanawī, Shams al-Dīn Muḥammad, *Zīj al-muḥaqqaq al-sulṭānī ʿalā uṣūl al-raṣad al-Īlkhānī (The testified zīj for the sultan on the basis of the parameter values of the Īlkhanid observations)*, MSS. T: Turkey, Aya Sophia Library, No. 2694; MS. Y: Iran, Yazd, Library of ʿUlūmī, no. 546 (its microfilm is available in the central library of the University of Tehran, no. 2546), P: Iran, Library of Parliament, no. 6435.

3

WĀBKANAWĪ'S OBSERVATION AND CALCULATIONS OF THE ANNULAR SOLAR ECLIPSE OF JANUARY 30, 1283[1]

3.1 Introduction

This chapter deals with the observation and computation of the annular solar eclipse of January 30, 1283, by Shams al-Dīn Muḥammad al-Wābkanawī al-Bukhārī (*fl. ca.* 1270–1320 CE), an Iranian astronomer at the Marāgha observatory (northwestern Iran).[2] This is apparently the third confirmed observation of an annular eclipse in history and the first "prediction" of a solar eclipse of this type on the basis of the medieval theories and methods grounded in the Ptolemaic context.[3] It is also interesting from both astronomical and mathematical aspects.

Within the Ptolemaic astronomical context, the solar angular diameter is constant and equal to the minimum lunar angular diameter,[4] and therefore, for Ptolemy, the annular solar eclipse is an impossible phenomenon. On the other hand, the Indian hypotheses for determining the solar and lunar angular diameters made the Moon's minimum apparent diameter less than the Sun's and thus indicated the possibility of the phenomenon.[5] In the first part of the tenth century, Abū al-Rayḥān al-Bīrūnī (973–1048) referred to Abu 'l-ʿAbbās al-Īrānshahrī's observation of the annular eclipse of July 28, 873, in Nīshāpūr and to the fact that, based on the principles of Indian astronomy, the Sun's angular diameter may be less than

---

1 Original publication: S. M. Mozaffari, "Wābkanawī's observation and calculations of the annular solar eclipse of 30 January 1283," *Historia Mathematica* **40** (2013), pp. 235–261. © 2013 Elsevier, and republished by permission.

2 The report and numeral data were first introduced in Mozaffari 2009.

3 It is worth noting at the outset that a "prediction" can be either *retroactive* or *proactive*, but this distinction makes no difference in cases like ours, where the underlying theories deployed for predicting natural phenomena were available at that time. Many of the quantitative predictions of celestial events documented in the medieval astronomical corpus might have been retroactive.

4 Toomer 1998, pp. 252–253.

5 *Sūrya Siddhánta* IV.1–5, 41. For the Indian hypotheses and formulas, see Mozaffari 2015.

DOI: 10.4324/9781003481966-6

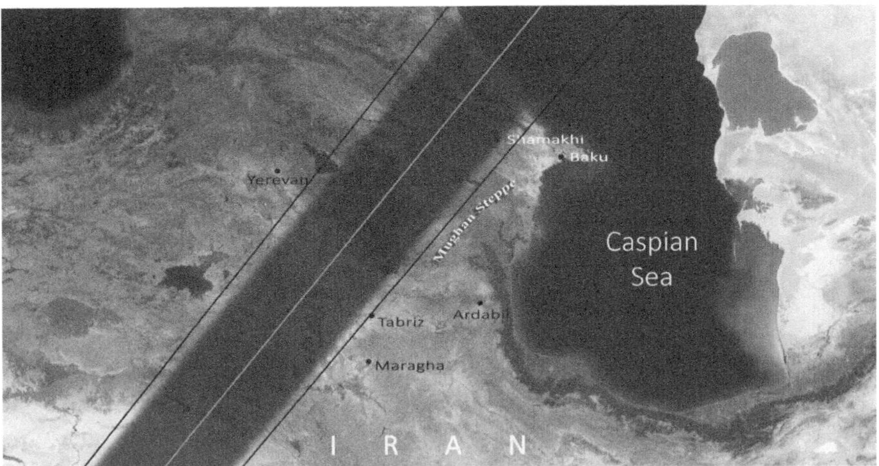

Figure 3.1 Path of the annular eclipse of 1283 (demarcated by the dark lines) in the vicinity of the Mughān steppe.

Source: Underlying map from © mapcarta.com.

the Moon's.[6] Since then, the annular (*ḥalqat al-nūr*, "bright ring") eclipse became a justified phenomenon. A clear account of it appeared in treatises and *zījes* written in the eastern Islamic lands:[7] for example, the *Īlkhānī zīj*; al-Ṭūsī's *Memoir* and his commentary on the *Almagest*; Shīrāzī, *Nihāya*, *Ikhtiyārāt*, and *Tuḥfa*; Nīshābūrī, *Kashf*, and *Sharḥ*; al-Maghribī's *Talkhīṣ al-majisṭī*; Wābkanawī's *Muḥaqqaq zīj*; Kamālī's *Ashrafī zīj*; and Kāshī's *Khāqānī zīj*.[8] It was in such a context that Wābkanawī had the opportunity to predict and observe the annular solar eclipse of 1283 in the vicinity of the Mughān steppe (Figure 3.1).

Wābkanawī's calculations involve an application of an iterative process for computing the lunar parallax and the time of occurrence of a solar eclipse. Iterative solutions were used in various problems of astronomy, for example, spherical astronomy,[9] planetary equations,[10] the sizes and distances of celestial bodies,[11]

---

6 Bīrūnī, *al-Qānūn al-masʿūdī*, vol. 2, p. 632. Bīrūnī used it as evidence for rejecting the homocentric solar model by Abū Jaʿfar al-Khāzin (d. *ca.* 961–971); see Samsó 1977, p. 274, Stephenson 1997, p. 467.

7 Their authors were well acquainted with Bīrūnī and/or his report of Īrānshahrī's observation. See Shīrāzī, *Ikhtiyārāt* II.6: f. 49v and *Tuḥfa*, f. 37r; Wābkanawī, T: ff. 103v, 121v.

8 al-Ṭūsī, *Īlkhānī zīj* II.9: C: f. 24r; P: f. 24r; *Memoir* II.13: pp. 238–239; *Taḥrīr*, P1: p. 167, P2: f. 43v, P3: f. 65v, B: f. 66r, AS: f. 48r; Shīrāzī, *Nihāya*, f. 134r; *Ikhtiyārāt* II.12.3: f. 102v; *Tuḥfa* II.15: f. 96r; Nīshābūrī, *Kashf*, f. 177v; *Sharḥ*, f. 69v; al-Maghribī, *Talkhīṣ* VI.17: f. 106r; Wābkanawī, *Zīj* III.13.22: T: ff. 70r–v, Y: ff. 126v–127r, P: ff. 106v–107r; Kamālī, *Zīj* V.18: ff. 152v–153r; Kāshī, *Zīj* III.14: IO: f. 89r.

9 Plofker 2002, esp. pp. 173–176.

10 Plofker 2002, esp. pp. 170–172.

11 Goldstein and Swerdlow 1970–1971, p. 144, Mancha 1998, pp. 35–39.

eclipses, parallax, and so on. Even if they were utilized in pure mathematics, the results were obtained for astronomical purposes. For instance, al-Kāshī used an iterative process to obtain a more accurate value for sin 1°. The result was employed in the calculation of a table of sines, which was applied in numerous types of astronomical computations or in constructing tables of planetary equations, which are sinusoidal functions.[12]

The most important applications of iterative procedures in ancient and medieval astronomy are the following:

1. The determination of the planets' orbital elements (eccentricity and longitude of the apsidal line) in the Ptolemaic context as explained in *Almagest* X.7 and XI.1&5.[13] Probably, the only surviving account of the determination of the planets' eccentricity from observational data in the Islamic period is that of Muḥyī al-Dīn al-Maghribī (d. 1283 CE) at the Marāgha observatory during the 1260s and 1270s. He repeated the process at least three times for each superior planet.[14]

2. The computation of lunar parallax in a solar eclipse. In Indian and early Islamic astronomy, an iterative method for computing the lunar parallax boiled down to an equation in the form

$$a = b + c \cdot \sin b.$$

Here, a is the projection on the celestial equator of an estimated arc of the ecliptic between the degree of the solar–lunar conjunction and the culminating point of the ecliptic at the time of the conjunction; c is a constant (equal to $-1;36$); and the term $c \cdot \sin b$ is the parallax in longitude. One first computes $b_1 = a - c \cdot \sin a$, then determines $b_2 = a - c \cdot \sin b_1$, $b_3 = a - c \cdot \sin b_2 \ldots$ up to finding a value b_n which satisfies the preceding equation to the desired precision.[15] The aforementioned expression was applicable in other astronomical computations as well.[16] It appeared later in connection with elliptical orbits and is now known as the *Kepler equation*, where a is the mean anomaly, b the eccentric anomaly, and c the planet's eccentricity. Iterative procedures for solving the equation were proposed by Kepler himself and by others.[17]

The computation of the lunar parallax in late Islamic astronomy was based on Ptolemy's *Almagest* and *Handy Tables*. The two are identical in their principles, but the input data and the necessary operations are different, as shall be described

---

12 See Aaboe 1954, Rosenfeld and Hogendijk 2002–2003. For Kāshī's method of the interpolation in the table of sines, see Hamadanizadeh 1980.
13 Toomer 1998, pp. 484–499, 507–519, 525–537, Neugebauer 1975, vol. 1, pp. 174–179. Levi b. Gerson also used iterative processes in order to obtain the parameters of his own lunar and planetary models (see Mancha 1998, pp. 22–30, 40–43).
14 *Talkhīṣ* VIII.3.7&11: ff. 124r–126v, 129r–131v, 133v–134v; see Saliba 1986, Mozaffari 2018–2019.
15 See Kennedy 1956b, pp. 50–52, Kennedy and Transue 1956, Plofker 2002, pp. 178–179.
16 See Kennedy 1969.
17 Therkel 1974, Dutka 1997, Swerdlow 2000.

later in Section 3.3, step 2. To work with these tables, one needs an iterative procedure in which a series of calculations of the longitudes, times, and the longitudinal and latitudinal components of parallax must be repeated in order to achieve the time of the apparent conjunction (mid-eclipse) and the magnitude of the eclipse (as shall be exemplified in the case of the annular eclipse of 1283, later in Section 3.3, step 4). The process is evidently not as simple as in Kepler's equation.

The iterative procedure for finding the parameters of a solar eclipse is one of the most complicated computations in medieval astronomy. Perhaps this is the reason that worked examples of solar eclipses are rarely found in the medieval astronomical literature.[18] One may well ask if a general description of the method without computational examples would be sufficient for understanding the whole procedure. From the late Islamic period, I know two other worked examples for the solar eclipses containing the iterative process: July 5, 1293, and October 28, 1296, calculated by Shams al-Bukhārī (i.e., Wābkanawī himself; see Section 3.2) for the latitude of Tabriz ($\varphi \approx 38°$) and preserved in Byzantine Greek sources.[19] In the case of lunar eclipses, where the computation does not require any iterative process, at least five worked examples may be found in *zījes* contemporary with Wābkanawī.[20]

Section 3.2 concerns the life and the career of Wābkanawī. Section 3.3 introduces and comments at length on our author's procedure, calculations, and numerical details, both in his astronomical context and with regard to modern data. The report may also be employed in order to elucidate, as a case study, the quality of observational astronomy in the late Islamic period in terms of the extent to which the parameters adopted in the Islamic *zījes* could lead to the results that agreed with the observational data and the degree to which the astronomers were aware of the accuracy of their parameters. This will be discussed in Section 3.4.

3.2 Wābkanawī's Life and Career

Shams al-Dīn Muḥammad al-Wābkanawī al-Bukhārī (known as Shams al-Munajjim)[21] was a member of the Marāgha observatory, which was founded by

18 See also Levi ben Gerson's computations and observational reports of the four solar eclipses from 1321 to 1337 in Orange, Southern France (see Goldstein 1979).

19 Some computational examples concerning the various parameters of solar eclipses for the latitude of Shīrāz may be found in the *Ashrafī zīj* V.18, but the dates were not indicated. Some of them, however, appear to be related to the solar eclipse of April 3, 1307 (based on the given solar longitude in the beginning and end of the eclipse, around 20°–22°, and the longitude of the ascending node, 203;36°; f. 149r). In addition, there is the computation of the eclipse of October 28, 1296, in the anonymous *Sulṭānī zīj* for the latitude of Yazd (ff. 138v–140r). No iterative process is explained or embedded in these examples.

20 The lunar eclipses of May 30, 1295, and November 23, 1295, in the *Sulṭānī zīj* (ff. 137r–138r) and those of January 4, 1303, May 9, 1305, and December 14, 1312, in the *Ashrafī zīj* (ff. 133v–134r, 145v–146r).

21 For a biographical outline of him, see van Dalen 2007. On his *zīj*, see Kennedy 1956a, no. 35; King and Samsó 2001, p. 46, Kennedy 1958, p. 251, Haddad and Kennedy, p. 91, Kennedy 1964, p. 443, King 1986, pp. 138–140, Kunitzsch 1964, pp. 398–399. Kennedy (1960, p. 211) employed the explanations given by Wābkanawī as regards the Marāgha observatory to verify some remarks

Hülegü (the first ruler of the Ilkhanid dynasty of Iran, d. 1265) and directed by Naṣīr al-Dīn al-Ṭūsī (d. 1274). His sons continued to direct it during the second period of its activities, in the period 1283–1320.[22] If Wābkanawī's identification with Σάμψ Πουχαρής (= Shams al-Bukhārī) in Greek sources is correct (see later text), then the date June 11, 1254, can be taken as his date of birth.

Wābkanawī was the royal astronomer of Ghāzān Khān (the seventh ruler of the Ilkhanid dynasty, b. December 4, 1271; reign: October 21, 1295–May 17, 1304), upon whose order he began to compile a *zīj* and a calendar with a new chronology.[23]

The *Zīj al-muḥaqqaq al-sulṭānī* ("Testified *zīj* for the sultan") was completed, according to our author, over an observational period of more than 40 years[24] in Marāgha and Tabriz, where Ghāzān had founded a new observatory. The first observation cited in this *zīj* was that of the Moon in Marāgha on December 3, 1272.[25] The last observation was that of the triple great conjunction of Jupiter and Saturn in 1305–1306.[26] Besides the annular eclipse of 1283, our author also observed the great conjunction of 1286.[27] Other observations were also mentioned in the work in order to show the inaccuracies in the preceding *zīj*es (see later text). The process of writing the *zīj* continued during the reign of Uljāytu (the eighth ruler of the Ilkhanid dynasty), and the work was finally dedicated to Sulṭān Abū Saʿīd Bahādur (the ninth ruler of the Ilkhanid dynasty; reign: 1316–1335 CE).[28] Therefore, the evidence confirms our author's claim that the task took around 40 years.

The new chronology established by Wābkanawī was called *Khānī*.[29] It became so widespread that Abū Muḥammad ʿAṭā al-Samarqandī, a fourteenth-century Muslim astronomer in the service of the Mongolian Yüan dynasty of China, also utilized it in his *zīj*.[30]

The Marāgha observatory was exceptional in many aspects, for example, the financial system applied in its directorship, the variety of astronomical instruments, the enormous costs of work of education and observation, and the number of scholars. The result of these activities in observational and mathematical astronomy was the *Īlkhānī zīj*, which was completed around 1270. The work was disappointing because its planetary parameters (e.g., mean motions, eccentricities)

---

by al-Kāshī in a letter to his father and, in another paper (1962, p. 24), quoted a section of the *zīj* related to chronology and astrology.

22 See Sayılı 1960, pp. 224–232.

23 Wābkanawī, T: f. 2v, Y: f. 3v, P: f. 3v.

24 Wābkanawī, T: f. 1v, Y: f. 2r, P: f. 2r.

25 Wābkanawī, T: ff. 89v–90r, Y: f. 155r, P: ff. 135r–136r.

26 Wābkanawī, T: f. 125r, Y: f. 235r, P: ff. 205r–v.

27 Wābkanawī, T: f. 2r, Y: ff. 2v–3r, P: ff. 2v–3r.

28 Wābkanawī, T: f. 4r, Y: f. 6r, P: f. 5v. He states that a preliminary edition of the *zīj* had been dedicated to Uljāytu, who had disseminated it by commanding to distribute the copies of it.

29 Wābkanawī, II.6: T: ff. 28r–30r, Y: ff. 49v–54r, P: ff. 42r–45v.

30 Kennedy 1987–1988, pp. 63, 74. However, he substituted the Mongolian months used by Wābkanawī for the Persian months.

were based on the preceding *zīj*es. In addition, some glaring deficiencies characterized various parts of the work. It seems that scholars soon realized those clear defects and were displeased with the application of the *zīj*. The historical sources contain some apocryphal stories on the astronomical activities in Marāgha and the corrections made in the *Īlkhānī zīj* after al-Ṭūsī, the first director, died in 1274.[31] Despite this, and probably out of respect for its main author, the *Īlkhānī zīj* continued to be used, especially for astrological prognostications, or with further corrections.[32]

A characteristic of Wābkanawī's *zīj* is its criticism of the preceding *zīj*es, especially the *Īlkhānī zīj*, on the grounds that the *calculated positions* of the seven planets were never in agreement with the *observed positions*. Wābkanawī enumerated the *zīj*es he had used during his career and criticized them for their deficiencies in the observational aspect of astronomy.[33] That he had sufficient knowledge of the preceding *zīj*es is easily demonstrable from his comparative discussions of the mathematical methods in the various *zīj*es. For example, he explained the different procedures in the *Almagest*, the Indian traditions, and four Islamic *zīj*es for calculating the time of the maximum phase of solar eclipses.[34] Later, in the process of calculating the parameters of the annular eclipse of January 30, 1283, he produced evidence to the effect that Khāzinī's procedure reduced the calculation steps (see Section 3.3, step 4). Throughout the *zīj*, our author emphasized that "the *calculated positions* of the planets must be in agreement with their *observed ones*." He directed his critiques especially at the *Īlkhānī zīj*. The following quotation may serve as an example:

||T: 2r|| ||Y: 2v|| ||P: 2v|| Therefore, the positions of planets calculated on the basis of the *zīj*es which are fashionable and current among people in this day do not agree with the observed positions of the planets. Because [in the case of] those great men who constructed those Tables, despite

---

31 For example, Rukn al-Dīn al-Āmulī (T: ff. 1v–2r, P: ff. 1v–2r) said that Aṣīl al-Dīn, one of al-Ṭūsī's sons, appealed to Shīrāzī to correct the errors, but the latter only made some minor corrections, mostly with respect to the planetary mean motions, some of which were reported by Āmulī. Āmulī went on to say that some members of the observatory, including al-Maghribī, corrected the *zīj* by means of observations in a 30-year period. This statement is in contradiction with the historical facts; for example, al-Maghribī worked in a solitary manner in the observatory and died in 1283 before the so-called 30-year period was finished. Āmulī's statement that the *Īlkhānī zīj* is different from the *Īlkhānī observations* also appears to be a misinterpretation of Wābkanawī's statements (see later text). Incidentally, it is worth mentioning that Āmulī was the main source of Sayılı's account of the activities in the Marāgha observatory in the second period (1283–*ca.* 1320) (Sayılı 1960, pp. 214ff). Consequently, some of Sayılı's concluding remarks have to be treated with caution.

32 al-Kāshī (d. 1432), for example, calls his *zīj* as *Zīj-i Khāqānī dar takmīl-i Zīj-i Īlkhānī* ("Khāqānī *zīj* for completing the Īlkhānī *zīj*") and gives in it a long list of the necessary corrections in the *Īlkhānī zīj*. See al-Kāshī, IO: ff. 3r–4r, P: pp. 22–24.

33 Wābkanawī, T: f. 3r, Y: ff. 4r–v, P: f. 4r.

34 Wābkanawī, III.13.14: T: ff. 68r–69v, Y: ff. 123r–126r, Y: ff. 104r–105r.

their perfect knowledge and abundant properties and the order of the king, their life failed them to attempt to complete [i.e., they died before completing] those important affairs. For this reason, as the occasion arose, they appealed to (the results of) the old observations; and in the course of time, those necessary fractions added up to integers.[35] And [as a result], notable divergences in the positions of the planets have appeared to such an extent that in the case of the conjunctions of the two superior planets [i.e., Jupiter and Saturn] – on which the world's commandments depend, at the two times when they were in conjunction with each other, some obvious divergences were observed. For example, in the year 684 H, the conjunction took place in the ninth degree of Aquarius [i.e., at the longitude of 309°]. The difference between the calculated [time of the conjunction] based on the *zīj* that is the most famous and reliable in these regions as well as in common use among people [apparently, our author means the *Īlkhānī Zīj*] and the [time when it was] observed was close to fifteen days. I mean, according to that *zīj*, the conjunction should have occurred ||P: 3r|| in the ninth hour of daylight on Wednesday, the twentieth [day] of [the month] Shawwāl in that year [19 December 1286], but according to observation, it took place on the night of the fifth day of the month Dhu al-Qaʿda [2 January 1286]. ||Y: 3r|| Again, according to the same *zīj*,[36] in the months of the year 705 H, the conjunction should have taken place at the end of Libra [viz. at the longitude of 209°], but according to the observation, it occurred during daylight on Friday, the thirteenth day of the Jumādā al-Ākhir [31 December 1305] in the second degree of Scorpio [viz. at the longitude of 212°]. There was a difference between the calculated [time] and the observed [time] of about eighteen days, and the degree of conjunction (*juzw-i qirān*), fell under another sign. Since then, close to that date, they [i.e., Saturn and Jupiter] formed two other conjunctions, which [the calculations] based on that *zīj* did not predict. The first conjunction: on the night of Wednesday, the fifth day of [the month] Shawwāl in that year [20 April 1306], in the situation (*ḥālat*) of retrograde motion [of the two planets]. The second conjunction: on the night of Monday, the fifth day of the month Muḥarram in the year 706 H [18 July 1306], again at the end of ||T: 2v|| Libra in the period (*zāmān*) of direct motion [of the two planets].

---

35 Here, our author alludes to a possible cause of the observational deficiencies of the *zīj*es. The values for the mean motions of the planets per day or per hour in sexagesimal fractions were determined from observations. While the mean motions were arranged in the entries of the tables of *zīj*es for greater periods of time (e.g., 1 year, 30 years), they were given at most to arc seconds and naturally were subject to rounding. As a result, in longer periods, small errors in fractions accumulated to several degrees.

36 The text in the three manuscripts reads *Tā'rīkh*, "date," which seems to have no meaning here, because the date is mentioned immediately. Rather, *Tā'rīkh* appears to be a scribal error of the word *zīj*.

The quote illustrates how our author dealt with the problem of reconciling theory and observation, an issue that only rarely arose in the medieval astronomical literature. After a period of testing the various *zījes*, he was finally convinced that Muḥyī al-Dīn al-Maghribī's basic parameter values (e.g., mean motions, equations, etc.) from his *Talkhīṣ al-majisṭī* and his last *zīj*, *Adwār al-anwār*, produced results in agreement with observations. Accordingly, Wābkanawī adopted all these parameter values in his own *zīj*. However, Wābkanawī says that he corrected minor disagreements between "the principles of the Ilkhānīd observations" (i.e., al-Maghribī's parameter values) and his own observations during the years of his career. These corrections concerned (1) the mean longitude of the Moon and Mars, (2) the mean anomaly of Venus, and (3) the latitude of the two inferior planets[37] (see Section 3.3, step 1, and Section 3.4, no. 3). Wābkanawī also arranged the entries of al-Maghribī's tables of planetary equations in a different way.

In parallel with the official activities at the Marāgha observatory, which apparently had little (if anything) to do with observations, al-Maghribī planned an extensive program to renew the determination of all the planetary parameter values (including planetary eccentricities and mean motions, the solar and lunar parallaxes, and so on). His values for the majority of the parameters differ from those in the *Almagest* and the *zījes* of his Islamic predecessors. The program took more than a decade to complete, and the results appeared in his last *zīj*, *Adwār al-anwār*. Interestingly, in a treatise named *Talkhīṣ al-majisṭī*, he explained his observations and methods for determining the parameter values on the basis of the observational data. According to the *Talkhīṣ*, the observations from which Muḥyī al-Dīn obtained his parameter values were made between March 8, 1262 (lunar eclipse), and August 12, 1274 (Jupiter).[38] Wābkanawī referred to al-Maghribī's *Adwār* and *Talkhīṣ* as "based on the New Ilkhanid Observations" (i.e., al-Maghribī's own observations) in order to distinguish them from the *Īlkhānī zīj*, which was assumed to have been based on the "Ilkhanid Observations" (i.e., the presumed observations supervised by al-Ṭūsī and performed by his colleagues).[39] Wābkanawī contends that the *Īlkhānī zīj* was based on the preceding astronomical tables rather than independent observations, and sometimes he only referred to al-Maghribī's *Adwār* as based on the "Ilkhanid Observations."[40] Although Wābkanawī had been in Marāgha since 1272, called al-Maghribī, "our master" (*mawlānā*, a prevalent title to honor great scientists of the day), and showed great confidence in the latter's astronomical heritage, there is no evidence that he had been directly instructed by al-Maghribī. Al-Maghribī died in June 1283;[41] thus, he lived to hear of the observational report of Wābkanawī. According to

37 Wābkanawī, T: f. 3r; Y: f. 4v; P: ff. 4r–v.
38 On Muḥyī al-Dīn, see Chapter 9.
39 Wābkanawī, III.3.1: T: f. 53r, Y: f. 96r; III.9.5: T: f. 60r, Y: f. 108v; III.13.6: T: f. 67r, Y: f. 120v and other places.
40 Wābkanawī, T: f. 3r, Y: f. 4v, P: f. 4r.
41 Ibn al-Fuwaṭī, vol. 5, p. 117.

the prologue of his *zīj*, Wābkanawī paid considerable attention to comparing observations of eclipse magnitudes and phases with calculations based on the *zījes* available to him.[42]

During his career, a Byzantine scholar named Gregory Chioniades was in Tabriz, where he translated the *Abridgement* (*Wajīz*) of al-Khāzinī's *Zīj al-mu'tabar al-sanjarī*,[43] Ibn al-Fahhād's *Zīj al-'alā'ī*,[44] and a text on *'ilm al-hay'a* ("cosmography")[45] into Greek. According to Chioniades, he received oral instructions from a person named Σάμψ Πουχαρής, who was born at Bukhārā on June 11, 1254.[46] Wābkanawī was an expert in the field of mathematical astronomy, made exact observations, and was so familiar with the Islamic *zījes* that he could have taught them to Chioniades. When Chioniades arrived in Tabriz, Wābkanawī was there, working on the *Khānī* calendar. Chioniades cited the parameters that "Shams" had calculated for the solar eclipses of July 5, 1293, and October 28, 1296, and the lunar eclipse of May 29, 1295, based on both the *Zīj al-'alā'ī* and the *Īlkhānī zīj*.[47] It thus seems reasonable to identify Σάμψ Πουχαρής with Wābkanawī. However, such an identification certainly needs further inquiry. According to Chioniades, Shams wrote a treatise about the astrolabe dedicated to Emperor Andronicus II (1282–1328 CE). Although manuscripts of this treatise have not survived in Arabic or Persian, but only one manuscript in Greek, it now appears likely that this treatise was the one attributed to Wābkanawī. Therefore, an exact comparison between Wābkanawī's treatise on the astrolabe and the Greek manuscript could clarify this identification.[48]

For the present study, three manuscripts of Wābkanawī's *zīj* were used:[49]

MS T: Turkey, Istanbul, Library of Aya Sophia, no. 2694. This manuscript is the most complete and has both the canons and the tables, which are missing in the other two manuscripts. The manuscript is undated, but there is a horoscope on folio 184v dated 22 Muḥarram 854 (= March 7, 1450), so the manuscript was written prior to this date. According to its first leaf, in Rabī' I 882 H (= June/July

---

42 Wābkanawī, T: f. 2v, Y: f. 3v, P: f. 3v.
43 See Leichter 2004.
44 See Pingree 1985.
45 See Paschos and Sotiroudis 1998.
46 Pingree 1985, pp. 16–17.
47 Pingree 1985, pp. 151–157, 321–333. I note that Wābkanawī's *zīj* is the only *zīj* of that period that has the table of solar parallax for the latitude of Tabriz. This shows that the author probably spent some time in Tabriz.
48 Ghāzān Khān received envoys of Emperor Andronicus II on September 1302. Wābkanawī's *Book on the knowledge on the northern astrolabe* (MS. Turkey, Topkapı Saray, no. 3327/4) was written in 1303–1304 (Storey 1958, p. 65). We know for sure that Chioniades spent several years between 1295–1297 and 1310–1314 in Tabriz (Westerink 1980, pp. 233–245).
49 Two other manuscripts are preserved in Tashkent, no. 3608 and Mumbai, Cama, Rÿl.51 (= Mullā Fīrūz Astr. 51). I thank Dr. Benno van Dalen for this information, which is to appear in his *A New Survey of Islamic Astronomical Handbooks*. Two fragments of the *zīj* are preserved in University of Tehran: no. 2452 (pp. 122–128, selected sections for the determination of hours), and Theology Faculty, no. 190 D (ff. 163v–175v).

1407), it was in the possession of a ʿAbd al-Raḥmān b. ʿAlī b. Muʾayyad.[50] After that, the manuscript was in possession of the rulers of Mecca, through which it seems to have been transferred to the library of the Ottoman Empire.

MS Y: Iran, Yazd, Library of ʿUlūmī, no. 546; a microfilm is available in Tehran University Central Library, no. 2546. The manuscript is incomplete, and the date of writing is not indicated; nevertheless, it appears to have been written around the fourteenth and fifteenth centuries. As indicated on its first leaf, a certain ʿAbd al-Karīm b. ʿAbd al-Wakīl owned the manuscript.

MS P: Iran, Tehran, Library of Parliament, no. 6435, written in 848 H/1444–1445 CE. This manuscript was in the possession of a certain Zayyan al-Dīn al-Lārī and, later, his son, Niẓām al-Dīn, whose seals can be seen on the latest folio. Zayyan al-Dīn (born in 995 H/1587 CE, working in his native town, Lār, in central Iran) was one of the first Islamic scholars who became acquainted with Renaissance European astronomy (e.g., the Tychonic and Keplerian world systems) through a treatise in Persian written by Pietro della Valle (1586–1625).[51]

The references to Wābkanawī's *zīj* in other treatises[52] and the manuscript tradition of this work evidently show that it enjoyed a wide dissemination in a good part of the Middle East (northern and central Iran, the Indian subcontinent, the Arabian Peninsula, and the Ottoman Empire).

3.3 The Report and Calculations

Wābkanawī's account and calculations of the annular eclipse of 1283 appear in Book 3, Section (*bāb*) 14,[53] which bears the title "On the knowledge of the solar eclipse according to the table." The whole section serves as a tutorial to acquaint the reader with the difficult procedures of calculating an eclipse. It consists of six chapters with the following titles:

Cap. 1: On mentioning of (some) preliminaries and determining the possibility and approximate occurrence of a (solar) eclipse.

Preliminary 1: On the knowledge of the parallax of the Moon in longitude and latitude according to the tables of Theon of Alexandria,

---

50 He seems most likely to be Muʾayyadzāda ʿAbd al-Raḥmān Efendi, a student of Jalāl al-Dīn Dawānī, who, according to Dawānī's *ijāza* to Muʾayyadzāda, studied with Dawānī in Shiraz for about seven years, including Qāḍī-zāda al-Rūmī's commentary on Jaghmīnī's *Mulakhkhaṣ fī 'l-hayʾa al-basīṭa* and al-Ṭūsī's *Taḥrīr Uqlīdis*. It is also noteworthy that Muʾayyadzāda returned to the Ottoman Empire, where he established a large, 7,000-volume library, which contained other works from the Marāgha circle as well. I thank Judith Pfeiffer (2015) for this information.

51 See Ben-Zaken 2009.

52 For example, in the *Laṭāʾif al-kalām*, written by a certain Sayyid Muḥammad (a local Astrologer from Lāhījān, a city in the north of Iran, completed in March 1421), pp. 322–324, in Farīd al-Dīn Masʿūd b. Ibrāhīm al-Dihlawī's (d. 1629) *Sirāj al-istikhrāj* (written in Lahore around 1598, as it includes the calculation of the lunar eclipse of August 16/17, 1598, on f. 39r), f. 17v: line 3, and the prolegomenon of his *Shāhjahānī zīj*, f. 4r: line 8.

53 Wābkanawī, T: ff. 71r–74v, Y: ff. 128r–132v, P: ff. 108r–113r.

who determined them for the latitudes of the Seven Climata under the assumption that the Moon is both at its greatest distance (to the Earth) and at the beginning of the (ecliptic) signs.

Preliminary 2: On the knowledge of the possibility and approximate occurrence of a (solar) eclipse.

Preliminary 3: On the correction of the parallax and its investigation.

Cap. 2: On the knowledge of the Fine Table (*jadwal-i laṭīf*) and its composition for (extracting) the parallax.

Cap. 3: On the determination of the apparent relative velocity (*sabaq-i mar'ī*) based on the Fine Table.

Cap. 4: On the correction of the time of mid-eclipse (*sā'āt-i wasaṭ-i kusūf*) with the aid of the Fine Table.

Cap. 5: On the determination of the apparent latitude (*'arḍ-i mar'ī*) of the Moon

Cap. 6: On the knowledge of the magnitude and time of the phases of the eclipse based on the (Fine) Table.

In what follows, the whole procedure for the calculation of the annular solar eclipse of January 30, 1283, is presented in the order in which it was described by our author. The calculations are expressed in modern symbols and concepts, which are introduced whenever the corresponding parameters are used for the first time. The numerical values mentioned throughout the following steps are our author's. Whenever the values are recomputed based on al-Maghribī's or our author's tables, they are given in brackets. In steps 1 and 6, some of the results will be compared with the recomputed modern values, which will be indicated explicitly. Critical comments, notes, and additional explanations will be given whenever needed. Since we are dealing with the calculation of a solar eclipse in the framework of Ptolemy's *Almagest*, the reader needs to be familiar with the following concepts: true and apparent ecliptical longitude and latitude, altitude, elongation, lunar anomaly, parallax, true and apparent conjunction, true and apparent angular velocity, equation of time, phases of a solar eclipse, magnitude of eclipse, and so on.[54]

Step 1: (Cap. 1, Pre. 1; Cap. 2) The Data for λ and T

The first step in the calculation of a solar eclipse consists in finding the longitude λ of the point on the ecliptic at which the Sun and the Moon are in conjunction and the time t (counted in days) + T (counted in fractions of a day or in hours) when the conjunction occurs. When λ and T have been determined based on the longitudes of the Sun, $\lambda_{\odot}$, and of the Moon, λ_{D}, they are called, respectively, the longitude and

---

54 See Neugebauer 1975, Pedersen 1974 [2010].

the time of true conjunction (*juz'-i ḥaqīqī-i ijtimā'* and *sā āt-i ḥaqīqī-i ijtimā'*). That is, at the time $t + T$, the lines connecting the Earth's center to the centers of the Sun and the Moon coincide with each other and, when extended, meet the ecliptic at the point with longitude λ. To determine λ and T, $\lambda_\odot$ and $\lambda_\jupiter$ are found from the tables of mean motions and equations, for the mean or true noon of the two successive days in which $\lambda_{\odot 1} > \lambda_{\jupiter 1}$ and $\lambda_{\odot 2} < \lambda_{\jupiter 2}$, respectively. Then, using the true angular motion of the luminaries in one day (i.e., $v_\odot = \lambda_{\odot 2} - \lambda_{\odot 1}$ and $v_\jupiter = \lambda_{\jupiter 2} - \lambda_{\jupiter 1}$), λ and T are easily determined. Wābkanawī presents the data for the solar eclipse of 1283 as follows:

For Mughān, $\varphi = 39°$ $L = 83°$ (from the Canary Islands):

t = Saturday, 29 Shawwāl 681 H
  = 26 Farwardīn 652 Yazdigird era[55] [= January 30, 1283, JDN 2189703] (3.1)
$T = 0;4^h$ (after noon in Mughān).
$\lambda = 317;59,38°$.

The true angular velocities (*buht*) of the luminaries are:

$$v_\odot = 0; 2,31°/h$$
$$v_\jupiter = 0;31,23°/h. \tag{3.2}$$

As a result, the true relative angular velocity (*sabaq*) is:

$$v = 0;28,52°/h, \tag{3.3}$$

which our author rounded to 0;29°/h. Since, as mentioned earlier, our author relied on Muhyī al-Dīn's *Adwār* and applied all his parameter values in his own *zīj*, it initially seems reasonable to assume that he had utilized them for his calculations of the solar eclipse of 1283. In the table of the geographical coordinates of cities and regions (T: f. 149r), the longitudes of Marāgha and Mughān are the same: $L = 83°$ (that is, both locations were assumed to be on the same meridian). However, in the *Īlkhānī zīj* and the *Adwār*, the longitude of Marāgha is 82°. Wābkanawī himself indicated $L = 82°$ in the prologue of his *zīj*. Thus, the solar and lunar longitudes calculated on the basis of the Marāgha *zīj*es must be adjusted for the longitude of Mughān by subtracting the amount of solar and lunar mean motions in 4 minutes of time. Both the *Adwār* and the *Talkhīṣ* give the following values for the true longitude of the luminaries on the two successive days, 26 and 27 Farwārdīn 652 Y (= January 30 and 31, 1283):

---

55 The Yazdigird date is also given in the marginal notes of MS. P: f. 110r.

	Day 1: 26th	Day 2: 27th		v (°/h)	
$\lambda_\odot$	317;59,27°	318;59,58°	$\rightarrow$	$\approx 0;\ 2,31$	(3.4)
λ_D	317;44,30	330;24,10	$\rightarrow$	$\approx 0;31,39$	

From (3.4), we have

$$v = 0;29,8°/h \tag{3.5}$$

Therefore, T and λ are determined as follows:

$$T = 0;31^\mathrm{h} \text{ after mean noon in Mughān} \tag{3.6}$$
$$\lambda = 318;0,44°$$

The recomputed values in (3.6) are obviously different from our author's values in (3.1). This is due to the fact that Wābkanawī's epoch mean longitude of the Moon (for March 13, 1266) is greater than al-Maghribī's value for that date by $0;13,11°$.[56] Taking this difference into account, we obtain the following values:

	Day 1: 26th	Day 2: 27th		v (°/h)	
λ_D	317;56,51°	330;37,18°	$\rightarrow$	$\approx 0;31,41$	(3.7)

From (3.7) and the solar longitudes given in (3.4), our author's values in (3.1) are obtained. However, judging from the values our author gives later for the lunar longitude for the times $T = 0;46^\mathrm{h}$ (following, in (3.15)) and $T = 0;48^\mathrm{h}$ (cf. (3.18)) and his value of the lunar velocity, (3.2), it seems he had computed $\lambda_{\mathrm{D}1} = 317;57,25°$ for the lunar longitude at noon on January 30. This small difference is probably due to rounding errors and does not affect the rest of the computation.

The modern values for Mughān, that is, for $L \approx 47°$, are

$$T = 0;14^\mathrm{h} \text{ after mean noon in Mughān} \tag{3.8}$$
$$\lambda = 317;54,52°$$

A comparison of (3.1) and (3.6) with (3.8) shows that Wābkanawī's modification made the time of the true Sun–Moon conjunction of January 1283 seven minutes more accurate than the time computed from al-Maghribī's original mean longitudes. We shall trace the effect of this difference on the computed time of the apparent conjunction (maximum eclipse) in Section 3.4 (no. 3).

---

56 The other two differences between Wābkanawī's epoch mean positions and the values given by al-Maghribī for that time are Mars's mean longitude (increased by $1;5°$) and Venus's mean anomaly (increased by $2;30°$).

Equation of Time.[57] In order to calculate the longitudes of the luminaries for a given time, the equation of time must be taken into account:

$$E_\odot(\lambda_\odot) = \lambda_\odot + c(\lambda_\odot, \lambda_{ap0}, e) - RA(\lambda_\odot) + RA(\lambda_{\odot 0}) - \overline{\lambda}_{\odot 0} \qquad (3.9)$$

in which RA stands for the right ascension, λ for the true longitude, $\overline{\lambda}$ for the mean longitude, c for the solar equation of canter, e for the solar eccentricity, and λ_{ap} for the longitude of the solar apogee. The subscript "0" indicates the previous parameters for the epoch.

It comes as a surprise that, according to Wābkanawī, Muḥyī al-Dīn believed that the equation of time does not exist (i.e., is zero), and he had given "proof" (*burhān*) of his claim in his *Majisṭī*.[58] Wābkanawī, however, did not accept al-Maghribī's false opinion and prepared his own tables for the correction of the longitude of the Sun and the Moon due to the equation of time, $E'_\odot(\lambda_\odot)$ and $E'_\text{☽}(\lambda_\odot)$ (preserved in T: f. 156v), from a "principal" (*aṣlī*) table of the equation of time, $E(\lambda_\odot)$ (not surviving, but mentioned in V.2.2),[59] apparently, on the basis of al-Maghribī's parameter values. The radix values in his *Talkhīṣ* (ff. 64r–v) and *Adwār* (M: f. 76r, CB: f. 74r) for 1 Farwardīn 601 Y (= January 18, 1232) are:

$\overline{\lambda}_{\odot 0}$	303;59,22°
λ_{ap0}	88;20,47 (modern: **89;45,35°**)
$\lambda_{\odot 0}$	305;11,31 (modern: **305;8,16°**)
$RA(\lambda_{\odot 0})$	307;33,36
$RA(\lambda_{\odot 0}) - \overline{\lambda}_{\odot 0}$	3;34,14

Muḥyī al-Dīn has $e = 2;6$ ($c_{max} = 2;0,21°$)[60] and $\varepsilon = 23;30°$.[61] Figure 3.2 displays the graph of $E(\lambda_\odot)$. On January 30, 1283, the equation of time was about its minimum and could be ignored; for $\lambda_\odot = 317;59,27°$ (3.4), Wābkanawī's tables give $E'_\odot = 0°$ and $E'_\text{☽} = 0;0,2° \approx 0°$.

---

57 "Equation of day" (Ar. *ta'dīl al-ayyām*, La. *equatio dierum*) in the medieval astronomical context. The modern term "equation of time" (*ta'dīl al-zamān*) is found in Ibn Yūnus's *Zīj al-kabīr*, p. 92 (line 13) and Bīrūnī's *al-Qānūn al-mas'ūdī*, vol. 2, p. 720; cf. Neugebauer 1975, vol. 1, p. 61, n. 2.
58 Wābkanawī, II.1.1: T: f. 16r, Y: f. 26v, P: f. 22r, V.2.1: T: f. 135v, Y: f. 237r, P: f. 206v. The *Majisṭī* ("Almagest") to which Wābkanawī refers seems to be either Muḥyī al-Dīn's *Khulāṣat al-majisṭī* (*Epitome of the* Almagest) or his *Talkhīṣ al-majisṭī* (*Compendium of the* Almagest), but nothing about his idea on the equation of time can be found in the extant manuscripts of either work: in the first, Ptolemy's remarks in *Almagest* III.9 are summarized (III.6: ff. 47v–48v), and in the latter, the equation of time is completely neglected, without any explanation. The opinion that the equation of time does not exist is found, at least, in another work from the thirteenth century: Kamālī (f. 52v) reports from the *Zīj al-mughnī* written by a certain Muntakhab al-Dīn Sakkāk (or Hakkāk) al-Yazdī that its author did not make any difference between the mean and true solar days (this work seems to be different from a *zīj* with the same name compiled by 'Abd al-Karīm al-Fahhād, twelfth century; see Kennedy 1956a, no. 64, Mozaffari 2019, p. 524).
59 Wābkanawī, T: f. 136r, Y: f. 237v, P: f. 207r.
60 *Talkhīṣ*, IV.5; see Saliba 1985.
61 See Chapter 9.

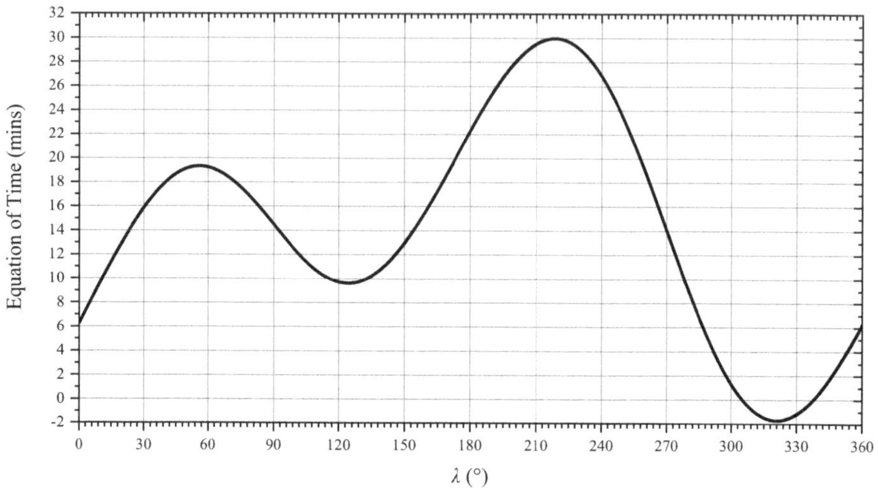

Figure 3.2 Graph of the equation of time on the basis of the solar parameters of Muḥyī al-Dīn.

Step 2: (Cap. 1, Pre. 3; Cap. 2) Parallax

For a celestial body sufficiently near the Earth, there is a vertical displacement (i.e., in the circle of altitude) between its true position with respect to the center of the Earth and its apparent position at the surface of the Earth (Figure 3.3 (left)). This is called parallax, and the value of its angle (angle *PSE*) depends on the altitude of the body and its distance from the Earth's center. The table of parallaxes in *Almagest* (V.18)[62] gives the values of the parallaxes of the luminaries from the true altitude, the true lunar anomaly α, and the mean elongation $\bar{\eta}$ between them. Since, in a true conjunction of the Moon with the Sun, the luminaries have the same true altitude, one can take into account the difference of the values of their parallaxes, which is called the *adjusted parallax* and which is used to compute the longitudinal and latitudinal components of the parallax (Figure 3.3 (right)).

Theon's tables of parallax give the values of the longitudinal and latitudinal components of the adjusted parallax of the luminaries, Π_λ and Π_β, respectively, for the Moon at maximum distance (i.e., $\alpha = \bar{\eta} = 0$) as functions of the geographical latitude φ (for the seven climates), the ecliptical longitude λ (= $n \cdot 30$, $n = 1, 2 \dots$ 12), and the time T in equinoctial hours before or after noon.[63] Π_λ and Π_β are used

---

62　Toomer 1998, p. 265, Pedersen 1974, Chapter 7, Neugebauer 1975, vol. 1, pp. 100–118.

63　The tables are completely based on *Almagest* V.18. They dominated medieval astronomy (see Neugebauer 1975, vol. 2, pp. 990–999). The standard study on parallax in Islamic times is Kennedy 1956b. In the Islamic *zījes*, the parallax is tabulated for different steps of λ and T. For example, in the *Ashrafī Zīj*, there are two sorts of Theonic parallax tables: In the first type (ff. 159r–164v), T is

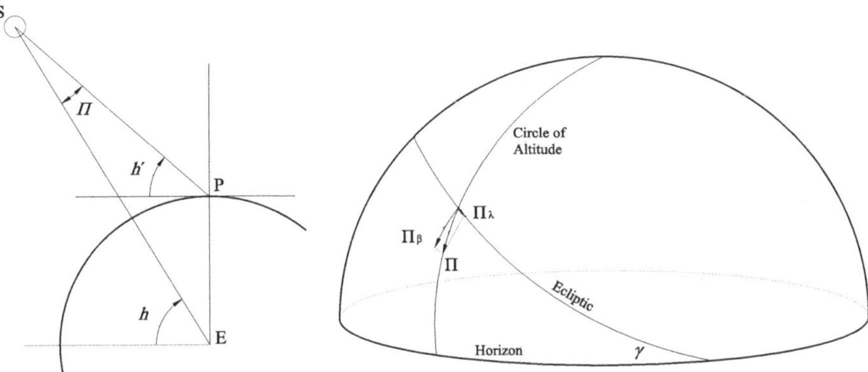

Figure 3.3 (Left) Parallax in the circle of altitude and (right) its longitudinal and latitudinal components.

to compute the apparent ecliptical coordinates of the luminaries and the time of their apparent conjunction. The procedure is as follows. First, for the time T when the Moon and Sun are in conjunction, α, η, and $\lambda = \lambda_{\odot 0} = \lambda_{\mathbb{D}0}$ are calculated. One then interpolates with φ and $[\lambda]$ in the tables of the parallax for the two neighboring climates to find the values of Π_λ and Π_β for the equinoctial hours before or after noon (the possible fractional part of λ has a very small effect on the parallax). Next, all the values are adjusted by the correction coefficients due to α and $\bar{\eta}$, symbolized, respectively, as $c_{\Pi1}$ and $c_{\Pi2}$. (Since in a solar eclipse $\bar{\eta}$ is very small, $c_{\Pi2} = 1$.) Due to the parallax, the Moon appears to be at $\lambda'_{\mathbb{D}0}$ in the time T, so that $|\lambda'_{\mathbb{D}0} - \lambda_{\mathbb{D}0}| = \Pi_{\lambda0}$. The apparent conjunction or the mid-eclipse, that is, the maximum phase of a solar eclipse, occurs when $\lambda'_{\mathbb{D}} = \lambda_\odot$. Therefore, the problem is to find the time T' when $|\lambda_{\mathbb{D}} - \lambda_\odot| = \Pi_\lambda$. In doing so, it is necessary to find the interval Δt_1, during which the two luminaries travel the distance $\Pi_{\lambda0}$. At this point, one can employ either the true relative velocity of the luminaries, that is, $v = v_{\mathbb{D}} - v_\odot$, or their apparent relative velocity, that is, the amount of v adjusted for parallax (cf. step 3). Then the longitudes $\lambda_{\mathbb{D}1}$ and $\lambda_{\odot 1}$ and the amount of parallax $\Pi_{\lambda1}$ for the time $T + \Delta t_1$ are computed. If $|\lambda_{\mathbb{D}1} - \lambda_{\odot 1}| = \Pi_{\lambda1}$, then $T' = T + \Delta t_1$. Otherwise, the procedure will be repeated to produce a time $T' = T + \Delta t_n$ when $|\lambda_{\mathbb{D}n} - \lambda_{\odot n}| = \Pi_{\lambda n}$. ($c_{\Pi1}$ changes from 1 for $\alpha = 0°$

---

given for each 10 minutes and λ proceeds in steps of 3° with 120 entries from 3° to 360°. Although it is not mentioned in the canons of the *zīj*, this table is for $\varphi = 36°$. In the second type, there are seven tables for the seven climates, in which the values of T are for each half an hour, and the values of λ for each 30°. The intermediate entries were calculated by means of linear interpolation between the two neighboring entries. Wābkanawī has the seven tables for the seven climates and two other tables for Marāgha, $\varphi = 37;20°$ (T: f. 168v), and Tabriz, $\varphi = 38°$ (T: f. 169r). (The latter is identical to al-Khāzinī's table of parallax for the latitude of Marw. See Kennedy 1956b, p. 46.) The tables for the intermediate values of φ were produced with the aid of the linear interpolation in the Theonic tables of the parallax for the seven climates, according to the simple procedure explained later, in no. 1.

to 1;14 for $\alpha = 180°$; since during one or two hours α changes only around 1°, one does not need to consider this effect.) As we shall see later, Wābkanawī iterated in computations while taking into account the two options at his disposal, that is, using the true and apparent velocities. Thus, he empirically discovered a way to reduce the steps of computation, as shall be explained in step 4.

Returning to Wābkanawī's calculation of the solar eclipse of January 30, 1283, our author explained (Cap. 1, Pre. 3) how the astronomer should work with the Theonic parallax tables and perform the necessary interpolations and corrections to obtain the desired value. Cap. 2 contains his worked example of the eclipse of January 30, 1283. Our input data are (3.1): $\varphi = 39°$, $\lambda = 317;59,38° \approx 318°$, and $T = 0;4^{h}$ after noon in Mughān.

The interpolations are as follows:

1. The interpolation for $\varphi = 39°$ between the two parallax tables for the latitude $\varphi_1 = 36°$ (the fourth climate) and $\varphi_2 = 41°$ (the fifth climate). We find all the amounts given in the tables of the parallax for the fourth climate, $\varphi_1 = 36°$, and for the fifth climate, $\varphi_2 = 41°$ (Wābkanawī, T: ff. 168v–169r), for the equinoctial hours after noon, and for $\lambda = 300°$ (0°♒) and $\lambda = 330°$ (0°♓), and write them in a separate table. Then we interpolate between the two relevant amounts by

$$\Pi = \Pi_1 + \frac{\phi - \phi_1}{\phi_2 - \phi_1}(\Pi_2 - \Pi_1). \tag{3.10}$$

Our author called $\phi - \phi_1$ *faḍl*, meaning "residue," and $\phi_2 - \phi_1$ *tafāḍul*, meaning "difference." The result is as follows (T: f. 73r):[64]

	$\varphi_1 = 36°$				$\varphi_2 = 41°$				$\varphi = 39°$			
	♒0°		♓0°		♒0°		♓0°		♒0°		♓0°	
	Π_λ	Π_β	Π_λ	Π_β	Π_λ	Π_β	Π_λ	Π_β	Π_λ	Π_β	Π_λ	Π_β
0	9	42	14	*34* [25]	10 [*13*]	43	15	38	10	43	15	*37* (33)
1	19	38	23	31	19	41	24	34	19	40	24	33
2	29	35	32	27	28	38	32	30	28	37	32	29
3	37	31	40	22	36	34	39	27	36	33	39	25
4	42	27	45	18	40	30	44	23	41	29	44	21
5	45	23	48	16	43	27	47 [*46*]	19	44	26	*45* [47]	18

Note: All the values are in arc minutes. The errors are indicated in italics. The amounts in square brackets are mentioned in the parallax tables in Wābkanawī's *zīj* (T: ff. 168v–169r). To eliminate scribal errors, they were compared with the similar tables in the *Ashrafī zīj* (ff. 166v–167r) and the *Īlkhānī zīj* (C: pp. 91–92). This showed that, in two cases, the scribal errors belonged to the parallax tables in Wābkanawī's *zīj*, that is, [*13'*] and [*46'*]. In one case, that is, *34'*, the error belonged to our author's table on f. 73r. It seems that our author made a mistake in copying the correct value from the tables of the parallax. This caused a mistake in calculating $\Pi_\beta(\text{♓}0°)$ for noon: 37' instead of 33'. An error also occurred in the calculation of $\Pi_\lambda(\text{♓}0°)$ for the fifth hour after noon. Both errors do not affect the subsequent steps.

64 The auxiliary tables for calculating the parallax (Table 3.3) are missing in MSS Y and P.

2. The interpolation for $\lambda = 317;59,38°$ between parallax for $\lambda = 300°$ ($0°\approx$) and $\lambda = 330°$ ($0°\mathcal{H}$). For this purpose, we again interpolate between the two relevant entries in the table, which was produced in the previous step by

$$\Pi = \Pi_1 + \frac{\lambda - \lambda_1}{\lambda_2 - \lambda_1}(\Pi_2 - \Pi_1) \tag{3.11}$$

Our author again called $\lambda - \lambda_1$ *faḍl*, or "residue," and $\lambda_2 - \lambda_1$ *tafāḍul*, or "difference." Thus, we obtain the values of the parallax for the equinoctial hours elapsed after noon for $\varphi = 39°$ and $\lambda = 318°$ (Table 3.1; T: f. 73r).

3. The interpolation for α. For $T = 0;4^h$ after noon, our author gave $\alpha = 74°$ (recomputed from the tables, *Talkhīṣ*, f. 73v; *Adwār*, M: f. 77r: $\alpha = 73;59,11°$). According to the table of the solar eclipses (Wābkanawī, T: f. 170r), we have

$$c_{\Pi 1}(74°) = 64/60 = 1;4 \tag{3.12}$$

4. The interpolation for $\bar{\eta}$. Wābkanawī explained this step but did not mention the resulting value because, as mentioned earlier, $\bar{\eta}$ is very small in the solar eclipses, so $c_{\Pi 2}(12°) = 1$, and this kind of interpolation is not necessary here. (The mean motion tables give $2\,\bar{\eta} = 11;55,54° \approx 12°$ for $T = 0;4^h$. Then from the table of the solar eclipses, we have $c_{\Pi 2}(12°) = 60/60 = 1$.

All the amounts obtained in steps 1 and 2 (the last two columns of Table 3.1) are then multiplied twice by $c_{\Pi 1}$ (3.12) and $c_{\Pi 2}$ (which is, of course, equal to 1). This produces what our author called *jadwal-i laṭīf*, the "Fine Table" (Table 3.2; T: f. 73r).[65]

Table 3.1 Computation of the parallax for the equinoctial hours elapsed after noon for $\varphi = 39°$ and $\lambda = 318°$ under the assumption $\alpha = \bar{\eta} = 0°$ (step 2, nos. 1 and 2)

	$\varphi_1 = 36°$				$\varphi_2 = 41°$				$\varphi = 39°$				$\varphi = 39°$	
	$\approx 0°$		$\mathcal{H}0°$		$\approx 0°$		$\mathcal{H}0°$		$\approx 0°$		$\mathcal{H}0°$		$\approx 18°$	
	Π_λ	Π_β	Π_λ	Π_β	Π_λ	Π_β	Π_λ	Π_β	Π_λ	Π_β	Π_λ	Π_β	Π_λ	Π_β
0	9	42	14	34 [25]	10 [13]	43	15	38	10	43	15	37 [33]	13	40 [37]
1	19	38	23	31	19	41	24	34	19	40	24	33	22	36
2	29	35	32	27	28	38	32	30	28	37	32	29	29 [30]	33 [32]
3	37	31	40	22	36	34	39	27	36	33	39	25	37 [38]	34 [28]
4	42	27	45	18	40	30	44	23	41	29	44	21	42 [43]	37 [24]
5	45	23	48	16	43	27	47 [46]	19	44	26	45 [47]	18	44 [46]	37 [21]

Note: For the two last columns, the error in $\Pi_\beta(\approx 18°)$ resulted from the error in the previous step. All the amounts of parallax for 2–5 hours after noon are in error; of course, none of them seem to be scribal. Nevertheless, since $T = 0;4^h$, none of these errors have a negative effect on the following steps.

---

65 This method and the name of "*jadwal-i laṭīf*" can be found in some preceding *zīj*es; for example, al-Khāzinī, *Wajīz* X.3.1: ff. 25rf, which appears to be the direct source of Wābkanawī (see step 6).

Table 3.2 Parallax for the equinoctial hours elapsed after noon for $\varphi = 39°$ and $\lambda = 318°$ with considering the computed values of α and $\bar{\eta}$ (step 2, nos. 3 and 4)

		$\varphi = 39°$	
		$18°\,\aquarius$	
		Π_λ	Π_β
	0	14	43 [39]
	1	23	39
	2	31 [32]	35 [34]
	3	39 [41]	34 [30]
	4	45 [46]	39 [26]
	5	47 [49]	39 [22]

Note: All the deviations here are produced from the noted errors in the previous steps.

Table 3.3 Apparent relative angular velocity of the luminaries for the equinoctial hours elapsed after noon

	$v'^{\,(°/h)} =$
1	$29 - (23 - 14) = 20$
2	$29 - (31 - 23) = 21 \; [29 - (32 - 23) = 20]$
3	$29 - (39 - 31) = 21 \; [29 - (41 - 32) = 20]$
4	$29 - (45 - 39) = 23 \; [29 - (46 - 41) = 24]$
5	$29 - (47 - 45) = 27 \; [29 - (49 - 46) = 26]$

5. The interpolation for T is connected to the procedure for finding λ' and T' (see step 4).

Step 3: (Cap. 3) The Apparent Angular Velocity of the Luminaries

The true relative angular velocities of the Sun and the Moon have already been determined ((3.2) and (3.7)). In order to calculate the apparent relative angular velocity (*sabaq-i mar'ī*) of the luminaries, we need to take the values of Π_λ into account. Figure 3.4 shows the relative positions of the Sun and the Moon in the two successive hours after noon. At these two instants, the vector of Π_λ is in a reverse direction with respect to that of v. Thus, the hourly apparent relative velocity is

$$v'^{\,(°/h)} = v^{\,(°/h)} - (\Pi_{\lambda 2} - \Pi_{\lambda 1})^{66} \tag{3.13}$$

From $v = 29°/h$ and Table 3.2, we can determine the amount of v' for each hour elapsed after noon (Table 3.3).

---

66 Note that the values of Π_λ contain the amounts of the longitudinal components of the parallaxes of both the Sun and the Moon.

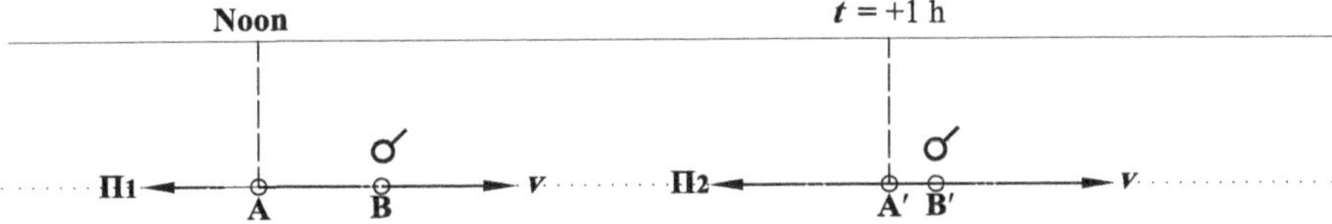

Figure 3.4 *A* and *A'*: the true position in the two times; *B* and *B'*: the apparent positions in the two times. $AA' > BB' \leftrightarrow v > v'$.

Step 4: (Cap. 4) The Time of the Apparent Conjunction
(= the Time of the Mid-Eclipse)

In this step, we determine the time T' of the apparent conjunction from the time T of the true conjunction from the amounts of the parallax calculated in the previous steps (Tables 3.2 and 3.3). T' is the time when the apparent longitude of the Moon, $\lambda'_{\mathbb{D}}$, is equal to the apparent longitude of the Sun, $\lambda'_{\odot}$ – that is, it is the time of the maximum phase of the eclipse (mid-eclipse) for the given location. In step 2, we worked with the values of the adjusted parallax found from the Theonic tables, in which, as mentioned before, the solar parallaxes had already been applied. Thus, we should take the amounts of the parallax into account to modify only the true longitude of the Moon in order to obtain its apparent longitude. Therefore, T' is the time when $\lambda'_{\mathbb{D}}$ is equal to the true longitude of the Sun, $\lambda_{\odot}$.[67]

In order to determine T', we can use an iterative procedure as follows. The time T of the true conjunction falls between the times t_1 and t_2, which are the two successive equinoctial hours before or after noon. In our example, $T = 0;4^h$, $t_1 = 0^h$, and $t_2 = 1^h$. First, the value of Π_λ for the time T is determined, which gives the angular distance between λ and $\lambda'_{\mathbb{D}}$. The time interval taken by the luminaries to travel from λ to $\lambda'_{\mathbb{D}}$ or, in reverse, from $\lambda'_{\mathbb{D}}$ to λ, is

$$\Delta t = \Pi_\lambda/v' \text{ or } \Pi_\lambda/v \tag{3.14}$$

which is called *sā 'āt-i ikhtilāf*. Then, if, in the time T, $|\eta| = |\lambda_{\odot} - \lambda_{\mathbb{D}}| = |\Pi_\lambda|$, then $T' = T + \Delta t$. If not, we compute $\Pi_{\lambda 1}$ (with the interpolation in the Fine Table, that is, Table 3.2), Δt_2 (from (3.14)), $\lambda_{\odot 1}$, and $\lambda_{\mathbb{D}1}$ for the time $T_1 = T + \Delta t_1$. If $|\eta| = |\lambda_{\odot 1} - \lambda_{\mathbb{D}1}| = \Pi_{\lambda 1}$, then $T' = T + \Delta t_1$. If not, we repeat the process until the condition $|\eta| = |\lambda_{\odot n} - \lambda_{\mathbb{D}n}| = \Pi_{\lambda n}$ is established for the time $T_n = T + \Delta t_n$, so that $t_1 < T_n < t_2$:

t_1	Π_λ/v or Π_λ/v'			T'
T	$\Pi_\lambda \rightarrow$	$\Delta t_1 \rightarrow$	$\|\lambda_{\odot} - \lambda_{\mathbb{D}}\| = \Pi_\lambda \rightarrow$	$T + \Delta t_1$
			$\|\lambda_{\odot} - \lambda_{\mathbb{D}}\| \neq \Pi_\lambda \rightarrow$	Continue the process
$T_1 = T + \Delta t_1$	$\Pi_{\lambda 1} \rightarrow$	$\Delta t_2 \rightarrow$	$\|\lambda_{\odot 1} - \lambda_{\mathbb{D}1}\| = \Pi_{\lambda 1} \rightarrow$	$T + \Delta t_2$
			$\|\lambda_{\odot 1} - \lambda_{\mathbb{D}1}\| \neq \Pi_{\lambda 1} \rightarrow$	Continue the process

67 As a result, from here onward, the values of $\lambda'_{\mathbb{D}}$ are no longer exactly the lunar apparent longitude.

$$T_2 = T + \Delta t_2 \quad \Pi_{\lambda 2} \quad \rightarrow \qquad \Delta t_3 \quad \rightarrow \quad |\lambda_{\odot 2} - \lambda_{\mathbb{D}2}| = \Pi_{\lambda 2} \quad \rightarrow \quad T + \Delta t_2$$
$$|\lambda_{\odot 2} - \lambda_{\mathbb{D}2}| \neq \Pi_{\lambda 2} \quad \rightarrow \quad \text{Continue the process}$$

$$\cdots \qquad \cdots \quad \cdots \qquad \cdots \quad \cdots \quad \cdots \qquad \qquad \cdots \quad \cdots$$

$$T_n = T + \Delta t_n \quad \Pi_{\lambda n} \quad \rightarrow \qquad \Delta t_{n+1} \quad \rightarrow \quad |\lambda_{\odot n} - \lambda_{\mathbb{D}n}| = \Pi_{\lambda n} \quad \rightarrow \quad T + \Delta t_{n+1}$$
$$|\lambda_{\odot n} - \lambda_{\mathbb{D}n}| \neq \Pi_{\lambda n} \quad \rightarrow \quad \text{Continue the process}$$

t_2

According to Wābkanawī, the majority of the medieval Islamic astronomers, except al-Khāzinī, utilized v instead of v' in (3.14). The numerical results of the preceding algorithm for the iterative procedure are summarized in the two auxiliary tables named *dastūr* (each of their rows is called a *muqaddama*, "preliminary step"). The first *dastūr* is based on $v = 29°/h$ ((3.3) and (3.5)), and the second one on $v' = 20°/h$ (calculated in step 3 for $t = 1$ hour after noon; cf. the first row in Table 3.3).

First *dastūr*			Second *dastūr*		
	Π_λ	$\Delta t = \Pi_\lambda/29$		Π_λ	$\Delta t = \Pi_\lambda/20$
$t_1 = 0$	14′		$t_1 = 0$	14′	
$T = 0;\ 4$	14 [15]	0;29 [0;31]	$T = 0;\ 4$	14 [15]	0;42 [0;45]
$T_1 = 0;33\ [0;35]$	19	0;39			
$T_2 = 0;43$	20	0;42 [0;41]			
$T_3 = 0;46\ [0;45]$	21	0;44 [0;43]			
$T_4 = 0;48\ [0;47]$	21	0;44[1]	$T_1 = 0;46\ [0;49]$	21	$T' = 0;48$
$t_2 = 1$	23		$t_2 = 1$	23	

[1] Text: 0;46 (scribal error: مو → مد)

Working with the second *dastūr*, for $T_1 = 0;46^\text{h}$, our author gives

$$\lambda_\odot = 318;\ 1,23°$$
$$\lambda_{\mathbb{D}} = 318;21,29°$$
$$(3.15)$$

All MSS used give $\lambda_\odot = 318;1,43°$ (Y: f. 131v, T: f. 73v, P: f. 111v), which is clearly an error, simply because in the following steps, our author gives $\lambda_\odot = 318;1,28°$ for $T = +0;48^\text{h}$ (see (3.18)), from which $\lambda_\odot = 318;1,28° - 0;2^\text{h} \times 0;2,31°/h = 318;1,23°$ may conveniently be obtained for $T = +0;46^\text{h}$.[68] The text also immediately gives $\eta = 20'\ 6''$. One may conclude that our author or the scribe committed an error in writing the value in abjad numerals (كح → كج). From the second *dastūr*, we have $\Pi_\lambda = 21'$ for the given time. Hence, $\eta \neq \Pi_\lambda$. Nevertheless, the difference $\Pi_\lambda - \eta$

---

68 This is also the case if one starts, in reverse, from $\lambda_\odot = 317;59,27°$ at noon (3.4) to obtain $\lambda_\odot = 318;1,23°$ for $T = 0;46^\text{h}$.

$= 54''$ is negligible. In addition, we know that Π_λ remains constant in the short time intervals; the first *dastūr* shows that, for the time interval $0;46^h - 0;50^h$, $\Pi_\lambda = 21'$. Therefore, the apparent conjunction occurs a very short time, that is,

$$(\eta - \Pi_\lambda)/v = 0;0,54°/0;29^{°/h} = 0;1,52^h \approx 2 \text{ minutes,} \tag{3.16}$$

after $T_1 = 0;46^h$. Therefore,

$$T' = 0;46^h + 0;2^h = 0;48^h \tag{3.17}$$

which our author wrote in the last cell of the column for Δt.

Using the first *dastūr*, we calculate $\lambda_\odot$ and λ_D for the time $T_1 = 0;33^h$ $[0;35^h]$ after noon. Our author only mentioned $\eta = 13'40''$ for the given time. From the first *dastūr*, we have for T_1, $\Pi_\lambda = 19'$. The difference $\Pi_\lambda - \eta = 5'20''$ is not negligible. Thus, we repeat the process for the times T_2, T_3, and T_4. Our author did not give the longitudes or the values of η for T_2 and T_3. For $T_4 = 0;48^h$, he gave

$$\lambda_\odot = 318; 1,28° \tag{3.18}$$
$$\lambda_\mathrm{D} = 318;22,31°$$

Thus, $\eta = 21' 3''$, which is approximately equal to $\Pi_\lambda = 21'$ for the given time. Therefore, $T' = T_4 = 0;48^h$, which is nearly the same value determined based on the second *dastūr*. Figure 3.5 depicts schematically the aforementioned steps.

Our author notices that the majority of astronomers would have used the relative true angular velocity v (an iterative process according to the first *dastūr*), while al-Khāzinī utilized the relative apparent angular velocity v', and al-Battānī both v and v'. In the end of Cap. 4, our author mentions, based on his calculations, that utilizing v' would decrease the steps of calculation.

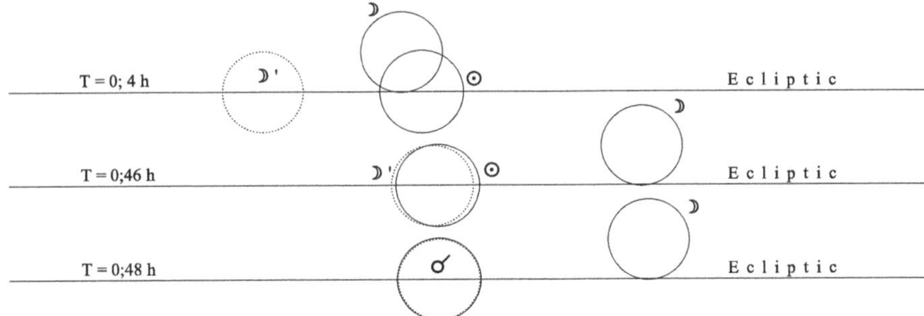

Figure 3.5 The continuous circles show the luminaries, each indicated with its own symbol. The dotted circles are the apparent Moon, drawn to scale (in order to avoid confusion, the angular diameters of the luminaries were reduced to 10% of their true amounts).

In the last step, our author gives the time of the apparent conjunction, which is equal to the time of the maximum phase (the middle) of the eclipse, counted from the instant of sunrise on January 30, 1283, in Mughān:

$$T' + \tfrac{1}{2} \text{ the hours of daylight} = 0;48^h + 5;6,48^h \approx 5;55^h \text{ (text: } 5;54^h) \qquad (3.19)$$

Step 5: (Cap 5) The Apparent Latitude of the Moon

The "mean longitude" (*wasaṭ*) of the orbital node of the Moon is defined as $\bar{\lambda}_\Omega = 360° - \lambda_\Omega$. For $T = 0;48^h$ after noon, our author gives $\bar{\lambda}_\Omega = 49;19,30°$ (recomputed from *Talkhīṣ*, f. 74r, and *Adwār*, f. 78r: $49;9,32°$). In order to determine the argument of latitude, λ_β, we have to add this amount to λ_D (3.18). As a result, $\lambda_\beta = 7;42,1°$ [$7;32,3°$]. Then, by interpolation in the table of the lunar latitude (Wābkanawī, T: f. 106r; *Talkhīṣ*, f. 83v), we have $\beta_\mathrm{D}(7;42,1°) = 0;40,11°$ [$\beta_\mathrm{D}(7;32,3°) = 0;39,20°$]. By interpolation in Table 3.2, $\Pi_\beta = -40'$ [$-39'$]. Therefore, the apparent latitude of the Moon would be

$$\beta'_\mathrm{D} = \beta_\mathrm{D} + \Pi_\beta = +0;0,11° \; [+0;0,20°] \qquad (3.20)$$

In fact, all the errors have finally produced only a minor error of $-9''$.

Step 6: (Cap. 6) The Duration of the Eclipse, the Times of the First and Last Contacts, and the Magnitude and Shape of the Eclipse

We have already determined the time T' of the apparent conjunction (*sāʿat-i ijtimāʿ-i marʾī*) (3.17). T' is also the time of the maximum phase (mid-eclipse) of the eclipse (*sāʿat-i wasaṭ-i kusūf*). The duration $\Delta\tau$ of the eclipse is the time interval between the first and last contacts, by which the instants of the first and last contacts (*sāʿat-i badw-i kusūf, sāʿat-i badw-i injilāʾ*) are determined as, respectively, $T - \tfrac{1}{2}\Delta\tau$ and $T + \tfrac{1}{2}\Delta\tau$. The distance $\Delta\lambda$ that the Moon travels during $\tfrac{1}{2}\Delta\tau$ is called *daqāʾiq-i suqūṭ*. The magnitude m of the eclipse is called *daqāʾiq-i kusūf* ("minutes of eclipse")[69] when counted in degree; it is called *aṣābiʿ-i quṭr* ("digits of diameter") when expressed in digits that assume that the solar apparent diameter consists of 12 digits (*iṣbaʿ*). In his table of the solar eclipses (T: f. 170r),[70] our

---

69 Cf. *minuta casuf* in some Latin sources (Pedersen 1974, p. 233).

70 The table is of the type of Ptolemy's solar eclipse table in the *Handy Tables* (Neugebauer 1975, vol. 2, pp. 990–1001), preserved, for example, in al-Battānī's *Zīj al-Ṣābiʾ* (Nallino 1899–1907, vol. 2, p. 89). However, there exist some remarkable variants in the entries of Wābkanawī's table. For example, the correction values for the parallax, $c_{\Pi 1}$ and $c_{\Pi 2}$, were increased by 60; as a result, instead of multiplying the amount of the parallax by $c_{\Pi 1}/60$, adding the result to the amount of the parallax, and then multiplying the result by $c_{\Pi 2}/60 + 1$, it is only necessary to multiply one time the amount of the parallax by $c_{\Pi 1}$ and $c_{\Pi 2}$. In other words, if the tabulated values are shown as $T_{\Pi 1} = 60 + c_{\Pi 1}$ and $T_{\Pi 2} = 60 + c_{\Pi 2}$, then one calculates $T_{\Pi 1}/60 \cdot T_{\Pi 2}/60 \cdot \Pi$ instead of $((c_{\Pi 1}/60) \cdot \Pi + \Pi) + (c_{\Pi 2}/60) \cdot ((c_{\Pi 1}/60) \cdot \Pi + \Pi)$. See step 2, nos. 3 and 4.

author embedded al-Maghribī's table of the solar eclipses,[71] in which the amounts of m and $\frac{1}{2}\Delta\lambda$ are given as two-variable functions of the lunar apparent latitude, $\beta_{\mathbb{D}}'$, and the lunar velocity, $v_{\mathbb{D}}$; $\beta_{\mathbb{D}}'$ is given for each arc minute from 1′ to 34′ and $v_{\mathbb{D}}$, for the three values 0;29°/h, 0;33°/h, and 0;37°/h, corresponding, respectively, to the maximum, mean, and minimum distances of the Moon from the Earth at syzygies. Since $v_{\mathbb{D}} = 0;31,23°/h$ (cf. (3.2)), one has to interpolate in the table to obtain m and $\frac{1}{2}\Delta\lambda$ for the given β' (cf. (3.20)) between the columns for $v_{\mathbb{D}} = 0;29°/h$ and $v_{\mathbb{D}} = 0;33°/h$. The procedure is as follows[72] (the values obtained based on our author $\beta_{\mathbb{D}}' = 0;0,11°$ are within parentheses).

	$v_{\mathbb{D}} = 0;29°/h$	$\frac{1}{2}\Delta\lambda$	m
	0°	0;31,8°	12; 0
$\beta_{\mathbb{D}}'$	0;0,20 (0;0,11)	0;31,7,20 (0;31,7,38)	11;52,20 (11;55,47)
	0;2	0;31,4	11;14

(3.21)

	$v_{\mathbb{D}} = 0;33°/h$	$\frac{1}{2}\Delta\lambda$	m
	0°	0;32;47°	12; 0
$\beta_{\mathbb{D}}'$	0;0,20 (0;0,11)	0;32,46,30 (0;32,46,44)	11;54,40 (11;57,4)
	0;2	0;32,44	11;28

Now, for $v_{\mathbb{D}} = 0;31,23°/h$, we have with interpolation:

$$\frac{1}{2}\Delta\lambda = 0;32,6,25,13°$$
$$m = 11;53,43,25$$

(3.22)

Or working with our author's value $\beta_{\mathbb{D}}' = 0;0,11°$, we have:

$$\frac{1}{2}\Delta\lambda = 0;32, 6,40,50°$$
$$m = 11;56,32,53$$

In the manuscripts, the amount of $\frac{1}{2}\Delta\lambda$ is given as 0;31,57,32,30° (T: f. 74r) and 0;31,57,30,35° (Y: f. 132v and P: f. 112v), and that of m as 12. It seems that our author applied the rounded value $v_{\mathbb{D}} = 0;31°/h$, which gives the results around 0;31,57° for $\frac{1}{2}\Delta\lambda$.

---

71 *Adwār al-anwār*, M: f. 96v; CB: 94v. Note that the values employed by al-Maghribī for the under-lying parameters, i.e., the Luminaries' minimum angular diameters, are different from Ptolemy's: $\theta_{\odot min} = \theta_{\mathbb{D}min} = 0;31,8°$, instead of 0;31,20°; however, this is not the result of observations, but the result of a remedy on Ptolemy's computations in *Almagest* V.14 (Toomer 1998, pp. 253–254); see *Talkhīṣ* VI.6: ff. 94r–v.
72 As described in the *Adwār al-anwār* II.14: M: ff. 23r–v, CB: ff. 21v–22r.

In order to determine the half-duration of the eclipse, we use

$$\tfrac{1}{2}\Delta\tau = \tfrac{1}{2}\Delta\lambda/v' \qquad (3.23)$$

Using (3.22) and $v' = 21°/h$ for $t = 1$ hour after noon (Table 3.3), we have $\tfrac{1}{2}\Delta\tau = 1;32^h$. With our author's value for $\tfrac{1}{2}\Delta\lambda$, we have $1;31^h$, as mentioned in the text. The errors that occurred in the steps of computation again produced only a minor error of around 1 minute in the final result. Thus, the duration of the eclipse and the instants of the first and last contact counted from sunrise (see (3.19)) would be

$$\Delta\tau = 3;4^h \ (\text{Text: } 3;2^h)$$
$$T' - \tfrac{1}{2}\Delta\tau = 5;55^h - 1;32^h = 4;23^h \ (\text{Text: } 5;49^h - 1;31^h = 4;18^h) \qquad (3.24)$$
$$T' + \tfrac{1}{2}\Delta\tau = 5;55^h + 1;32^h = 7;27^h \ (\text{Text: } 5;49^h + 1;31^h = 7;20^h)$$

I do not know why our author used the amount $5;49^h$ for the time of the mid-eclipse instead of $5;54^h$, which he himself had already determined (cf. (3.19)).

In the table of the solar eclipses, the values of the angular apparent diameters are not listed. Nevertheless, these can be calculated by the aid of the one-variable interpolation function of the apparent velocity, $v'_{\jupiter}$ and $v'_{\odot}$, based on the instructions of our author in his $z\bar{\imath}j$ III.11.2. The values given by our author are as follows:

$$\theta_{\odot} = 32'59'' \ [33'7'']$$
$$\theta_{\jupiter} = 32'11'' \qquad (3.25)$$

The modern recomputed values are $\theta_{\odot} = 0;32,22°$ and $\theta_{\jupiter} = 0;30,49°$.[73] Therefore, since $\theta_{\odot} > \theta_{\jupiter}$, the solar eclipse of January 30, 1283, is determined to be annular for the latitude of Mughān with the parameter values listed in (3.18), (3.19), (3.24), and (3.25). Also, since $\beta_{\jupiter}' = +11'' \ [+20'']$, the bright ring would be thicker in the direction of the south.[74] As our author concluded:

> Because the (angular) diameter of the Moon was less than the (angular) diameter of the Sun, it was clear that a ring of light would remain from the circumference of the surface of the Sun's body. And because the apparent

73 Our author's ratio $\theta_{\odot}/\theta_{\jupiter}$ is approximately 2.5% more than the modern recomputation. Following al-Maghribī, our author accepted the Ptolemaic assumption that the angular diameters of the luminaries are equal at their apogees, while, according to the Indian relation $\theta_{\jupiter} = 0;2,26 \cdot v_{\jupiter}$, where $v_{\jupiter}$ is in degree per day, one may obtain, with (3.2), $\theta_{\jupiter} = 0;2,26 \cdot 0;31,23 \cdot 24 \approx 0;30,33°$.
74 Furthermore, our author determined the astrological significance using the rules given by Ptolemy in *Tetrabiblos* II.7 (pp. 82–84). He elaborated on these rules in his $z\bar{\imath}j$ (III.11.7, 13, and 24) to determine the duration of the effect of an eclipse on the "territory of generation and corruption," that is, the Earth. Thus, astrological considerations were interconnected with observational and astronomical activities in the medieval period and thereafter. (Tycho Brahe has a prognostication around ten times as large as Wābkanawī's for the lunar eclipse of December 8, 1573; see Brahe, vol. 1, pp. 62–64.)

latitude of the Moon was 11 seconds to the north, it was revealed that the southern side of the bright ring was thicker (than its northern side). I observed this eclipse in the vicinity of Mughān, and (my observation) was obviously in agreement with this (i.e., our author's results). And this (i.e., the agreement between calculation and observation) is like a proof of the correctness of the principles applied in this *zīj*.

3.4 Discussion and Conclusion

1. As we have already mentioned, Wābkanawī made a comparative study of the astronomical tables available to him and tried to cope with the iterative process. This led him to the empirical insight that the use of the apparent relative velocity (the true relative velocity adjusted with the amount of parallax), as proposed by al-Battānī and al-Khāzinī, instead of the relative true velocity, as instructed in the majority of *zījes*, reduced the numbers of steps and drastically shortened the whole procedure. In the first case, five iterations were needed, but in the second case, only one iteration. Wābkanawī emphasizes that for eclipses occurring near the horizon, using the true velocity may even cause the astronomer to iterate for eight times in order to obtain the final result.[75] This is to be expected because the variation of the altitude of the Moon and Sun, and thus of their parallaxes, is large when the two are close to the horizon. One may expect that any other astronomer who was practically engaged in such a lengthy procedure would have proposed the same method as Wābkanawī. Nonetheless, even in the two important *zījes* written in the Middle East after Wābkanawī, that is, al-Kāshī's *Khāqānī zīj*[76] and Ulugh Beg's *Sulṭānī zīj*,[77] the astronomer is instructed to work with the true velocities.

2. In order to make a comparison between our author's eclipse timings, listed in (3.19) and (3.24), and the times computed by modern astronomical theories, we need to know the exact location of the observation, or the location for which the data were calculated. The historical geographical coordinates of Mughān are $\varphi = 39°$ (3.1) and $L = 47°$ from Greenwich: note that there is no precise method for transferring the historical longitude of Mughān, $L = 83°$, reckoned from the Canary Islands to the modern longitude measured from Greenwich, but according to the Islamic *zījes* (Section 3.3, step 1), Mughān is located 1° to the east of Marāgha, whose modern longitude is about 46° (46°12′ for the central seat of the observatory). Mughān is a relatively small plane located on the northwestern boundary of Iran (approximately between 39°–40° N and 47°–49° E). It turns out that the central path of the eclipse passed through its northwestern region (see Figure 3.1).

Table 3.4 presents the times of the eclipse for $\varphi = 39°$ and $L = 47°$. Col. 2 gives the values mentioned in the text. The recalculated values in (3.19) and (3.24) are

75 Wābkanawī, T: f. 74r, Y: f. 132r, P: ff. 111v–112r.
76 Kāshī VI.I.14: IO: f. 87v.
77 Ulugh Beg III.10: P1: f. 114r, P2: f. 126r.

Table 3.4 Times (counted after sunrise)

	Wābkanawī	*Recomputed*	*Modern*	*Dif: Wab–Mod*	*Dif: Rec–Mod*
First contact	$4;18^h$	$4;23^h$	$4;17^h$	+1 min	+6 mins
Maximum phase	5;49	5;55	5;53	−4	+2
Last contact	7;20	7;27	7;24	−4	+3

quoted in Col. 3. Col. 4 gives the times based on the modern theories, computed by means of NASA's Five Millennium Catalogue of Solar and Lunar Eclipses, which uses the estimate of Morrison and Stephenson (2004) for the computation of Delta T (the difference between terrestrial time and universal time). Since our author gave all the times since sunrise, the modern values for the starting moments of the three phases of the eclipse are counted since the sunrise of January 30, 1283, in Mughān, that is, UT 3:56 or 7:4 mean local time.

Irrespective of seeing the eclipse as annular or partial, for any other location in Mughān, the time deviations did not exceed ±15 minutes.

The quotation at the end of the preceding section shows that our author was delighted by the fact that the eclipse looked as his theoretical computations suggested. It should be noted that it is very difficult for the unaided eye to distinguish annularity in general, and the southern part of the annulus being thicker than its northern part, since there is so much glare from the bright annular ring, unless the eclipsed Sun is observed, for example, through clouds or in pinhole images (either artificial or natural). Thus, the question arises whether it was possible for our author to draw a realistic conclusion. In his *zīj*, Wābkanawī described an instrument similar to a pinhole image device that specifically was used for eclipse observations.[78] Prior to him, al-ʿUrḍī (d. 1266 CE), the instrument maker of the Marāgha observatory, also described a dioptra having a fixed and a movable pinnula and the two auxiliary disks that made it useful for eclipse observations.[79] Therefore, our author probably had at his disposal the necessary optical aids in order to make a reliable observation.

In order to estimate to what extent our author could make an accurate determination of time, it is worth discussing the two methods which he proposed. The first one is a clepsydra to which the Persian name *"pangān"* was assigned; the other is the standard method of computing the time from the altitude of the Sun (in the case of a solar eclipse) or of reference stars (in the case of a lunar eclipse).[80] They together, as our author emphasized, constituted a method that might reduce the probable errors in time measurement. The *pangān* appears to

---

78 Wābkanawī, IV.15.8: T: ff. 92r–v, Y: ff. 159r–v, P: ff. 139r–v. It is also described in an anonymous treatise titled *Ghāzānīd treatise on the observational instruments*. See Section 11.4.12.
79 Seemann 1929, pp. 61–71.
80 Wābkanawī, IV.15.8–9: T: ff. 92r–v, Y: ff. 159r–160r, P: ff. 139r–140r.

Table 3.5 Lunar eclipses observed by al-Maghribī in Marāgha

			Time (counted after true noon in Marāgha)		
Nos.	*Date*		*al-Maghribī*	*Modern*	*Error*
1	March 7, 1262	(JDN 2182069)	8;18[h]	8;13[h]	+5 mins
2	April 7/8, 1270	(JDN 2185022–23)	10;13	10; 8	+5
3	January 24/25, 1274	(JDN 2186409–10)	14; 0	13;57	+3

have widely been used at the Marāgha observatory, and al-Maghribī frequently referred to its application in his systematic observations (see Chapter 9). We only know the general shape of the instrument,[81] and based on the information given by al-Maghribī in his *Talkhīṣ al-majisṭī*, one may speculate about its calibrations, but nothing more is known about its structure and the possible mechanical components embedded in it. However, thanks to the quantitative details recorded by al-Maghribī, an objective estimate of its accuracy is possible. Table 3.5 displays the accuracy of the *pangān* used by al-Maghribī in the observatory in the case of the observations of three lunar eclipses.[82] The instrument could establish time intervals up to a few minutes.[83] Although the passage quoted in the end of the preceding section does not imply that our author measured times in this observation, simple altitude measurements by an astrolabe with an accuracy of 1° could also produce measurements sufficiently close to our author's computed times.[84]

81 It was in shape of a floating bowl (*ṭās*), having a hole in its apex and two graduated scales (usually drawn with the aid of the astrolabe) for both equal and unequal hours on its surface. The bowl was placed into a vessel of water. As water drained into it, one could read off the amount of time that had elapsed. The first description of it in the Islamic Period appears in al-Ṣūfī's *Book on the Astrolabe* (d. 986 CE) (Caps. 354–357: pp. 299–302). The instrument dates back to Babylonian and Indian texts of the first millennium BC (Pingree 1973, pp. 3–4). Archaeological excavations have unearthed the earliest models in India, apparently belonging to the same period (Rao 2005, pp. 205–206).

82 al-Magrhibī, *Talkhīṣ* V.4: f. 69v.

83 Because of the accuracy achieved by al-Maghribī in measuring times by the *pangān*, it cannot have been a simple drainage clepsydra. The use of the clepsydras having compound mechanical components was well established in Chinese astronomy since at least the eleventh century (Needham 1981, p. 136). Due to the verified cultural relations between the two realms of the Mongolian Empire (i.e., Iran and China), and especially considering the fact that the Chinese astronomers (at least Fu Mengchi or Fu Muzhai) were working at the Marāgha observatory (van Dalen 2002a, p. 334, 2002b, 2004), there may have been a connection between the *pangān* of the Marāgha observatory and the Chinese technology of making time-measuring devices.

84 The solar altitudes in the three phases were: start, 34.1° ≈ 34°; middle, 34.6° ≈ 35°; and end, 27.4° ≈ 27°. Our author's value for the half-duration of daylight is 5;6,48 hours (cf. (3.19)). A simple computation based on his instructions in IV.13.1–3 (T: ff. 86v–87v, Y: ff. 150v–152r, P: ff. 130v–132r) produces the following times counted from sunrise: start, 4:11; middle, 5:39; and end, 7:22, whose differences from our author's times (Table 3.4) are, respectively, −7, −10, and +2 minutes.

3. If we compute this eclipse from al-Maghribī's parameter values in (3.4), (3.5), and (3.6), through the steps and iterative process explained in this chapter, the maximum phase of the eclipse occurs at $T' = 1;25^h$ after noon or 6;21 hours after sunrise in Mughān. The difference between T' and the modern values in Table 3.4 is +28 minutes, seven times as large as the difference between the modern values and the time recomputed from Wābkanwī's initial values in (3.1), (3.2), and (3.3). This gives the impression that Wābkanawī's correction of al-Maghribī's lunar mean longitudes by $+0;13,11°$ was the result of observations, as he himself claims. This issue, as well as the two other corrections made in the epoch mean positions of Mars in longitude and Venus in anomaly, merits further study.

4. As we have seen in the previous chapter, the annular eclipse became widely known in the late Islamic period, because a good number of astronomers were familiar with Īrānshahrī's observation of the annular eclipse of July 28, 873, although there is no firm evidence that astronomers were aware of this event prior to the eleventh century. The Indian hypotheses on the apparent diameters of the luminaries which had already been incorporated in the Ptolemaic context of Islamic astronomy were supported by Īrānshahrī's observational evidence. The combination of the two made the annular eclipse the third type of solar eclipse that was mentioned in nearly all treatises or in the canons of *zījes* written in the Middle East after the eleventh century. It was in this context that Wābkanawī presented the first prediction of an annular solar eclipse.

References

Aaboe, A., 1954, "Al-Kāshī's iteration method for the determination of sin 1°", *Scripta Mathematica* **20**, pp. 24–29.

al-Āmulī, Rukn al-Dīn, *Zīj-i jāmi'-i Būsa'īdī*, MS. P: Iran, Parliament Library, no. 183/1, T: Tehran University Central Library, no. 2558.

Anonymous, *Sulṭānī zīj*, MS. Iran, Library of Parliament, no. 184.

Anonymous, 1860, *Sūrya Siddhánta: A Textbook of Hindu Astronomy* (Reprinted in 1997), Gangooly, P. and Burgess, E. (eds.), Burgess, E. and Gangooly, P. (trs.), Delhi: Motilal Banarsidass.

Ben-Zaken, A., 2009, "From Naples to Goa and back: A secretive Galilean messenger and a radical hermeneutist", *History of Science* **47**, 147–174.

al-Bīrūnī, Abū al-Rayḥān, 1954–1956, *al-Qānūn al-mas'ūdī* (*Mas'ūdīc canons*), 3 Vols., Hyderabad-Dn: Osmania Oriental Publication Bureau.

Brahe, T., 1913–1929, *Scripta Astronomica* or *Opera Omnia*, Dreyer, I. L. E. (ed.), 15 Vols., Copenhagen: Libraria Gyldendaliana.

van Dalen, B., 2002a, "Islamic and Chinese astronomy under the mongols: A little-known case of transmission", in: Dold-Samplonius, Y., *et al.* (eds.), *From China to Paris: 2000 Years Transmission of Mathematical Ideas*, Stuttgart: Franz Steiner, pp. 327–356.

van Dalen, B., 2002b, "Islamic astronomical tables in China: The sources for the Huihui li", in: Ansari, R. (ed.), *History of Oriental Astronomy* (Proceedings of the Joint Discussion 17 at the 23rd General Assembly of the International Astronomical Union, Organised by the Commission 41 (History of Astronomy), Held in Kyoto, August 25–26, 1997), Dordrecht: Kluwer, pp. 19–30.

van Dalen, B., 2004, "The activities of Iranian astronomers in Mongol China", in: Pourjavady, N. and Vesel, Ž. (eds.), *Science, Techniques et Instruments Dans le Monde Iranien*, pp. 17–28, Téhéran: Institut Français de Recherche en Iran.

van Dalen, B., 2007, "Wābkanawī", in: Hockey, T., *et al.* (eds.), *The Biographical Encyclopedia of Astronomers*, London: Springer, pp. 1187–1188.

al-Dihlawī, Farīd al-Dīn Masʿūd b. Ibrāhīm, *Shāhjahānī zīj*, MS London, British Library, Or. 372 (17th century).

al-Dihlawī, Farīd al-Dīn Masʿūd b. Ibrāhīm, *Sirāj al-istikhrāj*, MS Tashkent, Abū al-Rayḥān al-Bīrūnī Institute of Oriental Studies of the Uzbek Academy of Sciences, no. 702/2: ff. 16v–40v.

Dutka, J., 1997, "A note on 'Kepler's Equation'", *Archives for History of Exact Sciences* **51**, pp. 59–65.

Goldstein, B. R., 1979, "Medieval observations of solar and lunar eclipses", *Archives Internationales d'Histoire des Sciences* **29**, pp. 101–156.

Goldstein, B. R. and Swerdlow, N., 1970/71, "Planetary distances and sizes in an anonymous Arabic treatise preserved in Bodleian Ms. Marsh 621", *Centaurus* **15**, pp. 135–170.

Haddad, F. I. and Kennedy, E. S., 1971, "Geographical tables of medieval Islam", *Al-Abhath* **24**, pp. 87–102. Reprinted in Kennedy 1983, pp. 636–651.

Hamadanizadeh, J., 1980, "The trigonometric tables of al-Kāshī in his *Zīj-i Khāqānī*", *Historia Mathematica* **7**, pp. 38–45.

Ibn al-Fuwaṭī, Kamāl al-Dīn ʿAbd al-Razzāq b. Muḥammad, *Majmaʿ al-ādāb fī muʿjam al-alqāb*, Muḥamamd Kāẓim (ed.), Tehran, 1416 H.

Ibn Yūnus, Abu al-Ḥasan ʿAlī b. ʿAbd al-Raḥmān b. Aḥmad, *al-Zīj al-kabīr al-Ḥākimī*, Ms. Leiden, no. Or. 143.

al-Kamālī, Muḥamamd b. Abī ʿAbd-Allāh Sanjar, *Zīj-i Ashrafī*, MS. Paris, Biblithèque Nationale, no. suppl. Pers. 1488.

al-Kāshī, Ghiyāth al-Dīn Jamshīd, *Khāqānī zīj*, IO: England, India Office, no. Persian 430 (Ethé 2232), P: Iran, Parliament Library, no. 6198.

Kennedy, E. S., 1956a, *A Survey of Islamic Astronomical Tables*, Philadelphia: American Philosophical Society.

Kennedy, E. S., 1956b, "Parallax theory in Islamic astronomy", *Isis* **47**, pp. 33–53. Reprinted in Kennedy, pp. 164–184.

Kennedy, E. S., 1958, "The Sasanian Astronomical Handbook *Zīj-i Shāh* and the astronomical doctrine of Transit (*Mamarr*)", *Journal of American Oriental Society* **78**, pp. 246–262. Reprinted in Kennedy 1983, pp. 319–335.

Kennedy, E. S., 1960, "A letter of Jamshīd al-Kāshī to his father: Scientific research at a fifteen century court", *Orientalia* **29**, pp. 191–213. Reprinted in Kennedy 1983, pp. 722–744.

Kennedy, E. S., 1962, "The world-year concept in Islamic astronomy", *International Congress of History of Science*, pp. 23–43. Reprinted in Kennedy 1983, pp. 351–371.

Kennedy, E. S., 1964, "The Chinese–Uighur Calendar as described in the Islamic sources", *Isis* **55**, pp. 435–443. Reprinted in Kennedy 1983, pp. 652–660.

Kennedy, E. S., 1969, "An early method of successive approximations", *Centaurus* **13**, pp. 248–250.

Kennedy, E. S., 1983, *Studies in the Islamic Exact Sciences*, Beirut: American University of Beirut.

Kennedy, E. S., 1987–1988, "Eclipse prediction in Arabic astronomical tables prepared for the Mongol viceroy of Tibet", *Zritschrift für Geschichte der Arabisch-Islamischen Wissenschaften* **4**, pp. 60–80.

Kennedy, E. S. and Transue, W. R., 1956, "A medieval iterative algorithm", *American Mathematical Monthly* **63**, pp. 80–83.

al-Khāzinī, ʿAbd al-Raḥmān, *Wajīz* [Abridgment of] *al-Zīj al-Muʿtabar al-Sanjarī* [Considered *Zīj* of Sultan Sanjar]. MS. Istanbul, Suleymaniye Library, Hamadiye collection, no. 859.

King, D., 1986, "The earliest Islamic mathematical methods and tables for finding the direction of Mecca", *Zeitschrift fur Geschichte der Arabisch-Islamischen Wissenschaften* **3**, pp. 82–149.

King, D. and Samsó, J., 2001, "Astronomical handbooks and tables from the Islamic World", *Suhayl* **2**, pp. 9–105.

Kunitzsch, P., 1964, "Das Fixsterverzeichnis in der 'Persischen Syntaxis' des Georgios Chrysokokkes", *Byzantinische Zeitschrift* **57**, pp. 382–411.

Leichter, J. G., 2004, *The Zīj as-Sanjarī of Gregory Chioniades: Text, Translation and Greek to Arabic Glossary*. Ph.D. thesis. Brown University. Under supervision of D. Pingree (unpublished, private communication).

al-Maghribī, Muḥyī al-Dīn, *Adwār al-anwār*, MSS. M: Iran, Mashhad, Shrine Library, no. 332; CB: Ireland, Dublin, Chester Beatty, no. 3665.

al-Maghribī, Muḥyī al-Dīn, *Khulāṣat al-Majisṭī*, MS. Doha: Mathaf al-fann al-Islāmī, 791 (copied from a copy of an autograph, which was written by Ḥajjāj b. Khwāja ʿAzīz b. Sharaf al-Dahhān al-Munajjim al-Khwārizmī al-Ḥanūfī in Marāgha on Sunday, 6 July 1287).

al-Maghribī, Muḥyī al-Dīn, *Talkhīṣ al-majisṭī*, MS. Leiden, Or. 110.

Mancha, J. L., 1998, "Heuristic reasoning: Approximation procedures in Levi ben Gerson's astronomy", *Archives for History of Exact Science* **52**, pp. 13–50.

Morrison, L. V. and Stephenson, F. R., 2004, "Historical values of the Earth's clock error ΔT and the calculation of eclipses", *Journal for the History of Astronomy* **35**, pp. 327–326.

Mozaffari, S. M., 2009, "Wābkanawī and the first scientific observation of an annular eclipse", *The Observatory* **129**, pp. 144–146.

Mozaffari, S. M., 2015, "Annular eclipses and the considerations about the solar and lunar angular diameters in the medieval astronomy", in: Orchiston, W., Green, D. A., and Strom, R. (eds.), *StephensonFest: Studies in Applied Historical Astronomy and Amateur Astronomy*, New York: Springer, pp. 119–142.

Mozaffari, S. M., 2018–2019, "Muḥyī al-Dīn al-Maghribī's measurements of Mars at the Maragha observatory", *Suhayl* **16**, pp. 149–249.

Mozaffari, S. M., 2019, "Ibn al-Fahhād and the great conjunction of 1166 AD", *Archive for History of Exact Sciences*, **73**, pp. 517–549.

Nallino, C. A. (ed.), 1899–1907, *Al-Battani Sive Albatenii Opus Astronomicum*, 3 Vols., Milan: Milan University Press.

Needham, J., 1981, *Science in Traditional China: A Comparative Perspective*, Cambridge, MA: Harvard University Press.

Neugebauer, O., 1975. *A History of Ancient Mathematical Astronomy*, Berlin-Heidelberg-New York: Springer.

al-Nīshābūrī, Niẓām al-Dīn Aʿraj, *Kashf al-ḥaqāʾiq al-Zīj al-Īlkhānī* (a commentary on the *Īlkhānī zīj*), MS. Iran, Parliament Library, no. 1210 (copied at Jumādā I 825 H/April-May 1422).

al-Nīshābūrī, Niẓām al-Dīn Aʿraj, *Sharḥ-i Tadhkira* ("Explanation of *Memoir*", a commentary on al-Ṭūsī's *Memoir*), MS. University of Tehran, no. 488.

Paschos, E. A. and Sotiroudis, P., 1998, *The Schemata of the Stars: Byzantine Astronomy from A.D. 1300*, Singapore: River Edge & N.J. World Scientific.

Pedersen, O., 1974, *A Survey of Almagest*, Odense: Odense University Press; with annotation and new commentary by A. Jones, New York: Springer, 2010.

Pfeiffer, J., 2015, "Teaching the learned: Jalāl al-Dīn al-Dawānī's *Ijāza* to Muʾayyadzāda ʿAbd al-Raḥmān Efendi and the circulation of knowledge between Fārs and the Ottoman Empire at the turn of the sixteenth century", in: Pomerantz, M. and Shahin, A. (eds.), *The Heritage of Arabo-Islamic Learning*, Leiden: Brill, pp. 284–332.

Pingree, D., 1973, "The mesopotamian origin of early Indian mathematical astronomy", *Journal for the History of Astronomy* **4**, pp. 1–12.

Pingree, D. (ed.), 1985, *Astronomical Works of Gregory Chioniades*, Vol. 1: *Zīj al-ʿalāʾī*, Amsterdam: Gieben.

Plofker, K., 2002, "Use and transmission of iterative approximations in India and the Islamic World", in: dold-Samplonius, Y., *et al.* (eds.), *From China to Paris: 2000 Years Transmission of Mathematical Ideas*, pp. 167–186, Stuttgart: Franz Steiner Verlag.

Ptolemy, 1822, *Tetrabiblos*, Ashmand, J. M. (ed.), London.

Rao, N. K., 2005, "Aspects of prehistoric astronomy in India", *Bulletin of the Astronomical Society of India* **30**, 499–511.

Rosenfeld, B. A. and Hogendijk, J. P., 2002–2003, "A mathematical treatise written in the Samarqand observatory of Ulugh Beg", *Zeitschrift für Geschichte der Arabisch-IslamischenWissenschaften* **15**, pp. 25–65.

Saliba, G., 1985, "Solar observations at Maragha observatory", *Journal for the History of Astronomy* **16**, pp. 113–122. Reprinted in Saliba 1994, pp. 177–186.

Saliba, G., 1986, "The determination of new planetary parameters at the Maragha observatory", *Centaurus* **29**, pp. 249–271. Reprinted in Saliba 1994, pp. 208–230.

Saliba, G., 1994, *A History of Arabic Astronomy: Planetary Theories during the Golden Age of Islam*, New York-London: New York University Press.

Samsó, J., 1977. "A homocentric solar model by Abu Jaʿfar al-Khāzin", *Journal for the History of Arabic Science* **1**, pp. 268–275. Reprinted in Samsó, J., 1994, *Islamic Astronomy and Medieval Spain*, Trace XI, Aldershot, Variorum.

Sayılı, A., 1960, *The Observatory in Islam*, Ankara: Turk Tarih Kurumu Basimevi.

Sayyid Muḥammad the Astrologer, *Laṭāʾif al-kalām fī aḥkām al-aʿwām*, MS. Iran, Parliament Library, no. 6347.

Seemann, H. J., 1929, "Die Instrumente der Sternwarte zu Maragha nach den Mitteilungen von al-ʿUrḍī", in: Schulz, O. (ed.), *Sitzungsberichte der Physikalisch-medizinischen Sozietät zu Erlangen*, **60** (1928), pp. 15–126, Erlangen: Kommissionsverlag von Max Mencke.

al-Shīrāzī, Quṭb al-Dīn, *Ikhtiyārāt-i muẓaffarī* (*Muẓaffarid Selections*, dedicated to Muẓaffar al-Dīn Bulāq Arsalān, d. 1305), MS. Iran, National Library, no. 3074F, up to f. 176r (copying completed on Monday, 20 Jumādā I 682/16 August 1283).

al-Shīrāzī, Quṭb al-Dīn, *Nihāyat al-idrāk fī dirāyat al-aflāk* (*Limit of Comprehension in the Knowledge of Celestial Spheres*), MSS. B: Berlin, no. Ahlwart 5682 = Petermann I 674 (towards the end of Dhi al-qaʿda 726/October 1326), P1: Iran, Parliament Library, no. 6457 (copied at Ulugh Beg's school at Samarqand one month during Dhi al-qaʿda and Dhi al-ḥijja 844/April–May 1441), P2: Iran, Parliament Library, no. 16008 (5 Muḥarram 1120/27 March 1708).

al-Shīrāzī, Quṭb al-Dīn, *Tuḥfa al-shāhiyya fī ʾl-hayʾa* (*Gift to the King on Astronomy*), MS. Iran, Parliament Library, no. 6130 (copied in Rabīʿ I 703 H/October 1303 CE).

Stephenson, F. R., 1997, *Historical Eclipses and Earth's Rotation*, Cambridge: Cambridge University Press.

Storey, C. A., 1958, *Persian Literature*, Vol. 2, Part 1, London: Luzac.

al-Ṣūfī, ʿAbd al-Raḥmān, 1995, *Al-ʿamal bi-ʾl-Asṭurlāb*, Morocco: ISESCO.

Swerdlow, N. M., 2000, "Kepler's iterative solution to Kepler's equation", *Journal for the History of Astronomy* **31**, pp. 339–341.

Therkel, J. N., 1974, "Peder Horrebow's solution of the Kepler-problem", *Centaurus* **18**, pp. 173–183.

Toomer, G. J., 1998, *Ptolemy's Almagest*, Princeton: Princeton University Press.

al-Ṭūsī, Naṣīr al-Dīn Muḥammad, *al-Tadhkira fī [ʿilm] al-hayʾa* (*Memoir on [the Science of] Cosmography*), 1993, *Naṣīr al-Dīn al-Ṭūsī's* Memoir on Astronomy, Ragep, F. J. (ed. & En. tr.), 2 Vols., New York: Springer.

al-Ṭūsī, Naṣīr al-Dīn Muḥammad, *Īlkhānī zīj*, MSS. C: University of California, Caro Minasian Collection, no. 1462, P: Iran, Parliament Library, no. 181.

al-Ṭūsī, Naṣīr al-Dīn Muḥammad, *Taḥrīr al-majisṭī* (*Exposition of the* Almagest), MSS. Iran, Parliament Library, P1: no. 3853 (commented by ʿAbd al-ʿAlī al-Bīrjandī, d. 1525/1526 CE), P2: no. 6357, P3: no. 6395, B: Staatsbibliothek zu Berlin, Sprenger 1838 = Ahlwardt 5655, ff. 1v–152r (copying finished at Marāgha on Wednesday, 22 Ramaḍān 655 [H.] = 3 October 1257 CE, for a certain Maḥmūd al-Khāṭīb: f. 152r), AS: Istanbul, Aya Sofia, no. 2583, ff. 1v–112v (copying completed on Thursday, 5 Shawwāl 686 [H.] = 13 November 1287 CE: f. 112v).

Ulugh Beg, *Sulṭānī zīj*, MS. P1: Iran, Library of Parliament, no. 72, MS. P2: Iran, Library of Parliament, no. 6027.

Wābkanawī, Shams al-Dīn Muḥammad, *Zīj al-muḥaqqaq al-sulṭānī ʿalā uṣūl al-raṣad al-Īlkhānī* (*The testified zīj for the sultan on the basis of the parameter values of the Īlkhanid observations*), MSS. T: Turkey, Aya Sophia Library, No. 2694; MS. Y: Iran, Yazd, Library of ʿUlūmī, no. 546 (its microfilm is available in the central library of the University of Tehran, no. 2546), P: Iran, Library of Parliament, no. 6435.

Westerink, L., 1980, "La Profession de foi de Gregoire Chioniades", *Revue des études Byzantines* **38**, pp. 233–245.

4

BĪRŪNĪ'S EXAMINATION OF
THE PATH OF THE CENTER
OF THE EPICYCLE IN
PTOLEMY'S LUNAR MODEL[1]

Ptolemy's complete lunar model in *Almagest* V produces a large variation of distance of the center of the epicycle, between its maximum at mean conjunction and opposition to the mean position of the Sun and its minimum at mean quadrature, so that the resulting path of the center of the epicycle about the Earth is an oval figure. In Figure 4.1, with the Earth at T, the center of the epicycle L moves uniformly about T on an eccentric of radius R with center E and apogee A, through the mean elongation from the mean position of the Sun S in the direction of increasing longitude, while the apsidal line of the eccentric rotates in the opposite direction through the same mean elongation. Consequently, L reaches the apogee A and perigee B of the eccentric twice in each mean synodic month, which produces the oval path, farthest from the Earth at mean conjunction and opposition, closest at mean quadrature. The Moon M moves on the epicycle of radius r in the direction opposite to the motion of L through the mean anomaly, completed in an anomalistic month, uniformly with respect to the mean apogee F, which has an "inclination" (*prosneusis*) toward a point P, opposite to the direction of E from the Earth and with the same eccentricity, $e = PT = TE$.[2] The true apogee G lies on the line TLG from the Earth. Our concern here is the path described by the variable distance of the lunar epicycle from the Earth $\rho = TL$. At mean conjunction and opposition, $\rho = EA + ET = R + e$, and at mean quadrature, $\rho = EA - ET = R - e$. Using historical values by Muḥyī al-Dīn al-Maghribī (d. 1283) of the Marāgha observatory in northwestern Iran,[3] where $R + e = 60$, $R = 51$, $e = 9$, and $r = 5;12$. Thus, at conjunction and opposition, $\rho = R + e = 60$, and at quadrature, $\rho = R - e = 42$.

---

1 Original publication: S. M. Mozaffari, "Bīrūnī's examination of the path of the centre of the epicycle in Ptolemy lunar model," *Journal for the History of Astronomy* **45** (2014), pp. 123–127. © 2014 Sage, and republished by permission.
2 See Toomer 1998, p. 220f, Pedersen 1974, chapter 6, Neugebauer 1975, vol. 1, p. 53f.
3 See Chapters 7 and 9.

DOI: 10.4324/9781003481966-7

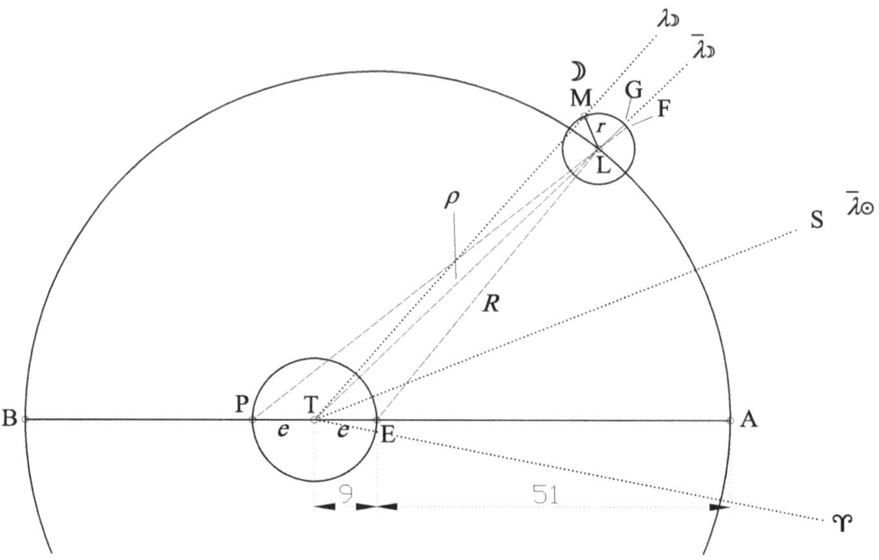

Figure 4.1 Ptolemy's lunar model.

Ptolemy, as Pedersen remarks, "always conceives the motion of the lunar epicycle center as a circular motion around the moving center of the deferent. He
never asks for the orbit described by the lunar epicycle center relative to the center of the Earth."[4] Abū al-Rayḥān al-Bīrūnī (973–1048) explores the issue more
fully in a passage written in the form of catechism in the end of his *al-Qānūn
al-masʿūdī* VII.7.1.[5] The translation of the relevant passage runs as follows:

> Q: What [shape] does its [i.e., the Moon's] epicycle center describe by
> this motion [i.e., according to Ptolemy's model of a movable eccentric]?
>
> A: If it is assumed that the Sun is at rest and if the lunar epicycle center
> is at its orb's apogee in its [mean] conjunction or opposition [with/to
> the Sun] and is at the perigee in its [mean] quadrature, it will describe a
> rounded rectangular shape by its motion. It might be thought that it [i.e.,
> the path of the epicycle center] is an ellipse of the [right circular] conic
> or cylindrical sections. [But,] it is not so.
>
> Take the apogee of the Moon's orb being *A*, its center being at *E*, at the
> time of the [mean] conjunction, and the circle that this center revolves on

4 Pedersen 1974, p. 188.
5 Bīrūnī, *al-Qānūn*, F: f. 116v, B: f. 248r, C: f. 143r, I: f. 215r, K: p. 344, L: f. 143r, R: p. 307;
   1954–1956, vol. 2, pp. 794–795. The contents of this work are introduced in Kennedy 1971.

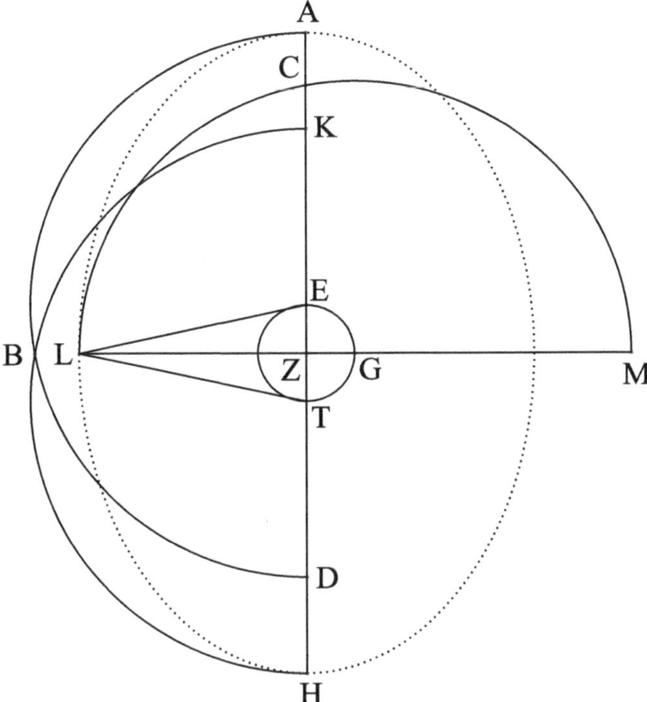

Figure 4.2 Trajectory of the lunar epicycle's center.

its periphery being *EGT* [cf. Figure 4.2][6]; the position of the eccentric orb is then *ABD*. At the time of the [first mean] quadrature, the center of the eccentric orb is at *G* and its position is *MCL* in which *L* is the perigee. At the time of the [mean] opposition, the center of the eccentric orb is at *T* and its position is *HBK*. If the points *A*, *L*, and *H* are on the periphery of the ellipse with two foci (*markaz*, "center") *T* and *E*, drawing the triangle *ELT*, the sum of *TA* and *AE* will be equal to the sum of the two sides *TL* and *LE*. Half of either of these two sums is *ZA*, which is the greatest distance [of the epicycle center from the Earth at the point *Z*]. Its square is equal to [the sum of] the squares of *AE* and *EZ* plus twice the multiplication of *AE* by *EZ*. The side *EL* is the sum of the squares of *ZL*, which is the least distance [of the epicycle center from the Earth *Z* in the first mean quadrature], and *ZE*. Thus, *EL* is less than *ZA*. Therefore, *E* and *T*

6 The original lettering is transcribed according to the standard proposed by E. S. Kennedy (1991–1992).

are not the foci of the ellipse. Moreover, the Sun is in motion. Therefore, what the [epicycle] center describes is not an exact shape.

Commentary: Bīrūnī's intention is to prove that the three points A, L, and H are *not* on the periphery of an ellipse with foci E and T, that is, E and T are not the foci of the dotted ellipse, not drawn in the original. His argument is a *reductio ad absurdum*. If the three points A, L, and H are on the periphery of the ellipse with E and T as foci, then $\frac{1}{2}(TA + AE) = \frac{1}{2}(TL + LE) = ZA$. Since triangles ELZ and TLZ are congruent, $LE = TL$, and thus, $EL = ZA$. But $ZA = AE + EZ$, so $ZA^2 = AE^2 + EZ^2 + 2\,AE \cdot EZ$. And $EL^2 = ZL^2 + EZ^2$. It is thus evident that $EL < ZA$. (Note that EZ^2 is common to both relations and AE is the radius of the eccentric, while ZL is the least distance of the epicycle center; thus, $AE^2 + 2AE \cdot EZ > ZL^2$.) Therefore, our first assumption is false, and it follows that the path of the epicycle center is *not* an ellipse whose foci are the center of the eccentric E and T in mean conjunction and opposition to the Sun, and the three points A, L, and H are not on its periphery. QED.

There are some further points worth mentioning. Ptolemy's model for Mercury uses a movable eccentric like his lunar model. In this model, the center of the eccentric revolves in a small circle which is centered not at the Earth but at a point displaced from the Earth twice as far as the distance between the Earth and the equant point, which is thus located between the center of the small circle and the Earth. In a justly famous study published in 1955, W. Hartner showed that with Ptolemy's parameters, the curve described by the center of the epicycle is nearly indistinguishable from an ellipse.[7] Bīrūnī does not investigate Mercury, perhaps because he assumed that if the lunar epicycle did not describe an ellipse, still less would be the epicycle of Mercury. Ibn al-Zarqālluh (d. 1100) and al-Kāshī (1380–1429) describe the construction of oval, elliptically shaped plates for Mercury's deferent in equatoria for finding the longitude of the planet, very close to the curve in Ptolemy's model, and Ibn al-Zarqālluh identifies the curve as an ellipse (Kāshī's equatorium, named *Ṭabaq al-manāṭiq*, "Plate of zones," is described in his *Nuzhat al-ḥadāʾiq*, "A fruit-garden stroll," completed in 818 H/1416 CE).[8] But somewhat strangely, neither of them did take the locus of the lunar epicycle's center as an ellipse, not least because neither in Ibn al-Zarqālluh's graphical method for finding lunar distances nor in the lunar plate of al-Kashī's instrument such a smart and useful auxiliary consideration was taken into account.[9] Albeit, about a decade later, when Kāshī was on the senior staff of the Samarqand observatory, he appended ten afterthought passages (*Ilḥāqāt al-ʿashara*, completed on 15 Shaʿbān 829/June 22, 1426) to his treatise, the first two of which are on the elliptical shapes of the paths of the centers of the epicycles of the Moon and Mercury; Kāshī remarks that from the geometrical proofs and computational procedures, we obtain that the shapes of both trajectories

---

7 Hartner 1955, pp. 109–122. Following Hartner's study, see Pedersen 1974, pp. 320–324, Neugebauer 1975, vol. 1, pp. 168–169.
8 See Kāshī, *Nuzha*, pp. 266–267; Kennedy 1952, pp. 46–47, 49–50, Kennedy 1960, pp. 40–41, Samsó and Mielgo 1994.
9 See Kennedy 1950, pp. 182–183, Puig 1989.

are neither elliptical/oval (*ihlīlajī*) to be drawn by a compass nor any (of the known conic) section(s) to be drawn by a "perfect compass" (*farjār al-tām*),[10] but they closely resemble an ellipse. However, no proof is given there, and he only puts forward an approximate method for drawing the oval-shaped deferents.[11]

It was because of the influence of Bīrūnī's *al-Qānūn* on the later medieval Middle Eastern astronomers that the oval trajectories of the centers of the epicycles of the Moon and Mercury are noted and schematically represented in the cosmographical works of the so-called Marāgha school. In Naṣīr al-Dīn al-Ṭūsī's (1201–1274) *Multaqaṭāt min kitāb al-Majisṭī ʿalā ḥasab iṣlāḥ baʿḍ al-mutaʾakhkhirīn* (*The collection of the corrections made by some modern [astronomers] to the* Almagest), which is a brief survey, and a dense assemblage, of the corrections made by Islamic astronomers in the first six books of the *Almagest*, the first topic related to *Almagest* V is the word-by-word quotation of Bīrūnī's aforesaid passage in *al-Qānūn* VII.7.1 on the oval shape of the trajectory of the lunar epicycle's center.[12] In his first notable treatise on cosmography, titled the *Muʿīniyya* (1235 CE), he also correctly extended the same idea to the trajectory of the center of Mercury's epicycle around the center of the planet's hypocycle in Ptolemy's model.[13] In his *Appendix* to the same work (the *Ḥall* or *Dhayl*), finished a decade later, in October 1245 CE,[14] he devotes chapter 4 to the locus of the epicycles' centers of the Moon and Mercury; of course, the description is brief, without any geometrical proof.[15] The same figures in the *Muʿīniyya* also appear, about two decades and a half later, in his most important treatise, the *Tadhkira* (1261 CE).[16] Analogously, in his three cosmographical treatises, Quṭb al-Dīn al-Shīrāzī (d. 1311) mentions that the path of the center of the lunar epicycle is elliptical/oval (*ihlīlajī biyḍī*), not an ellipse (*qaṭʿ nāqiṣ*, "deficient [conic] section"), but without giving any geometrical proof or identifying his source; in the case of Mercury, he also solely refers to the elliptical shape of the trajectory as drawn in a figure.[17] Similarly, in his preliminary cosmographical work, *Lubāb* I.5 and 7, Kāshī briefly notes the oval paths of the epicycle centers of the Moon and Mercury.[18]

Georg von Peurbach (1423–1461), in his *Theoricae Novae Planetarvm*, maintains the oval trajectory of the epicycle center only in the case of Mercury, without referring to the existence of a similar situation for the Moon.[19] In his commentary

---

10 A specialized compass for drawing the conic sections; for example, see Sezgin and Neubauer 2010, vol. 3, pp. 149–151.

11 Kāshī, Nuzha, pp. 289–291 (the fact that when writing the appendixes Kāshī was at Samarqand is deduced from the colophon on p. 299); Kennedy 1960, pp. 170–174.

12 Ṭūsī, *Multaqaṭāt*, f. 63v: lines 20–29.

13 Ṭūsī, *Muʿīniyya*, P: pp. 55, 73, N: pp. 37, 48; 2020, pp. 79, 96.

14 On the dates of completion of Ṭūsī's works cited here, see Saliba 1987, pp. 146–147, 151–152, 155, Ragep 2020.

15 Ṭūsī, *Appendix* to the *Muʿīniyya*, N: p. 119; 2020, pp. 226–227.

16 Ṭūsī, *Tadhkira*, ff. 23r, 27r; 1993, vol. 1, pp. 162–163, 176–177.

17 Shīrāzī, *Ikhtiyārāt*, ff. 66r–v, 84r–v; *Nihāya*, B: ff. 77r–v, 95r, P1: ff. 34r, 40v, 41v, P2: ff. 82r–v, 103v; *Tuḥfa*, ff. 50v–51r, 61v.

18 Kāshī, *Lubāb*, N: ff. 7r, 9r–v, P1: pp. 187, 191–2, P2: ff. 8r, 10v.

19 See Hartner 1955, pp. 130, 132–134, Aiton 1987, p. 26, his figure, captioned "Figure 16," is on p. 27.

on Peuerbach's *Theoricae*, Erasmus Reinhold (1511–1553) describes, and illustrates through graphical constructions, the paths of the centers of the epicycles of the Moon and Mercury.[20] In both cases, the figures are determined by the exaggerated eccentricities used for the drawing. For the Moon, he illustrates what he calls an "oval or rather lenticular figure," like a lentil. For Mercury, he describes the figure as "like an egg," with the broader end at apogee and the narrow at perigee, as indeed it appears, but only because the eccentricity is taken so large.

References

Aiton, E. J., 1987, "Peurbach's *Theoricae Novae Planetarvm*: A translation with commentary", *Osiris*, 2nd Series, **3**, pp. 4–43.

al-Bīrūnī, Abū al-Rayḥān, 1954–1956, *al-Qānūn al-mas'ūdī* (*Mas'ūdīc canons*), 3 Vols., Hyderabad: Osmania Bureau; MSS. F*: Paris, Bibliothèque Nationale de France, Ar. 6840 (the end of Ramaḍān 501 h, about the middle of May 1108 CE), B: Berlin, Staatsbibliothek zu Berlin, no. Or. Oct. 275 = Ahlwardt 5667, C*: Cairo, Dār al-Kutub, Mīqāt collection, no. 866 (673 h), I*: Istanbul, Süleymaniye, Veliyüddin, no. 2277 (before 536 h), K: Konya, Yusuf Agha, no. 1797, L*: London, BL, no. Or. 1997 (570 h), R: India, Rampur, Raza, no. 3700 (asterisks indicate the manuscripts used in the Hyderabad edition).

Hartner, W., 1955, "The Mercury horoscope of Marcantonio Michiel of Venice", *Vistas in astronomy* **1**, pp. 84–138; reprinted (with additions) in: Hartner 1984, Vol. 2, pp. 440–495.

Hartner, W., 1984, *Oriens-Occidens*, 2 Vols., Hildesheim: Georg Olms Verlag.

Jones, A. and Carman, C. (eds.), 2020, *Instruments-Observations-Theories: Studies in the History of Early Astronomy in Honor of James Evans*. DOI: 10.5281/zenodo.3928498.

al-Kāshī, Jamshīd Ghiyāth al-Dīn, *Lubāb-i iskandarī*, MSS. N: Iran, National Library, no. 32066/5 (folios unnumbered, the first 32 pages), P1: Iran, Parliament Library, no. 17065/5, pp. 174–208 (copying finished on 1 Ramaḍān 1015/19 December 1606, P2: Iran, Parliament Library, no. 6312.

al-Kāshī, Jamshīd Ghiyāth al-Dīn, *Nuzhat al-ḥadā'iq* (*A fruit-garden stroll*), with *Ilḥāqāt al-'ashara* (*ten Appendixes*), published as lithographs with Kāshī's *Miftāḥ al-ḥisāb* at Tehran in 1306 H/1888–1889 CE, pp. 250–313; the digital scans from the courtesy of the Iran National Library, no. 2099/1.

Kennedy, E. S., 1950/1952, "A fifteenth-century planetary computer: al-Kāshī's 'Ṭabaq al-Manāteq' I. Motion of the Sun and the Moon in longitude"/"II. Longitudes, distances, and equations of the planets", *Isis*, Part 1: **41**, pp. 180–183, Part 2: **43**, pp. 42–50; reprinted in: Kennedy *et al.* 1983, pp. 452–455, 472–480.

Kennedy, E. S., 1960, *The Planetary Equatorium of Jamshīd Ghiyāth al-Dīn al-Kāshī*, Princeton: Princeton University Press.

Kennedy, E. S., 1971, "Al-Bīrūnī's Masudic Canon", *Al-Abhath* **24**, pp. 59–78; reprinted in: Kennedy *et al.* 1983, pp. 573–592.

Kennedy, E. S., colleagues, and former students, 1983, *Studies in the Islamic Exact Sciences*, Beirut: American University of Beirut.

---

20 There are many editions of Reinhold's commentary. In the Paris 1553 edition, the figure for the Moon is between ff. 38 and 39, and for Mercury between ff. 68 and 69.

Kennedy, E. S., 1991–1992, "Transcription of Arabic letters in geometric figures", *Zeitschrift fur Geschichte der Arabisch-Islamischen Wissenschaften* 7, pp. 21–22.

Neugebauer, O., 1975. *A History of Ancient Mathematical Astronomy*, Berlin-Heidelberg-New York: Springer.

Pedersen, O., 1974, *A Survey of Almagest*, Odense: Odense University Press, 1974; with annotation and new commentary by A. Jones, New York: Springer, 2010.

Puig, R., 1989, "Al-Zarqālluh's graphical method for finding lunar distances", *Centaurus* 32, pp. 294–309.

Ragep, F. J., 2020, "The origins of the Ṭūsī-couple revisited', in: Jones and Carman 2020, pp. 229–237.

Saliba, G., 1987, "The role of the *Almagest* commentaries in medieval Arabic astronomy: A preliminary survey of Ṭūsī's Redaction of Ptolemy's *Almagest*", *Archives Internationales d'Histoire des Sciences* 37, pp. 3–20; reprinted in Saliba 1994, pp. 143–160.

Saliba, G., 1994, *A History of Arabic Astronomy: Planetary Theories During the Golden Age of Islam*, New York: New York University Press.

Samsó, J., 2007, *Astronomy and Astrology in al-Andalus and the Maghrib*, Aldershot–Burlington: Ashgate (Variorum Collected Studies Series).

Samsó, J. and Mielgo, H., 1994, "Ibn al-Zarqālluh on Mercury", *Journal for the History of Astronomy* 25, pp. 289–296; reprinted in: Samsó 2007, Trace IV.

Sezgin, F. and Neubauer, E., 2010, *Science and Technology in Islam*, 5 Vols., Frankfurt: Institut für Geschichte der Arabisch-Islamischen Wissenschaften.

al-Shīrāzī, Quṭb al-Dīn, *Ikhtiyārāt-i muẓaffarī* (*Muẓaffarid Selections*, dedicated to Muẓaffar al-Dīn Bulāq Arsalān, d. 1305), MS. Iran, National Library, no. 3074F, up to f. 176r (copying completed on Monday, 20 Jumādā I 682/16 August 1283).

al-Shīrāzī, Quṭb al-Dīn, *Nihāyat al-idrāk fī dirāyat al-aflāk* (*Limit of Comprehension in the Knowledge of Celestial Spheres*), MSS. B: Berlin, no. Ahlwart 5682 = Petermann I 674 (towards the end of Dhi al-qaʿda 726/October 1326), P1: Iran, Parliament Library, no. 6457 (copied at Ulugh Beg's school at Samarqand one month during Dhi al-qaʿda and Dhi al-ḥijja 844/April–May 1441), P2: Iran, Parliament Library, no. 16008 (5 Muḥarram 1120/27 March 1708).

al-Shīrāzī, Quṭb al-Dīn, *Tuḥfa al-shāhiyya fī 'l-hay'a* (*Gift to the King on Astronomy*), MS. Iran, Parliament Library, no. 6130 (copied in Rabīʿ I 703 H/October 1303 CE).

Toomer, G. J., 1998, *Ptolemy's Almagest*, Princeton: Princeton University Press.

al-Ṭūsī, Naṣīr al-Dīn Muḥammad, *al-Risāla al-Muʿīniyya* (*al-Mughniya*) *dar ʿilm-i hay'a* (*the Muʿīniyya treatise on the science of the configuration* [of the celestial bodies]) and *Ḥall-i mushkilāt al-Muʿīniyya* (*The resolution of difficulties in the* Muʿīniyya)/*Dhayl-i Muʿīniyya* (*Appendix to the* Muʿīniyya), MSS. P: Tehran, Parliament Library, no. 6347 (only the first), N: Tehran, National Library, no. 21303 (both treatises); edition 2020 by Nikfahm-Khubravan, S., and Savadi, F., Tehran: Miras-e Maktoob.

al-Ṭūsī, Naṣīr al-Dīn Muḥammad, *al-Tadhkira fī* [*ʿilm*] *al-hay'a* (*Memoir on* [*the science of*] *cosmography*), MS. British Library, Or. 11209 (copying finished on Thursday, 11 Shawwāl 688 h, according to the astronomical Hijra calendar = 27 October 1289 CE: f. 74r); 1993, *Naṣīr al-Dīn al-Ṭūsī's* Memoir on Astronomy, Ragep, F.J. (ed. & En. tr.), 2 Vols., New York: Springer.

al-Ṭūsī, Naṣīr al-Dīn Muḥammad, *Multaqaṭāt min kitāb al-Majisṭī ʿalā ḥasab iṣlāḥ baʿd al-mutaʾakhkhirīn* (*The collection of the corrections made by the modern* [*astronomers*] *to the* Almagest), MS. Oxford, Bodleian Library, Thurston 3, ff. 59v–69v.

5

SOLAR AND LUNAR OBSERVATIONS AT ISTANBUL IN THE 1570s[1]

5.1 Introduction

Generally speaking, astronomical observations in the medieval period were made for one of two purposes:

1. Testing contemporaneous astronomical tables and ephemeris. Some instances of such tests contain interesting cases of reconciling theory and observations.[2]
2. The derivation of the fundamental parameters of astronomical theories, most notably the solar, lunar, and planetary orbital elements.

In the early Islamic period, the observation of heavenly phenomena such as planetary conjunctions and solar and lunar eclipses was directed at testing the accuracy of contemporary astronomical tables, the so-called *zījes*.[3] The most extensive collection of such observations is reported in Ibn Yūnus's *Ḥākimī zīj* (d. 1009). For example, from the observation of the conjunction of Jupiter with Regulus on September 6, 864, Ḥabash al-Ḥāsib (d. after 869) found that the mean longitude of the planet as computed from the tables of the mean motions

---

1 Original publication: S. M. Mozaffari and J. Steele, "Solar and lunar observations at Istanbul in the 1570s," *Archive for History of Exact Sciences* **69** (2015), pp. 343–362. © 2015 Springer, and republished by permission.

2 For example, Shams al-Dīn Muḥammad al-Wābkanawī's (1254?–after 1316) test of the times and longitudes computed from the *Īlkhānī zīj* for four conjunctions of Jupiter with Saturn in 1286 and 1305–1306 against observations (see Sections 3.2 and 9.2) and his observation of the annular eclipse of January 30, 1283 (see Chapters 2 and 3); the *Īlkhānī zīj* was the formal product of the first period of the activities in the Marāgha observatory (northwestern Iran, *ca*. 1259–1271) under the directorship of Naṣīr al-Dīn al-Ṭūsī (1201–1274; on it, see Kennedy 1956, pp. 161–162, Samsó *et al.* 2001, p. 46). Another example: Abraham Zacut's observation of the occultation of Venus by the Moon on July 24, 1476 (see Goldstein and Chabás 1999). For the previous studies on the role of observation and its relation to theory in medieval astronomy, see, for example, Hartner 1977, Goldstein 1985a, 1988, Saliba 1994. For the astronomers referred to here, see *DSB*, *NDSB*, *BEA*, *EI*$_2$, Sezgin 1978, Rosenfeld and İhsanoğlu 2003.

3 On Islamic astronomical tables, see Kennedy 1956, Samsó *et al.*, 2001.

DOI: 10.4324/9781003481966-8

in the *Mumtaḥan zīj* (complied at Baghdad about 830 under the observational program ordered by the Abbasid caliph, al-Maʾmūn, who reigned in 813–833) should be decreased by 0;47°.[4] Also, from the observation of the conjunction between Venus and Mars on October 22, 864, he argued that the mean anomaly of Venus as computed from that *zīj* should increase by 4;30°, and the mean anomaly of Mars should reduce by 0;30°.[5] Abū al-Rayḥān al-Bīrūnī's works (973–1048), most notably his *al-Qānūn al-masʿūdī*, are an important source for Islamic solar and lunar observations other than eclipses. For instance, he gathers and discusses in detail nearly all the solar observations made by his Islamic predecessors for the determination of the orbital elements of the Sun (eccentricity and tropical direction of the apsidal line)[6] and of the obliquity of the ecliptic. He also informs us of the Banū Mūsā's measurement of the lunar maximum latitude (the two non-Ptolemaic values of 4;45° and 5;3° are preserved for the inclination of lunar orbit from the medieval Islamic period).[7]

In the late Islamic period, post-1000 CE, the number of observational reports significantly decreases, despite the fact that the great Islamic observatories were constructed in the same period.[8] Muḥyī al-Dīn al-Maghribī (d. 1283) gives a summary of his own "extensive" observations carried out at the Marāgha observatory between 1262 and 1274.[9] A short while later, al-Kawāshī, a thirteenth-century Yemenite scholar, presents 13 random observations of planetary conjunctions, planetary appulses to stars, and occultations that he made in Yemen and Egypt during 1277–1284.[10] Nothing is known about the details of the observations made at the Samarqand observatory, although it is striking that Ulugh Beg's *Sulṭānī zīj*, the principal fruit of the astronomical program conducted there, shows the

---

4 Ibn Yūnus, L: pp. 108–109; Caussin 1804, p. 155. The date given in the text is "Wednesday, 30 Rajab 250 Hijra" (= September 6, 864) and "21 Murdād 23*8* Yazdigird," which is the equivalent of Monday, September 5, 86*9*. The year in the latter date is obviously incorrect and should be 23<u>3</u> (on September 5, 869, Jupiter and Regulus were more than 125° apart).

5 Ibn Yūnus, pp. 108–109, Caussin 1804, pp. 155, 157. The date given in the text is "Sunday, *6* Ramaḍān 250 H" (= October *11*, 864) and "7 Mihr 233 Y" (= October 22, 864). The first date is in error, simply because October 11, 864, was a *Wednesday*. Moreover, in the report it is stated: "Aḥmad b. ʿAbd-Allāh [Ḥabash] said: '. . . at daybreak (*ṭulūʿ al-fajr*; i.e., the start of the morning twilight), I saw Venus and Mars being close to (associated with) each other (*mutalāṣiqayn*) in [the zodiacal sign] Virgo, as if the two were one star'"; at the mentioned time on October 22, 864, the two planets were less than 5' apart, while 11 days earlier, they were about 6° apart (cf. Caussin 1804, p. 156).

6 See Chapter 1. For the solar meridian altitude observations, see Said and Stephenson 1995, Newton 1972, which deals specifically with the solar data recorded by Bīrūnī.

7 Bīrūnī, *al-Qānūn al-masʿūdī* VII.5: 1954–1956, vol. 2, p. 779, King 1999, pp. 502–503.

8 On the observatories founded in the Islamic period, Sayılı [1960] 1988 is still the only available study, although some of his argumentations and conclusions should be treated with caution; for example, in the case of the latter period of the Marāgha observatory, see some critical remarks in Mozaffari and Zotti 2013, pp. 61–62.

9 On Muḥyī al-Dīn, see Chapter 9.

10 See King and Gingerich 1982.

application of a good number of the unprecedented values for the solar, lunar, and planetary parameters in addition to the incorporation of the now well-known star catalogue into this work.[11] Over one century later, Taqī al-Dīn Muḥammad b. Maʿrūf (1526–1585) documented the solar and lunar observations made at the Ottoman territory, most notably in the short-lived observatory at Istanbul before its deconstruction, which are the main concern of this chapter. He was not provided with an opportunity and, more important, facilities to deal with the planets, and so his main contribution to observational astronomy was confined to the determination of the solar and lunar parameters.

Let us make another distinction between simple/random and purposed/systematic observations. Al-Kawāshī's observations are typical of the first category, while Taqī al-Dīn's and Muḥyī al-Dīn's observations fall into the second one, where an astronomer explains quantitatively how he has derived his own parameter values from direct observations, often a problematical task with unexpected difficulties requiring sufficient and reasonable justifications, so that other medieval astronomers show little intention to do so. Ibn Yūnus, for instance, never explains whether and how he derived his non-Ptolemaic parameter values from observations, although it cannot be far from the truth to assume that his new parameter values are actually based upon the data he gathered from his documented observations. For instance, the possibility exists that his non-Ptolemaic values for the radii

---

11 See Chapter 7. In his detailed commentary on Ulugh Beg's *Zīj*, ʿAlī Qūshčī (d. 1474), one of the contributors of this work, says nothing as to the details of the astronomical observations at Samarqand. This source is, of course, invaluable for checking the parameter values deduced from the tables in that *zīj*. Much less emphasis in the secondary, modern literature has been put on the parameter values underlying the tables in this work; the eccentricity of Venus is a good example: the majority of the early Islamic astronomers, including al-Battānī and Ibn Yūnus, influenced by Indian astronomy, took the eccentricity of this planet to be equal to that of the Sun (i.e., the Earth). Although some astronomers like Bīrūnī and ʿAbd-al-Raḥmān al-Khāzinī (*fl.* the first half of the twelfth century) kept the Ptolemaic distinction of the eccentricity of Venus from that of the Sun, this idea did not disappear completely in the Middle Eastern branch of Islamic astronomy until the foundation of the Marāgha observatory, as it can be traced back in some *zījes* until the mid-thirteenth century (e.g., in Muntakhab al-Dīn al-Yazdī's *Manẓūm zīj*, "Versified zīj," written in Yazd, central Iran, *ca.* 1252, f. 46v). Nevertheless, it can be found prevalently in the Western branch of Islamic astronomy (Spain and northwestern Africa) in the latter periods (e.g., in the *zīj* of Ibn al-Bannāʾ of Marrakech, d. 1321; see Samsó and Millás 1998, pp. 265, n. 19, 266) and was transmitted to the late medieval Latin and Jewish astronomy (e.g., see Swerdlow 1977, p. 205, Goldstein 1985b, p. 113, Chabás and Goldstein 1994, p. 33, Goldstein and Chabás 1999, p. 188, Goldstein 2003, pp. 160–161, Chabás and Goldstein 2003, pp. 253–254, Chabás 2004, p. 188, Chabás and Goldstein 2009, p. 34). By this idea, the double eccentricity of the planet (i.e., the distance of the equant point from the Earth) remains larger than 2 (the radius of the orbit = 60). But the geocentric eccentricity of Venus approximately remains equal to about 1.74 in the past two millennia. The maximum equation of center of Venus in Ulugh Beg's *Zīj* (P1: f. 144r; P2: f. 161v) is equal to 1;39,19°, which corresponds to a double eccentricity of 1.73, in agreement with the value 0;52 Qūshčī gives for the half of it (N: pp. 273–274, PN: f. 241v). See Mozaffari 2019a.

of the epicycles of the interior planets[12] were the fruit of his own observations of these planets.[13]

For lunar eclipses, a distinction between random and systematic observations is especially relevant. A good number of reports of observations of lunar eclipses survive from early Islamic astronomy, especially in Ibn Yūnus's *Ḥakimī zīj*. These observations have played a pivotal role in the modern estimation of the rate of the deceleration of the Earth's rotation about its axis (ΔT, the difference between terrestrial and universal times).[14] For a medieval astronomer, lunar eclipses were the only means by which the lunar orbital elements in the Ptolemaic model could be determined. In order to measure the size of the lunar epicycle, a trio of lunar eclipses is required. Observations of the Moon at quadratures are necessary for determining the eccentricity.

Some of the preserved reports that belong to the early Islamic period appear to be simple observations that, at best, only fulfill the first purpose posited in the beginning of this chapter, namely, to test available *zījes*, and do not show any clear relation to the second purpose, that is, the determination of the parameters of the lunar model:[15] Al-Māhānī observed a trio of lunar eclipses in 854–856 but only measured the times of the beginning of eclipses and/or immersions (i.e., first and second contacts), while for the measurement of the radius of the lunar epicycle, it is necessary to determine the times of the middle (maximum phase) of lunar eclipses. Ibn al-Amājūr's observations of five lunar eclipses in a decade from 923 to 933 were mainly directed at testing Ḥabash's *zīj*. Al-Battānī describes only two lunar eclipses that he observed in 883 and 901. By them, he shows the existence of glaring differences in magnitudes and timings of the eclipses between what are computed on the basis of the *Almagest* and what are observed. He also employs them to derive the apparent angular diameter of the Moon at mean distance.[16] Nevertheless, from this period, we have three values for the radius of the lunar epicycle; the first two are the Banū Mūsā's 5;22 and Ibn al-A'lam's 5;4.[17] No lunar

12 He has the value 22;52 for the radius of the epicycle of Mercury (Ptolemy: 22;30 in the *Almagest* and 22;15 in the *Planetary Hypotheses*) and 43;28 for that of Venus (Ptolemy: 43;10) if the radius of the geocentric orbit of the epicycle center, the deferent, is taken as 60 arbitrary units. These values are derived from the maximum value for the epicyclic equation of these planets at mean distance as tabulated in Ibn Yūnus's *zīj*, that is, 22;24° and 46;25°, respectively, for Mercury and Venus (Ibn Yūnus, L: pp. 121, 190, 192; Caussin 1804, p. 221).

13 He observed some conjunctions of the inferior planets with each other (e.g., the morning of June 22, 985; modern: the evening of June 18, 985), with stars (e.g., Venus and Regulus: 1 hour after sunset in Cairo on June 23, 990; modern: about 3 hours after midnight on June 24, 990), and with the other planets (e.g., Venus and Saturn: half an hour before the sunrise in Cairo on January 20, 988; modern: about 2 hours before the sunrise in Cairo on the given date); see Ibn Yūnus, *Zīj*, L: pp. 113–114; Caussin, pp. 179–184.

14 Stephenson 1997, Chapters 12 and 13, and Steele 2000, Chapter 4.

15 What follows is based upon Stephenson 1997, pp. 476–493, Steele 2000, pp. 107–124.

16 Nallino [1899–1907] 1969, vol. 3, p. 87, Swerdlow 1972.

17 The radius of the lunar orbit, the inclined eccentric deferent, is taken as 60 arbitrary units. These two values are derived respectively from the maximum values given for the first inequality of

eclipse is reported from the Banū Mūsā, and their own value for the maximum lunar first inequality is mentioned in a later source, namely, the thirteenth-century *Ashrafī zīj*, while Bīrūnī has nothing to say about it; nevertheless, on account of his clear evidence of other lunar observational data from the Banū Mūsā, it seems reasonable to accept the validity of this attribution and that their own value for the radius of the epicycle was actually an observational achievement. The same can also be true of Ibn al-A'lam, from whom a non-Ptolemaic table of the lunar equations has survived, though not any observation of a lunar eclipse. The third value is Ibn Yūnus's 5;1 as derived from a maximum lunar first inequality of 4;48° as tabulated in his own *zīj*. Ibn Yūnus observed ten lunar eclipses spread over a period from 979 until 1002; for half of them, the times of the first and last contact are given either directly or with reference to the altitudes of the Moon or of some luminous clock stars.

In the late medieval Islamic period, the situation drastically changed so that the astronomers of this period no longer seem intent on simply presenting the results of their own observations of eclipses for the purpose of testing astronomical tables against the obtained observational data;[18] rather, all the 12 lunar observations we

---

the Moon by al-Kamālī in his *Ashrafī zīj*, ff. 49r and 229v–230r: 5;8° and 4;51°. Muhyī al-Dīn al-Maghribī adopts Ibn al-A'lam's lunar equations in his first *zīj*, *Tāj al-azyāj* (*Crown of the zījes*), compiled in Damascus before his joining the Marāgha observatory (see Dorce 2002–2003, p. 203, 2003, pp. 127, 184).

18 A main factor appears to be the fair agreement between the computed and observed results, as a good number of such accounts scattered in the late Islamic *zījes* testify; in them, an astronomer explains his computation of the circumstances and parameters of an eclipse and then usually claims that they were in agreement with observation, which can easily be checked by aid of modern data. For example, in his *'Alā 'ī zīj* (preserved in a unique copy in India, Hyderabad, Salar Jung Library, no. H17; see van Dalen 2004), on pp. 32–35, Farīd al-Dīn Abu al-Hasan 'Alī b. 'Abd al-Karīm al-Fahhād of Shirwān or Bākū (in the northern Āzarbāijān state of Iran) presents at length his computation of the parameters of a solar and a lunar eclipse that were to take place, respectively, in the conjunction and opposition about the month Shawwāl of the year 571 H/April–May 1176. For the solar eclipse (which occurred on April 11, 1176), he computes the ecliptic longitude at the instant of the apparent conjunction (i.e., the topocentric longitude of the Sun and the Moon in the maximum phase of the eclipse) as $\lambda_\sigma = 27;32°$, the time of mid-eclipse as about $T = 4;40$ hours before noon, and its magnitude as 11;46 digits (the diameter of the solar disk is taken as 12 digits). He then states that he observed this eclipse and found its circumstances in agreement with the computed results. It is not precisely known whether the place of observation was Bākū or Shirwān; for the first, the modern values are: $\lambda_\sigma = 27;56°$, $T = 8:18$ MLT, and magnitude 0.996. For the lunar eclipse (which occurred on April 25, 1176), he gives the longitude of the Moon at the instant of the mid-eclipse as about $\lambda_\varphi = 222;31°$, $T = 3;53$ hours after sunset, and magnitude 6;51 digits (the diameter of the lunar disk is taken as 12 digits). The modern values are $\lambda_\varphi = 221;59°$, $T = 22:34$ MLT (sunset: 18:53 MLT), and magnitude 0.673. In both cases, the computed longitudes are of errors of about $^1/_2°$; the computed magnitude of the solar eclipse and the time of the lunar counterpart are of good accuracy. Such accuracies are not entirely matters of coincidence, since similar instances can be traced back in medieval Islamic astronomy (a notable case may be Wābkanawī's calculation of the circumstances of the annular solar eclipse of January 30, 1283; see note 1 earlier). Rather, this reflects our lack of knowledge about the quantitative precision of some Islamic *zījes* that were the fruits of undertaking the difficult task of continuous observations and derivations of

have at our disposal from the period in question pertain to the four extant accounts of the lunar measurements surviving from Islamic astronomy. In them, the four Islamic astronomers present their observational data of a trio of lunar eclipses and explain how they have computed their own values for the radius of the lunar epicycle from them:

1. Bīrūnī, in his *al-Qānūn al-masʿūdī* (*Masʿūdic canons*) VII.3: the lunar eclipses of February 19, 1003, August 14, 1003, and July 5, 1004; the first two observed at Jurjān (Gurgān, northern Iran), and the last in Jurjāniyya (Konye-Urgench, Turkmenistan).[19]
2. Muḥyī al-Dīn al-Maghribī, in his *Talkhīs al-majisṭī* V: the lunar eclipses of March 7, 1262, April 7, 1270, and January 24, 1274, observed from Marāgha.[20]
3. Jamshīd Ghiyāth al-Dīn al-Kāshī, in the prologue of his *Khāqānī zīj*: the lunar eclipses of June 2, 1406, November 26, 1406, and May 22, 1407, observed in Kāshān (central Iran).[21]
4. Taqī al-Dīn Muḥammad b. Maʿrūf, in *Sidrat muntaha al-afkar fī malakūt al-falak al-dawwār* (*The Lotus Tree in the Seventh Heaven of Reflection*) V.2: the lunar eclipses of 1576–1577 observed in Istanbul, Cairo, and Thessalonica.[22]

Bīrūnī and Muḥyī al-Dīn determined the value of 5;12 for the radius of the lunar epicycle; Kāshī reached the figure about 5;17; and Taqī al-Dīn derived the value of about 5;24. Of them, only Muḥyī al-Dīn and Taqī al-Dīn explain their observations of the Moon near quadrature for the sake of determination of the lunar eccentricity in Ptolemaic model; the first derives the value 9, and the latter a value a bit more than 9;46 (radius of orbit = 60).

The first three trios have already been studied. Here, Taqī al-Dīn's solar and lunar observations are presented and analyzed. The accuracy of his lunar observations is also compared with the precision attained in the three earlier sets of observations of the triple lunar eclipses from the late Islamic period as well as in the

---

parameters of Ptolemaic models, and the fact that if Ptolemaic models were quantified anew by the re-measurement of its fundamental parameters, it would be probable to predict eclipses with precisions within an hour, 1° in longitude, and one digit in magnitude. Wābkanawī replaced al-Fahhād's computations and eclipses by his calculation of the solar eclipses of July 5, 1293, and October 28, 1296 (for latitude of Tabriz, northwestern Iran), and the lunar eclipse of May 30, 1295, when he taught al-Fahhād's *zīj* to Gregory Chioniades (*ca.* 1240–1320), who translated it into Greek (see Pingree 1985, p. 352f). On Ibn al-Fahhād, see Mozaffari 2019b, 2023.

19 al-Bīrūnī 1954–1956, vol. 2, pp. 740–742. These eclipses are nos. 07224, 07225, and 07227 in NASA's Five Millennium Catalog of Lunar Eclipses (hereafter: 5MCLE). For the analysis of Bīrūnī's observations, see Said and Stephenson 1997, pp. 45–46, Stephenson 1997, pp. 491–492.
20 See Chapter 9; Mozaffari 2014, pp. 72–74. The eclipses nos. 07878, 07897, and 07907 in 5MCLE.
21 Kāshī, IO: ff. 4r–6r, P: pp. 24–28. The eclipses nos. 08220, 08221, and 08222 in 5MCLE. See Mozaffari 2020–2021.
22 Taqī al-Dīn, *Sidra*, K: ff. 42r–v, N: ff. 53r–v, V: f. 44r. The eclipses nos. 08610, 08611, and 08612 in 5MCLE.

extensive observations of the lunar eclipses both from the late medieval European and early Islamic periods.

5.2 Taqī al-Dīn's Observations

For the present study, we made use of three manuscripts of the *Sidra* (introduced in the bibliography at the end of this chapter; MS K is a collection of some works by Taqī al-Dīn copied in his own hand). The *Sidra* is the first *zīj* (astronomical tables with accompanying instructions to use them) he composed.[23] As usual in this genre of *zījes*, in the canons, our author presents the variant topics pertaining to theoretical, mathematical, and practical astronomy, such as the sections on chronology, trigonometry, spherical astronomy, the methods for the derivation of the fundamental parameters, and so on. In the parts related to the Sun and the Moon, detailed accounts of his observations and the instruments applied to them are given,[24] and then he explains how he derived his own values for the solar and lunar parameters from the date obtained in these observations. In what follows (Section 5.2.1), we first present Taqī al-Dīn's solar observations as translated from the original Arabic text, which are also summarized in Table 5.1, together with a brief commentary upon the accuracy of the unprecedented values he achieved for the solar parameters. This is followed by presenting the accounts of his four lunar observations in the same way (Section 5.2.2); the fourth observation is investigated there, but his first three lunar observations, that is, the trio of lunar eclipses, shall be analyzed at length in Section 5.3. We number his nine observations in the chronological order and indicate those of the Sun by the prefix *S*, and of the Moon by *M*. *M1* (1576) is the earliest documented observation, and *M4* (1579) the latest. As the contents of this work shows, Taqī al-Dīn's observations are limited only to the two luminaries; as he definitely says in the account of M4, at that time, he had not yet dealt with the stellar observations. He died in 1585 and apparently did not find any opportunity to deal with the stars and planets; moreover, in his later *zīj*, *Kharīda*, he strangely returns back to Ulugh Beg's values for the solar and lunar parameters (see note 38).

Two notes about the dates and a technical astronomical term in the following accounts merit consideration: the Alexander's era mentioned in the reports is in fact the Seleucid or Byzantine era (October 1, 311), to which Ptolemy refers as "according to the Chaldaeans"[25] ("Two-Horned," that is, Alexander, in Islamic astronomy), in which the years are Julian years of $365\frac{1}{4}^{\mathrm{d}}$; although our author repeatedly makes use of the alternative of "the death of Alexander" for this calendar, it has nothing to do with the Philip era, which is referred to as "Death of Alexander" (November 12, 323) throughout the *Almagest* and which is used with

---

23 King 2004–2005, vol. 1, p. 64.

24 For the illustration of the instruments of the Istanbul observatory, see Sezgin and Neubauer 2010, vol. 2, pp. 53–61.

25 *Almagest* IX.7 and XI.7: Toomer 1998, pp. 452–453, 541.

Table 5.1 Taqī al-Dīn's solar observations

Observation	Date and time	Noon altitude/solar longitude	Computed equinox time and solar longitude	Errors	Instrument
1	Tuesday 24 Rabīʿ I 985 June 11, 1577 JDN 2297219 true noon	h_{max} = 72;30,8,29°	72;29, 1°	~+1′	[Quadrant?]
2	Wednesday 1 Shawwāl 985 December 11, 1577 JDN 2297402 true noon	h_{min} = 25;32,20,14	25;31, 9	~+1′	[Quadrant?]
3	Wednesday 13 Muḥarram 987 March 11, 1579 JDN 2297857 2;34,47^h before true noon	$\lambda = 0°$ <u>Vernal equinox</u>	8:26:25 LT	~+1^h	*Dhāt al-awtār* ("Having the chords")
4	Saturday 22 Jumādā II 987 August 15, 1579 JDN 2298014 true noon	$\lambda = 151;21,15°$	151;21,17°	~−2″	Quadrant and armillary sphere
5	Monday 23 Rajab 987 September 14, 1579 JDN 2298044 true noon ↓ <u>Autumnal equinox:</u> Sunday September 13 9;22,36^h after true noon	$\lambda = 180;36,0°$	180;35,43° 21:20:31 LT	~+17″ ~+2^m	Quadrant and armillary sphere

the Egyptian years of 365^d. Also, The Hijra date in the first observation of the Sun is according to the *civil* reckoning (the epoch July 16, 622), but in the other four solar observations as well as in all the lunar observations, the Hijra dates are according to the *astronomical* reckoning (the epoch July 15, 622).[26] The terms *sā'āt al-bu'd* and *daqā'iq al-bu'd*, literally, "hours/minutes of the distance," as found in all the passages, refer to the interval of time remaining to or passed from the meridian passage/transit of a heavenly body (in the case of the Sun: true noon) counted in terms of equal hours or minutes.[27]

5.2.1 The Solar Observations

[S1] We observed the extremal [noon-altitudes of the Sun] at the two solstices in the same year. The second [first (?) observation] had been made at true noon (*nisf nahār*; lit. "middle of daylight/midday") on Tuesday, 24 Rabi' al-Awwal [3] . . . in the year 985 of Hijra and [11 Ḥazīrān [9]] 1888 of Alexander's era. After the correcting adjustment for making the true altitude and [i.e., deriving the summer solstice altitude from the noon-altitude on this solstice day by] considering the period/argument (*ḥiṣṣa*) [between noon and the time of occurring the summer solstice], the maximum altitude at the summer solstice was 72;30,8,29°.[28]

[S2] But, the first [second (?) observation] had been made at true noon on Wednesday, the first day (*ghurrat*) of Shawwāl [10] in the mentioned year [i.e., 985]. After doing the adjustments, the extremal altitude at the winter solstice was 25;32,20,14°.[29]

From these statements from Taqī al-Dīn as well as the precision to which the two values just mentioned are given, it is evidently understood that they are the product of some kind of adjustment. In fact, the solar noon altitudes in the solstitial days can be representative of the Sun's extremal meridian altitudes, if and only if the solstices take place exactly at true noon. Otherwise, medieval astronomers undertook some methods for extrapolating the extreme solstitial noon altitudes. Taqī al-Dīn does not explain his adopted method in order to do this, but the practical procedures for such adjustments can be addressed in the works of his predecessors, for example, in Bīrūnī's *Taḥdīd nahayāt al-amākin*.[30] Taqī al-Dīn should have computed in some way the period/argument from true noon on the given dates to the time when the solstice occurred and then extrapolated the extremal altitude from the rate of change in the Sun's declination about the solstices; this is, however, very minor, about 14″, and thus, it can be deduced that his observed values for the solar noon altitude at the summer and winter solstices should not differ

---

26 See B. Van Dalen's entry *Ta'rīkh* (date, chronology) in *EI*₂, vol. 10, pp. 259, 261.
27 Taqī al-Dīn, *Sidra*, K: f. 22v, N: f. 29r, V: ff. 28v–29r.
28 Taqī al-Dīn, *Sidra*, K: f. 17v (in the right margin), N: f. 22r, V: f. 25r.
29 Taqī al-Dīn, *Sidra*, K: f. 17v (in the right margin), N: f. 22r, V: f. 25r.
30 See al-Bīrūnī 1967, pp. 61–64, Kennedy 1973, pp. 34–38.

too much from 72;30° and 25;32,30°. From these two observations, he derives his own unique value ε = 23;28,54° for the obliquity of the ecliptic,[31] which is only about −0;1° in error, and φ = 40;58,46° for the latitude of Istanbul.

[S3]   [I]n order to derive the time of the vernal equinox, we installed the instrument having the chords (*dhāt al-awtār*), and observed the shadow-covering by means of it. Then, [we found that] it took place before true noon (*al-zawāl*) on Wednesday, 13 Muḥarram [1] in the year 987 of the noble Hijra, 20 Pharmouthi [8] in the year 2327 Nabonassar, 11 Ādhār [6] 1890 after the death of Alexander, at 2;34,47 equal hours from true noon (*sāʿāt buʿd muʿaddala*).[32]

[S4]   Then, on Saturday, 22 Jumādā al-Ākhira [6] in the year 987 of Hijra, 22 Thoth [1] in the year 2328 Nabonassar, 15 Āb [11] in the year 1890 from the death of Alexander, we observed the body of the Great Luminary [i.e., the Sun] by the armillary sphere some minutes before true noon (*al-zawāl*) and by the mural meridional quadrant (*lubna*) at it [i.e., true noon] for the examination of the correctness of the two observations. After the agreement of the two observations by taking into account the time [of the first observation] from true noon in minutes (*daqāʾiq al-buʿd*), [we found that] it was in the [ecliptic] sign Virgo 1;21,15° at the time of transiting the meridian (tawassuṭ).[33]

[S5]   After it, on Monday, 23 Rajab [7] in the year 987 of Hijra, the longitude (*mawḍiʿ*, lit. "position") of the Great Luminary was in the [ecliptic] sign Libra 0;36° in the time of passing the meridian as observed by the armillary sphere before true noon and the mural quadrant at it and the correct agreement of the two observations by the adjustment (*taʿdīl*) mentioned earlier. From the [Sun's] mean motion known from the New Observations and the derivation of its true daily motion (*al-buht*), it necessitates that the time of the Sun's entrance into the head of Libra, the autumnal equinox, in hours and their fractions from true noon (*sāʿāt al-buʿd wa kusūrihā*), was 9;22,36 hours on Sunday, 22 Rajab [987], the 21st of the month of Phaophi [2] in the year 2328 Nabonassar, 13 Aylūl [12] in the year 1890 after the death of Alexander.[34]

These three solar observations are related to the determination of the times of equinoxes of 1579 and the position of the Sun at an intermediary point, from which the basic parameters of the solar model are derived. The accuracy of Taqī

---

31 Taqī al-Dīn, *Sidra*, K: f. 17v, N: f. 22r, V: f. 25r; also, see King 2004–2005, vol. 1, pp. 57, 116, 123, 133, 151.
32 Taqī al-Dīn, *Sidra*, K: f. 35r, N: f. 46r, V: f. 39v.
33 Taqī al-Dīn, *Sidra*, K: f. 36r, N: f. 47r, V: f. 40r.
34 Taqī al-Dīn, *Sidra*, K: f. 36r, N: f. 47r, V: f. 40r.

al-Dīn's values for the solar noon altitudes and times of equinoxes is significant and comparable with that of the outstanding figures of the early Islamic period.[35] Owing to an error in counting the time between the vernal equinox of this year and Ptolemy's observation of the same equinox in 140, Taqī al-Dīn deduced a value about 365;14,38,34 days for the length of the solar year, which is about 3 minutes too long.[36] Then, having employed the general three-point method,[37] he obtained the value of about 2;0,34 for the solar eccentricity (radius of orbit = 60, or 0.01675, if the radius of orbit is taken as the unit) and 95;33° for the longitude of its apogee.[38] Taqī al-Dīn's documentation of his solar observations makes it possible to compute the true values for the eccentricity of the Earth and the longitude of the solar apogee in a circular orbit, which should be, respectively, 0.01686 and 95;10° for 1579 (in the elliptical orbit: 0.01688 and 95;43°).[39] His values for the solar orbital elements are highly accurate: for the longitude of the solar apogee, he has one of the most accurate values measured in Islamic astronomy (Chapter 1). Of course, for the eccentricity, his accuracy had already been reached by Ulugh Beg, the best of what was achieved in late Islamic astronomy, but not repeating the brilliant accuracy of the Mumtaḥan observers in the ninth-century Baghdad and Bīrūnī with errors, respectively, $\sim\!-1 \times 10^{-5}$ and $+5 \times 10^{-5}$.[40] It is noteworthy that his value for the eccentricity is remarkably better than that of his Danish contemporary, Tycho Brahe, who derived 0.01792 in 1588 (computed value for a circular orbit: 0.01690; true value in the elliptical orbit: 0.01688).[41]

5.2.2 The Lunar Observations

In what follows, Taqī al-Dīn's reports of his four lunar observations are presented, which are also summarized in Table 5.2. For the lunar eclipses, Taqī al-Dīn uses both types of the description of the date of a lunar eclipse as customary in medieval Islamic treatises: "on the night whose morning was [the day after eclipse]" and "after the meridian passage of the Sun on [the preceding day]."

> [M1] The first of the triple lunar eclipses we observed in the house of
> the great master . . . Saʿd al-Milla wa-ʾl-Dawla wa-ʾl-Dīn . . . ,[42]

---

35 See Said and Stephenson 1995.

36 See Mozaffari 2018, pp. 216–218.

37 See Chapter 1, pp. 30–32.

38 Taqī al-Dīn, *Sidra*, K: ff. 36r–v, N: ff. 47r–v, V: f. 40r (see Tekeli 1962, 2008). In his later *zīj* (*Kharīda*, B: B: ff. 26v–28v, 32v–38v, 44r–v, C1: ff. 48v–51r, 55v–63v, 71r–v, C2: ff. 34v–38r, 41v–48r, 55r–v, E: ff. 2v–9v, 10r–11r, 17r–v, K: ff. 46r–49r, 53r–61v, 69r–v), Taqī al-Dīn comes back to the solar and lunar theories of Ulugh Beg's *Sulṭānī zīj*.

39 See Mozaffari 2018, pp. 205–208.

40 See Mozaffari 2018, pp. 203, 206.

41 See Brahe, *Opera Omnia*, vol. 2, pp. 19–28; Dreyer 1890, p. 333, Moesgaard 1975, pp. 85–89, Thoren and Christianson 1990, p. 223–224, Swerdlow 2010, p. 155.

42 The vacant places only indicate the glorying titles Taqī al-Dīn ascribes to Saʿd al-Dīn Efendī.

Table 5.2 Taqī al-Dīn's lunar eclipse observations

Observation	Date	Time of mid-eclipse		Magnitude
1	Sunday/Monday 15/16 Rajab 984 October 7/8, 1576 JDN 2296972/3	12; 3,56^h **12;43,15** LT	Error ∼−39.3^m	$\frac{9}{12}$ **0.842**
2	Tuesday/Wednesday 14/15 Muḥarram 985 April 2/3, 1577 JDN 2297149/50	9; 8,46 **10; 1, 6** LT	Error ∼−52.3	Total **1.560**
3	Thursday/Friday 14/15 Rajab 985 September 26/27, 1577 JDN 2297326/7	13;36,36 **14;21,32** LT	Error ∼−44.9	Total **1.487**

Note: Modern computed values are given in bold for comparison (modern times given in terms of the apparent local time (LT)).

the distance of which from the observatory (*dār al-raṣad*) does not make any perceptible difference in seconds [of time]. The time of the middle [i.e., the maximum phase] of the eclipse was 12;3,56 hours after the meridian transit of the Sun [i.e., true noon] on Sunday, 15 Rajab [7] 984. . . . The Moon was eclipsed by 9 digits of its light.[43]

Saʿd al-Dīn Efendī (d. 1599) was one of Taqī al-Dīn's supporters, whom he praises in the prologue.[44] Observations used for the derivation of the mean motions and orbital elements of the Sun, Moon, and planets should be made in or converted to the local time of a specific meridian. The place of the observatory was representative of the principal longitude of Istanbul, from which its latitude was also measured. Taqī al-Dīn notes that the difference in longitude between Saʿd al-Dīn's house, where this first observation was made, and the observatory, where the second lunar eclipse was observed, is sufficiently small as to have no undesirable consequence in the use of these observations.

[M2] [T]he observation of the total lunar eclipse that took place on the night whose morning was Wednesday, 15 Muḥarram [1] in the year 985 of the noble Hijra. We found the time of its middle with the utmost investigation [in collaboration] with excellent masters of observations by the great instruments installed in the new royal observatory . . . to be 9;8,46 hours after the meridian passage of

43 Taqī al-Dīn, *Sidra*, K: ff. 42r–v, N: f. 53r, V: f. 44r.
44 Taqī al-Dīn, *Sidra*, K: f. 2v, N: f. 2v, V: f. 11r.

the Sun on Tuesday, 14 Muḥarram in the mentioned year. [This was] a total lunar eclipse with a perceptible duration (*makth*, lit. "staying").[45]

[M3] We were not able to observe the third eclipse, because of the entrance of the clouds. Our excellent brothers from Egypt told us of it and also Dāwūd al-Riyāḍī transmitted it to us with the measurement of [the altitude?] of [the star] Aldebaran [i.e., α Tau]. Then, we converted it to the longitude of Constantinople. Then, the time of its middle was 13;36,36 hours after the meridian transit of the Sun on Thursday, 14 Rajab [7] in the year 985. So, it occurred on the night whose morning was Friday 15 [Rajab].[46]

A. Ben-Zaken identifies Dāwūd al-Riyāḍī (the mathematician) from Thessalonica mentioned in the report of M3 as David Ben-Shushan, a Jewish scholar.[47] He appears to have measured the altitude of the star Aldebaran (α Tau) at the time of the maximum phase of the eclipse, since Taqī al-Dīn immediately mentions that he converted it to the meridian of Istanbul, and then gives the time of the middle of the eclipse. No information is given on what Taqī al-Dīn believes is the difference in longitude of Istanbul and whatever place in Egypt the observation was made (probably Cairo?).

For all three eclipses, Taqī al-Dīn reports only the time of the middle of the eclipse. Because the midpoint of an eclipse is difficult to determine directly from observation, it is likely that in all cases he has calculated the midpoint from observations of the times of the beginning and end of either the whole eclipse or the total phase of the eclipse. This suggests that the reports of the eclipse given by Taqī al-Dīn represent observations that have already been through a process of analysis, rather than the original raw data of the observations. A similar conclusion can be drawn from the lack of details about the altitude of Aldebaran in the final report; only the reduced time has been given.

The fourth observation is used for the measurement of the maximum value of the second inequality of the Moon, and hence its eccentricity in Ptolemy's lunar model. Such observations should fulfill some essential conditions: in the time of the observation, the Moon should be near quadrature and have the maximum distance from its mean longitude as well as it should culminate, so that its vertical circle of altitude is perpendicular to the ecliptic, which is to neutralize the effect of the longitudinal component of parallax.[48]

---

45 Taqī al-Dīn, *Sidra*, K: f. 42v, N: ff. 53r–v, V: f. 44r. The report of this eclipse is also given: *Sidra* V.1 (K: f. 41v, N: f. 52v, V: f. 43v).
46 Taqī al-Dīn, *Sidra*, K: f. 42v, N: f. 53v, V: f. 44r.
47 Ben-Zaken 2010, pp. 21–24.
48 See Neugebauer 1975, vol. 1, pp. 86–87.

[M4] God rendered those circumstances easy for us in the early morning of Friday, 18;48,46 hours after the meridian passage of the Sun on Thursday, 21 Shawwāl [10] in the year 987 of the noble Hijra. The Moon was nearly in the mentioned limits. It was not possible for us to observe it by the armillary sphere neither with the Sun, for it being below the horizon, nor with any of the fixed stars, since it was not previously possible for us to record any of them from a reliable observation. Thus, we purposed to observe the Moon by the [instrument] having the azimuth and altitude, and we derived the [oblique] ascension (*maṭāliʿ*) [of the Moon] at the time of the observation and endeavored to record the procedures with the extreme diligence. . . . Then, the longitude of the Moon was 176;27°.[49]

Taqī al-Dīn evidently states that he could not use the armillary sphere at the time of the observation because he had not yet measured the longitudes of some reference stars trustworthily – the task he apparently never found time to accomplish. He thus adhered to the methods of the spherical astronomy in order to derive the longitude of the Moon from its horizontal coordinates as observed by the altitude–azimuthal instrument, which is the same instrument called the Two Quadrants by Muʾayyad al-Dīn al-ʿUrḍī in his treatise *Fī kayfiyyat al-irṣād*, "On how to make [astronomical] observations," and constructed by him at the Marāgha observatory.[50] The intended time of this observation is when the vertical circle of the altitude of the Moon is perpendicular to the ecliptic; this occurred at 7:15 MLT on Friday, 22 Shawwal 987/December 11, 1579 (JDN 2298132). Like the other three lunar observations, which shall presently be discussed in the next section, the time Taqī al-Dīn computes for his fourth lunar observation is quite probably in error. Accordingly, the precise time of this observation cannot be determined with certitude, although it seems to have been made somewhere between 5:45 MLT (the meridian passage of the Moon) and 6:50 MLT (the start of civil twilight). At 6:49 MLT, apparent longitude $\lambda_\text{☽} \approx 176;15°$, so Taqī al-Dīn's longitude is about +12′ (less than the semidiameter of the lunar disk) in error.

5.3 Analysis of the Lunar Eclipse Observations

In Table 5.2, we summarize the trio of lunar eclipse observations reported by Taqī al-Dīn. It is evident that the times of mid-eclipse that Taqī al-Dīn derived from observations are considerably in error. Indeed, they are consistently earlier than the times of mid-eclipse computed using modern ephemerides by amounts ranging from just under 40 minutes to over 50 minutes. This poor level of accuracy

49 Taqī al-Dīn, *Sidra* V.7: K: f. 48r, N: ff. 61r–v, V: f. 48v.
50 See Seemann 1929, pp. 72–81.

is rather surprising and compares unfavorably with the observation of eclipses by other Islamic astronomers, both from the early and late period. For example, among the eclipses reported by al-Battānī, Ibn Yūnus (including also observations from Ḥabash al-Ḥāsib, al-Māhānī, and the Banū Mūsā), and al-Bīrūnī, no single eclipse timing is in error by more than about 36 minutes, and the vast majority have errors of less than 20 minutes,[51] irrespective of whether the time was determined using a clepsydra or by the observation of the altitude of either the eclipsed luminary or a fixed star.

Of the late Islamic astronomers, Muḥyī al-Dīn has significantly more accurate values for the times of the maximum phase of his trio of lunar eclipses than Taqī al-Dīn; the errors in Muḥyī al-Dīn's times do not exceed 5 minutes.[52] Muḥyī al-Dīn employed an accurate clepsydra which was probably implemented by some mechanical components and which made it possible for an operator to measure hour and minute separately.[53] By contrast, the accuracy of Kāshī's times for two of his trio lunar eclipses (June 2, 1406, and May 22, 1407) is similar to those of Taqī al-Dīn.[54] Neither Kāshī nor Taqī al-Dīn gives full details of how they determined the times. Taqī al-Dīn does not explicitly mention whether he used his own mechanical clocks, which were seemingly influenced by European sources and models,[55] in the observation of his first two lunar eclipses; he only refers to "the great instruments installed in the new observatory." For the last observation, he should have applied the method of spherical astronomy to convert the computed time to the meridian of Constantinople, as his reference to the star Aldebaran gives testimony to it. The use of this method with not highly accurate values for basic parameters (e.g., the geographical latitudes) might partly be responsible for the appearance of such great errors. The other contributing factor might have been the values applied for the difference in longitudes between Cairo/Thessalonica and Istanbul. The support comes from the fact that these lunar eclipses were also of geographical use for Taqī al-Dīn: in a cancelled passage in *Sidra* II.4 (preserved only in autograph K),[56] he explicitly asserts that from his observations of this trio of lunar eclipses, he derived the value 56;39,45° for the longitude of Istanbul from the Fortunate Islands; also, in *Sidra* V.1,[57] where he converts the time of one of the lunar eclipses which Ptolemy observed at Alexandria to the meridian of Istanbul, he takes the meridian of Istanbul equal to 56;40° and that of Alexandria as 61;54° and states that the then resulting difference of 5;14° in terrestrial longitude between

---

51 Steele 2000, pp. 112–124.
52 See Table 9.2(A) on p. 227.
53 See Section 9.5.2.
54 Of course, as explained in detail in Mozaffari 2020–2021, pp. 81–87, it is highly probable that Kāshī's timings of these two lunar eclipses were the *theoretical* (not *observational*) data in relation to the *zījes* of the Marāgha tradition (the *Īlkhānī zīj* and al-Maghribī's *Adwār al-anwār*).
55 Taqī al-Dīn, *Sidra* II.3: K: f. 16v, N: ff. 20v–21r, V: ff. 24r–v (with a figure and additional details in K: f. 90r); see Sezgin and Neubauer 2010, vol. 3, pp. 118–122.
56 Taqī al-Dīn, *Sidra*, K: f. 17v, N: f. 22r, V: f. 25r.
57 Taqī al-Dīn, *Sidra*, K: f. 41v, N: f. 52v, V: f. 43v.

the two cities corresponds to a difference of $0;20,56^h$ in local times between them. However, he presumably discarded the value $56;39,45°$ for the longitude of Istanbul later, since the relevant lines on f. 17v are blacked out; moreover, in the geographical table attributed to him,[58] the longitudes of Istanbul and Alexandria are given, respectively, as $60°$ and $61;55°$. Note that Istanbul (longitude $L = 28;57°$ E from Greenwich) is actually only about $1°$ west of Alexandria ($L = 29;55°$ E).

Taqī al-Dīn's and al-Kāshī's eclipse timings also compare unfavorably with European astronomers of that time. Regiomontanus and his colleague Bernard Walther, at the end of the fifteenth and beginning of the sixteenth centuries, observed many eclipses of both the Sun and the Moon and timed the eclipses with an accuracy of better than 15 minutes (in many cases, significantly better), and even Copernicus, an astronomer not generally regarded as a particularly accomplished observer, in the middle of the sixteenth century, observed the times of eclipses with errors of less than about 30 minutes.[59] And toward the end of the sixteenth century, at the same time as Taqī al-Dīn, Tycho Brahe was determining the time of eclipses to an accuracy of about 12 minutes.[60] Indeed, it is worth noting that two of the three eclipses reported by Taqī al-Dīn were also observed by Tycho: the eclipses of April 2/3, 1577, and September 26/27, 1577. Tycho observed the time of the four phases of the April 2/3, 1577, eclipse, each with an error of about -6, $+1$, -3, and $+3$ minutes, respectively, in contrast to Taqī al-Dīn's error of about -52 minutes in his determination of the time of mid-eclipse. For the eclipse of September 26/27, 1577, Tycho observed the time of the end of totality with an error of about -10 minutes in contrast to an error of -45 minutes in Taqī al-Dīn's time of mid-eclipse.

Of the three eclipses reported by Taqī al-Dīn, two were total and one partial. Taqī al-Dīn gives the magnitude in terms of the decrease in the brightness of the *lunar disk*, a term that is not encountered in the previous Islamic reports. It is not known whether Taqī al-Dīn refers to the eclipsed portion of the lunar diameter or surface; however, the naked eye estimation of the eclipsed area of the Sun and the Moon, or even the measurement of it with the aid of medieval optical aids, such as camera obscuras or pinhole image devices, is difficult, and consequently, we assume that Taqī al-Dīn refers to the eclipsed diameter of the Moon. However, the modern magnitude of 0.842 for the eclipse is equal to about 10 digits of the lunar diameter and nearly corresponds to the 10 digits of its surface as well, according to the Ptolemaic norm that the angular radius of umbra (i.e., the Earth's shadow in the distance of the Moon from the Earth) is 2.6 times as large as the apparent diameter of the Moon.[61] Thus, regardless of whether Taqī al-Dīn refers to the eclipsed diameter or surface of the Moon, his measured magnitude is -1 digit in error.

Of the other late Islamic astronomers, Muḥyī al-Dīn has exceptionally accurate values of the magnitudes for the two partial lunar eclipses he observed at Marāgha

---

58 See King 2004–2005, vol. 1, p. 449–450.
59 Steele 2000, p. 139–150.
60 Steele 2000, p. 151–154.
61 *Almagest* V.14: Toomer 1998, p. 254.

(April 7, 1270, and January 24, 1274). He expresses the magnitudes in more fractions than one may expect from the ancient and medieval normal unit of one-twelfth of the diameter of the lunar disk. His values might have been the results of doing some interpolations after observing the shape of the eclipses in the dioptra and pinhole image devices available to him at Marāgha.[62]

5.4 Conclusion

Taqī al-Dīn was among a small number of outstanding figures of Islamic astronomy that show the admirable intentions to give the full accounts of their observations and derivations of parameters. Although, unlike his solar observations, the accuracy of his lunar observations compares unfavorably both with earlier and contemporary astronomers, his work is important for studying the relationship between observation and theory in Islamic astronomy. It is curious that Taqī al-Dīn was among the possessors of the only surviving manuscript of Muḥyī al-Dīn's *Talkhīṣ al-majisṭī*,[63] the work that undoubtedly reflects the acme of observational astronomy in the thirteenth-century Middle East; it is not hard to imagine a probable positive influence that this work might have exercised on Taqī al-Dīn to document his observations and to explain the procedures of derivations of parameter values from them.

References

Bearman, P., Bianquis, T., Bosworth, C. E., van Donzel, E., and Heinrichs, W. P., 1960–2005, *Encyclopaedia of Islam*, 2nd edn., 12 Vols., Leiden: Brill.

Ben-Zaken, A., 2010, *Cross-Cultural Scientific Exchanges in the Eastern Mediterranean, 1560–1660*, Baltimore, MD: The John Hopkins University Press.

al-Bīrūnī, Abū al-Rayḥān, 1954–1956, *al-Qānūn al-mas'ūdī* (*Mas'ūdīc canons*), 3 Vols., Hyderabad: Osmania Bureau.

al-Bīrūnī, Abū al-Rayḥān, 1967, *Taḥdīd nahayāt al-amākin li-taṣḥīḥ masāfāt al-masākin* (*Determination of the Coordinates of Positions for the Correction of Distances between Cities*), Ali, J. (En. tr.), Beirut: American University of Beirut.

Brahe, T., 1913–1929, *Tychonis Brahe Dani Opera Omnia*, Dreyer, J. L. E. (ed.), 15 Vols., Copenhagen: Libraria Gyldendaliana.

Caussin de Perceval, J.-J.-A., 1804, "Le livre de la grande table hakémite, Observée par le Sheikh,..., ebn Iounis", *Notices et Extraits des Manuscrits de la Bibliothèque nationale* 7, pp. 16–240.

Chabás, J., 2004, "Astronomy for the court in the early sixteenth century, Alfonso de Córdoba and his *Tabule Astronomice Elisabeth Regine*", *Archive for History of Exact Sciences* **58**, pp. 183–217.

---

62 See Sections 9.5.3 and 11.4.12; Mozaffari and Zotti 2013, pp. 127–135.
63 See Mozaffari 2018–2019, pp. 160–162.

Chabás, J. and Goldstein, B. R., 1994, "Andalusian astronomy: *al-Zīj al-Muqtabis* of Ibn al-Kammād", *Archives for the History of Exact Sciences* **48**, pp. 1–44.

Chabás, J. and Goldstein, B. R., 2003, *The Alfonsine Tables of Toledo*, Dordrecht: Kluwer Academic Publishers.

Chabás, J. and Goldstein, B. R., 2009, *The Astronomical Tables of Giovanni Bianchini*, Leiden: Brill.

van Dalen, B., 2004, "The *Zīj-i Nāṣirī* by Maḥmūd ibn Umar: The earliest Indian Zij and its relation to the *'Alā'ī Zīj*", in: Burnett, C., *et. al.* (eds.), *Studies in the History of the Exact Sciences in Honour of David Pingree*, Leiden: Brill, pp. 825–862.

Dorce, C., 2002–2003, "The *Tāj al-azyāj* of Muḥyī al-Dīn al-Maghribī (d. 1283): Methods of computation", *Suhayl* **3**, pp. 193–212.

Dorce, C., 2003, "El Tāŷ al-azyāŷ de Muḥyī al-Dīn al-Maghribī", in: *Anuari de Filologia*, Vol. 25, Secció B, Número 5, Barcelona: University of Barcelona.

Dreyer, J. L. E., 1890, *Tycho Brahe: A Picture of Scientific Life and Work in the Sixteenth Century*, Edinburgh: Adam and Charles Black.

Farīd al-Dīn Abu al-Ḥasan ʿAlī b. ʿAbd al-Karīm al-Fahhād al-Shirwānī or al-Bākū'ī, *Zīj al-'Alā'ī*, MS. India, Salar Jung, no. H17.

Gillipsie, C. C., *et al.* (eds.), 1970–1980, *Dictionary of Scientific Biography*, 16 Vols., New York: Charles Scribner's Sons.

Goldstein, B. R., 1985a, *Theory and Observation in Ancient and Medieval Astronomy*, London: Variorum Reprints.

Goldstein, B. R., 1985b, *The Astronomy of Levi Ben Gerson (1288–1344), a Critical Edition of Chapters 1–20 with Translation and Commentary*, New York: Springer-Verlag.

Goldstein, B. R., 1988, "A new set of fourteenth century planetary observations", *Proceedings of the American Philosophical Society* **132**, 371–399.

Goldstein, B. R., 2003, "An anonymous Zij in Hebrew for 1400 A.D.: A preliminary report", *Archives for the History of Exact Sciences* **57**, pp. 151–171.

Goldstein, B. R. and Chabás J., 1999, "An occultation of Venus observed by Abraham Zacut in 1476", *Journal for the History of Astronomy* **30**, 187–200.

Hartner, W., 1977, "The role of observations in ancient and medieval astronomy", *Journal for the History of Astronomy* **8**, pp. 1–11.

Hockey, T., *et al.* (eds.), 2014, *The Biographical Encyclopedia of Astronomers*, 2nd edn., New York [etc.]: Springer.

Ibn Yūnus, ʿAlī b. ʿAbd al-Raḥmān b. Aḥmad, *Zīj al-kabīr al-Ḥākimī*, MS. L: Leiden, no. Or. 143.

al-Kamālī, Muḥammad b. Abī ʿAbd-Allāh Sanjar (Sayf-i munajjim), *Ashrafī zīj*, MS. Paris: Bibliothèque Nationale, no. 1488.

al-Kāshī, Jamshīd Ghiyāth al-Dīn, *Khāqānī zīj*, MS. IO: London: India Office, no. 430, MS. P: Iran: Parliament Library, no. 6198.

Kennedy, E. S., 1956, "A survey of Islamic astronomical tables", *Transactions of the American Philosophical Society*, New Series **46**, pp. 123–177.

Kennedy, E. S., 1973, *A Commentary Upon Bīrūnī's* Kitāb Taḥdīd al-Amākin, Beirut: American University of Beirut.

King, D. A., 1999, "Aspects of Fatimid astronomy: From hard-core mathematical astronomy to architectural orientations in cairo", in: Barrucand, M. (ed.), *L'Égypte Fatimide: son art et son histoire – Actes du colloqie organisé à Paris les 28, 29 et 30 mai 1998*, Paris: Presses de l'Université de Paris-Sorbonne, pp. 497–517.

King, D. A., 2004–2005, *In Synchrony with the Heavens: Studies in Astronomical Time-keeping and Instrumentation in Medieval Islamic Civilization*, 2 Vols., Leiden-Boston: Brill.

King, D. A. and Gingerich, O., 1982, "Some astronomical observations from thirteenth-century Egypt", *Journal for the History of Astronomy* **13**, pp. 121–128.

Koertge, N., 2008, *New Dictionary of Scientific Biography*, 8 Vols., Detroit: Charles Scribner's Sons.

al-Maghribī, Muḥyī al-Dīn, *Adwār al-anwār*, MSS. M: Iran, Mashhad, Holy Shrine Library, no. 332; CB: Ireland, Dublin, Chester Beatty, no. 3665.

al-Maghribī, Muḥyī al-Dīn, *Talkhīṣ al-majisṭī*, MS. Leiden: Universiteitsbibliotheek, Or. 110.

Moesgaard, K. P., 1975, "Tychonian observations, perfect numbers, and the date of creation: Longomontanus's solar and precessional theories", *Journal for the History of Astronomy* **6**, pp. 84–99.

Mozaffari, S. M., 2014, "Muḥyī al-Dīn al-Maghribī's lunar measurements at the Maragha observatory", *Archive for History of Exact Sciences* **68**, pp. 67–120.

Mozaffari, S. M., 2018, "An analysis of medieval solar theories", *Archive for History of Exact Sciences* **72**, pp. 191–243.

Mozaffari, S. M., 2018–2019, "Muḥyī al-Dīn al-Maghribī's measurements of Mars at the Maragha observatory", *Suhayl* **16**, pp. 149–249.

Mozaffari, S. M., 2019a, "The orbital elements of Venus in medieval Islamic astronomy; Interaction between traditions and the accuracy of observations", *Journal for the History of Astronomy* **50**, pp. 46–81.

Mozaffari, S. M., 2019b, "Ibn al-Fahhād and the great conjunction of 1166 AD", *Archive for History of Exact Sciences* **73**, pp. 517–549.

Mozaffari, S. M., 2020–2021, "Kāshī's lunar measurements", *Suhayl* **18**, pp. 69–127.

Mozaffari, S. M., 2023, "Sources of the planetary theories in Fahhād's *ʿAlāʾī zīj*: Solving a medieval case of intellectual fraud", *Suhayl* **20**, pp. 141–219.

Mozaffari, S. M. and Zotti, G., 2013, "The observational instruments at the Maragha observatory after AD 1300", *Suhayl* **12**, pp. 45–179.

Muntakhab al-Dīn al-Yazdī, *Manẓūm Zīj*, MS. Iran, Mashhad University, Theology Faculty, no. 674.

Nallino, C. A. (ed.), 1899–1907/1969, *Al-Battani sive Albatenii Opus Astronomicum*. Publicazioni del Reale osservatorio di Brera in Milano. n. XL, pte. I–III, Milan: Mediolani Insubrum. The Reprint of Nallino's edition: Minerva, Frankfurt, 1969.

Neugebauer, O., 1975, *A History of Ancient Mathematical Astronomy*, Berlin-Heidelberg-New York: Springer.

Newton, R. R., 1972, "The Earth's acceleration as deduced from al-Bīrūnī's solar data", *Memoirs of Royal Astronomical Society* **76**, pp. 99–128.

Pingree, D. (ed.), 1985, *Astronomical Works of Gregory Chioniades*, Vol. 1: *Zīj al-ʿAlāʾī*, Amsterdam: Gieben.

Qūshčī, ʿAlī b. Muḥammad, *Sharḥ-i Zīj-i Ulugh Beg* (*Commentary on the Zīj of Ulugh Beg*), MSS. N: Iran, National Library, no. 20127–5, P: Iran, Parliament Library, no. 6375/1, PN: USA, Rare Book & Manuscript Library of University of Pennsylvania, no. LJS 400.

Rosenfeld, B. A. and İhsanoğlu, E., 2003, *Mathematicians, Astronomers, and Other Scholars of Islamic Civilization and Their Works (7th-19th c.)*, Istanbul: IRCICA.

Said, S. S. and Stephenson, F. R., 1995, "Precision of medieval Islamic measurements of solar altitudes and equinox times", *Journal for the History of Astronomy* **26**, pp. 117–132.

Said, S. S. and Stephenson, F. R., 1997, "Solar and Lunar Eclipse measurements by medieval Muslim astronomers, II: Observations", *Journal for the History of Astronomy* **28**, pp. 29–48.

Saliba, G., 1994, *A History of Arabic Astronomy: Planetary Theories During the Golden Age of Islam*, New York: New York University Press.

Samsó, J., King, D. A., and Goldstein, B. R., 2001, "Astronomical handbooks and tables from the Islamic world (750–1900): An interim report," *Suhayl* **2**, pp. 9–105.

Samsó, J. and Millás, E., 1998, "The computation of planetary longitudes in the *zīj* of Ibn al-Bannā", *Arabic Science and Philosophy* **8**, pp. 259–286; reprinted in Samsó, J., *Astronomy and Astrology in al-Andalus and the Maghrib*, Aldershot: Ashgate, 2007, Trace VIII.

Sayılı, A., 1960/1988, *The Observatory in Islam*, 2nd edn., Ankara: Türk Tarih Kurumu Basimevi.

Seemann, H. J., 1928/1929, "Die Instrumente der Sternwarte zu Marāgha nach den Mitteilungen von al-'Urḍī", in: Schulz, O. (ed.), *Sitzungsberichte der Physikalisch-medizinischen Sozietät zu Erlangen*, Vol. 60), Erlangen: Kommissionsverlag von Max Mencke, pp. 15–126.

Sezgin, F., 1978, *Geschichte des arabischen Schrifttums, Band VI: Astronomie bis ca. 430 H.*, Leiden: Brill.

Sezgin, F. and Neubauer, E., 2010, *Science and Technology in Islam*, 5 Vols., Frankfurt: Institut für Geschichte der Arabisch–Islamischen Wissenschaften.

Steele, J., 2000, *Observations and Predictions of Eclipse Times by Early Astronomers*, Dordrecht-Boston-London: Kluwer Academic Publishers, reprinted by Springer.

Stephenson, F. R., 1997, *Historical Eclipses and Earth's Rotation*, Cambridge: Cambridge University Press.

Swerdlow, N. M., 1972, "Al-Battānī's determination of the solar distance", *Centaurus* **17**, pp. 97–105.

Swerdlow, N. M., 1977, "A summary of the derivation of the parameters in *Commentariolus* from the *Alfonsine Tables*", *Centaurus* **21**, pp. 201–213.

Swerdlow, N. M., 2010, "Tycho, Longomontanus, and Kepler on Ptolemy's solar observations and theory, precession of the equinoxes, and obliquity of the ecliptic", in: Jones, A. (ed.), *Ptolemy in perspective (Archimedes, 23)*, Dordrecht-Heidelberg-London-New York: Springer, pp. 151–202.

Taqī al-Dīn Muḥammad b. Ma'rūf, *Kharīdat al-durar wa jarīdat al-fikar (The non-bored pearls and the arrangement of ideas)*, MSS. B: Berlin, Staatsbibliothek zu Berlin, no. Ahlwardt 5699 = WE. 193; C1: Cairo, Dār al-Kutub, Ṭal'at Mīqāt Collection, no. 900; C2: Cairo, Dār al-Kutub, Ṭal'at Mīqāt Collection, no. 76; E: Istanbul, Süleymaniye, Esad Efendi Collection, no. 1976; K: Kandilli Observatory, no. 183.

Taqī al-Dīn Muḥammad b. Ma'rūf, *Sidrat muntaha 'l-afkār fī malakūt al-falak al-dawwār (The Lotus tree in the seventh heaven of reflection)* or *Shāhanshāhiyya Zīj*, MSS. K: Istanbul, Kandilli Observatory, no. 208/1 (up to f. 48v; autograph), N: Istanbul, Süleymaniye Library, Nuruosmaniye Collection, no. 2930, V: Istanbul, Süleymaniye Library, Veliyüddin Collection, no. 2308/2 (from f. 10v).

Tekeli, S., 1962, "Solar parameters and certain observational methods of Taqī al Dīn and Tycho Brahe", *Ithaca* 26 VIII-2 IX, Vol. 2, Paris: Hermann, pp. 623–626.

Tekeli, S., 2008, "Taqī al-Dīn", in: Selin, H. (ed.), *Encyclopaedia of the History of Science, Technology, and Medicine in Non-Western Cultures*, Netherlands: Springer, pp. 2080–2081.

Thoren, V. E. and Christianson, J. R., 1990, *The Lord of Uraniborg: A Biography of Tycho Brahe*, Cambridge: Cambridge University Press.

Toomer, G. J. (ed.), 1998, *Ptolemy's Almagest*, Princeton: Princeton University Press.

Ulugh Beg, *Sulṭānī Zīj*, MS. P1: Iran, Parliament Library, no. 72; MS. P2: Iran, Parliament Library, no. 6027.

Part III

PLANETARY ASTRONOMY

6

FOUR-POINT METHOD FOR DETERMINING THE ECCENTRICITY AND THE DIRECTION OF THE APSIDAL LINES OF THE SUN AND SUPERIOR PLANETS[1]

6.1 Introduction

In the Ptolemaic models for the superior planets and Venus, the center of the epicycle is located on a circle (the so-called deferent), the center of which is displaced from that of the Earth by an eccentricity e, but its motion is uniform with respect to the so-called equant point, removed from the Earth by an eccentricity $2e$ from the Earth on the side of the center of the deferent. The derivation of the planet's orbital elements (eccentricity and direction of the apsidal line) requires observations of three oppositions of the planet to the mean sun, when the planet points to the center of the epicycle. A difficulty arises since the angles of mean motion of the planet between two successive mean oppositions are known with respect to the equant point, while the angles of the true motions between two consecutive mean oppositions are measured with respect to the Earth, the center of the deferent being located exactly between the two. This causes the derivation of the eccentricity and the longitude of the apogee of a superior planet to require the solution of an equation of the eighth degree. Ptolemy, in the *Almagest* (X.7, XI.1, and XI.5), explains an iterative algorithmic solution for finding the two parameters.[2]

In 1987, N. M. Swerdlow published a detailed analysis of an alternative method to Ptolemy's, proposed by Jābir b. Aflaḥ of Seville (*fl.* the first quarter of the twelfth century), based on Gerard of Cremona's 1175 CE Latin translation of Jābir's *Iṣlāḥ al-majisṭī* (*Improvement of the Almagest*). This method is theoretically sound but

---

1 Original publication: S. M. Mozaffari, "Bīrūnī's four-point method for determining the eccentricity and the direction of the apsidal lines of the superior planets", *Journal for the History of Astronomy* **44** (2013), pp. 207–211. © 2013 Sage, and republished by permission.
2 Hill 1900, Toomer [1984] 1998, pp. 484–498, 507–519, 525–537, Pedersen 1974, pp. 271–283, Neugebauer 1975, vol. 1, pp. 172–179, Duke 2005, pp. 179–182.

DOI: 10.4324/9781003481966-10

practically infeasible, as demonstrated by Swerdlow.[3] I will refer to it as the "four-point method" due to its reliance on four oppositions of an outer planet to the mean sun. It can also be applied to measure the Sun's orbital elements.

What has been overlooked until 2014 was that this method appears in *al-Qānūn al-mas'ūdī* of Abū al-Rayḥān al-Bīrūnī (973–1048 CE), written roughly a century before Jābir, apparently, during the reign of Mas'ūd I, the ninth ruler of the Ghaznavid dynasty of Iran (1030–1041). More significantly, Bīrūnī's primary source is Ibrāhīm b. Sinān b. Thābit b. Qurra (909–946 CE). In the following sections, we first provide a thorough description of the method as explained by Bīrūnī supplied, complete with mathematical explanations (Section 6.2). We then delve into Ibrāhīm b. Sinān's account (Section 6.3), its transmission to the Western Islamic domain (Section 6.4), and other mentions of it in the medieval Islamic astronomical corpus (Section 6.5).

6.2 Bīrūnī's Account of the Method

At the end of *al-Qānūn* X.3.1,[4] Bīrūnī states that Ptolemy could alternatively determine the eccentricity and the direction of the apsidal line of a superior planet by seeking out four oppositions of the planet with the mean sun satisfying the following condition. Figure 6.1, as drawn in the edited text of *al-Qānūn* with the original lettering (except those bearing prime), transcribed according to the standard proposed by the late E. S. Kennedy,[5] shows the deferent *ABCGK* of a superior planet about center *D*; *E* represents the Earth, and *T* the center of uniform motion, that is, the equant point. *TE* is then the eccentricity of the equant point, and *DE* that of the deferent; in the Ptolemaic context, $TE = 2\,DE = 2e$. In the four mean oppositions, the planet and, thus, the center of the epicycle are located, respectively, at the points *A*, *B*, *G*, and *K*. Our author first explains the essential condition in the application of the method: the difference between the true longitudes of the planet in each pair of oppositions (*A–B* and *G–K*) should be identical: angle *AEB* = angle *GEK*. The trajectories travelled by the epicycle's center on the deferent in each pair of oppositions should also be identical (arc *AB* = arc *GK*) so that the mean motions of the planet in each pair of oppositions are equal: angle *ATB* = angle *KTG*. These conditions may be formulated as follows: in the two pairs of oppositions, the center of the planet's epicycle describes the equal arcs, *AB* and *GK*, in equal periods of time. Bīrūnī follows that "what we mentioned is the property of the two arcs on the deferent equidistant from the diameter of the deferent passing through the apogee and perigee." Then, point *C*, which marks one of the two apses (here, the apogee), is at the middle of the arc *BG* between the two arcs

---

3 Swerdlow 1987. On the astronomy of Jābir and his critical remarks concerning Ptolemy's *Almagest*, see, for example, Lorch 1995, Chapters VI, VIII, XVI; Samsó 2001, Bellver 2006, Bellver 2008.
4 al-Bīrūnī 1954–1956, vol. 3, pp. 1183–1184. The contents of Bīrūnī's *al-Qānūn* are introduced in Kennedy 1971.
5 Kennedy 1991–1992.

AB and *GK* (point *O*, diametrically opposed to *C*, is then the perigee). Our author had already proved (*al-Qānūn* VI.8)[6] the two particular theorems' treating of the relation between the true motion in longitude and the corresponding equation of centrum in the solar eccentric orbit. The second of these is that "retardation and acceleration occur according to the increase and the diminution of the difference in the equation." From this, the special consequence can then be derived that the aforementioned condition can occur only when the two points *A* and *K*, as well as *B* and *G*, are symmetrical to the apsidal line, because only in this case will the difference in equations of centrum, and thus the change in the angular velocity of the center of the epicycle in the two pairs of the oppositions, be identical.

Then, in order to determine the eccentricity *TE*, we drop the two perpendiculars *TL* (text: *GL*) and *DM* to *AE*. Since the angle *AET* (text: *ATE*) is half the difference in the true longitude between the first and fourth oppositions, the angles in the triangle *TLE* are known, and the sides are known in terms of *TE*. Angle *TAE* is

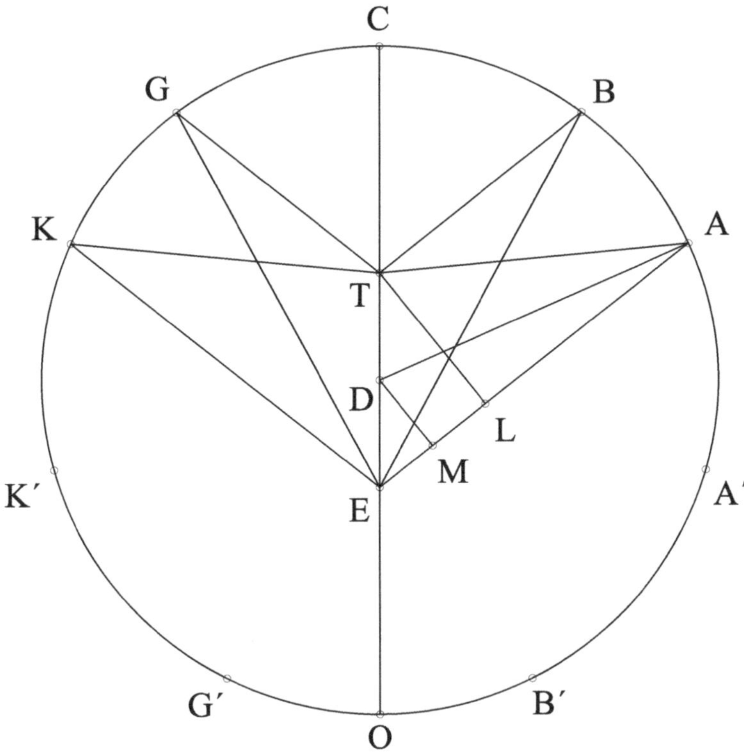

Figure 6.1 Four-point method in the case of the superior planets.

6 al-Bīrūnī 1954–1956, vol. 2, pp. 665–667; see Hartner and Schramm 1961, pp. 212–214.

the equation of centrum of the planet in the first opposition to the mean sun; it is equal to the difference between the two angles AEC and ATC, which are, respectively, half the true and mean motion of the planet between the two oppositions, that is, first and fourth. Thus, in triangle ATL, the angles are known, and the sides are known in terms of TL. And ML is half of LE, and MD is half of LT. Then, $AD = \sqrt{(AM^2 + MD^2)}$ is known. AD is either the radius of the deferent or may be considered as the radius of the trigonometric base circle, to both of which an arbitrary length $R = 60^\mathrm{p}$ is assigned; in Bīrūnī's terms, the length of AD is equal to the total sine, that is, $AD = \mathrm{Sin}\ 90^\circ = 60^\mathrm{p}$. Then, we express the length of the eccentricity TE in terms of $AD = R = 60^\mathrm{p}$. Therefore, the longitude of the apogee and the eccentricity are known from this method, and this is what we wished to show.

Let the true longitude of the planet in the four oppositions be λ_1, λ_2, λ_3, and λ_4, and its mean longitude λ_{m1}, λ_{m2}, λ_{m3}, and λ_{m4}. Then, angle $AEC = \frac{1}{2}(\lambda_4 - \lambda_1) = \eta$, and angle $TAE = \frac{1}{2}(\lambda_{m4} - \lambda_4 + \lambda_1 - \lambda_{m1}) = \zeta$. From triangle TEL, $LM = \frac{1}{2}\ LE = \frac{1}{2}\ TE \cos \eta$ and $MD = \frac{1}{2}\ LT = \frac{1}{2}\ TE \sin \eta$. From triangle ATL, $AL = LT/\tan \zeta$. Inserting these into the relation $(AL + LM)^2 + MD^2 = AD^2 = R^2$, one can easily obtain the ratio $2e/R$ as follows:

$$\frac{2e}{R} = \frac{TE}{AD} = \frac{2\tan\zeta}{\sqrt{4\sin^2\eta + 2\sin 2\eta \cdot \tan\zeta + \tan^2\zeta}}$$

This method enables one to compute TE and DE independently from each other. Bīrūnī does not, however, allude to this option, perhaps because he believed or took it for granted that Ptolemy's bisection of the eccentricity must be correct. The longitude of the point C may simply be obtained from $\lambda_{ap} = \lambda_1 + \frac{1}{2}(\lambda_4 - \lambda_1)$. A note that escaped our author's attention is that, although the two pairs of mean oppositions satisfying the already-mentioned condition are adequate to determine the direction of the apsidal line, they are, nevertheless, by no means sufficient to distinguish which point, C or O, is the apogee or perigee. For this, one needs two other pairs of mean oppositions to be compared with the first two. They must satisfy the already-mentioned condition as well as the other one saying that the time interval between the oppositions in the latter pairs is equal to that between the oppositions in the first pairs. Take the latter pairs of oppositions to be $A'\!-\!B'$ and $G'\!-\!K'$ in Figure 6.1, not drawn in the original: arc $A'B'$ = arc $G'K'$, angle $A'EB'$ = angle $G'EK'$, and angle $A'EB'$ = angle $G'EK'$ (the angles not shown). The center of the planet's epicycle moves faster when it is closer to the perigee, and slower when it is closer to the apogee. The two sets of the oppositions are then compared with each other: if arc $A'B'$ = arc $G'K'$ > arc AB = arc GK (such as the case here), then point C is the apogee, and point O the perigee, and *vice versa*. Of course, provided with the longitudes given for the planets' apogees in the *Almagest* or in an earlier *zīj*, one may distinguish the apses without needing the two additional pairs of the mean oppositions.

It is somewhat obvious that the method was inspired by Ptolemy's method for the derivation of the structural parameters of the inferior planets in *Almagest* IX.7 and X.1.

In *al-Qānūn* VI.8,[7] Bīrūnī explains the method in the case of the Sun. One observes the longitudes of the Sun at noon day by day during one year and then seeks out two equal arcs of the ecliptic, which the Sun has traversed in equal times (either angle *KTA* and angle *ATB* or angle *ZTK* and angle *BTG* in Figure 6.2); then it will be known that the apogee *A* is located at the midway of the two. The interval of time during which the Sun has travelled, for example, from *K* to *B*, is known. Its half is the time taken by the Sun to traverse either of arc *KA* and arc *AB* of the eccentric; thus, from the equation angle (q = angle *EKT*), that is, the difference between the solar true and mean motions in longitude (respectively, angle *KTA* and arc *KA*), one can readily find $EH = \mathrm{Sin}\, q$, and then $e = TE$, as all the angles in triangle *EHT* are already known (angle $H = 90°$).

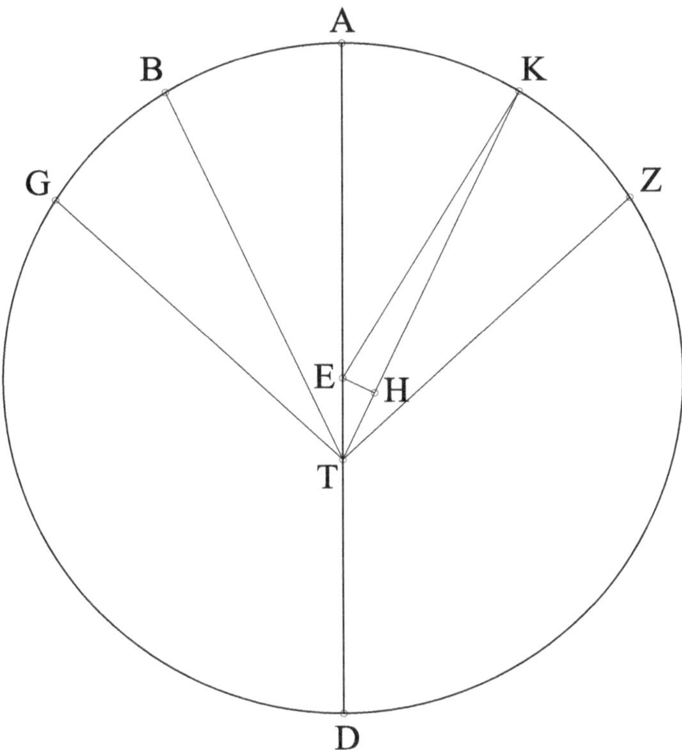

Figure 6.2 Four-point method in the case of the Sun.

7 al-Bīrūnī 1954–1956, vol. 2, pp. 684–685.

The method in the case of the Sun was essentially to refrain from serious observational difficulties in the determination of the times of the solstices which were used in the derivation of the solar orbital elements according to the seasons method put forward in *Almagest* III.4.

6.3 The Inventor of the Method

Ibrāhīm b. Sinān,[8] in his *Fī ḥarakāt al-shams* (*On the solar motions*; completed in 933 CE), explains in full the four-point method in the case of the Sun as a close parallel to another method of his own, which he

> had already elucidated in [his now lost] book on the correction of the chapter[s in the *Almagest*], in which Ptolemy explained [the method for the determination of] the eccentricity of Saturn, Jupiter, Mars, and the other planets which have two anomalies in their [longitudinal] motions accounted for by the epicyclic and eccentric hypotheses.[9]

Therefore, there is no doubt that he is the inventor of the four-point method, both for the superior planets and for the Sun.

Bīrūnī was well acquainted with Ibrāhīm's works, not least because he directly refers to him in his works in connection with the model for the variable precession described in the *On the solar motions* (the model was hypothesized by Abū Jaʿfar al-Khāzin, 900–after 970 CE, in collaboration with Ibrāhīm).[10] So he almost certainly learned the method from Ibrāhīm.

6.4 Transmission of the Method to Western Islamic Astronomy: Jābir B. Aflaḥ

The four-point method in the case of the outer planets as described by Bīrūnī is principally that posed by Jābir b. Aflaḥ in his *al-Kitāb fī al-hay'a* (also known as the *Iṣlāḥ al-majisṭī*)[11] about one century later, although the mathematical procedures applied by the two astronomers are substantially different: Bīrūnī's is obviously more straightforward and simpler, which is partly due to the fact that the two eccentricities are computed separately in Jābir's, while such an option, as already mentioned, is not included in Bīrūnī's.

It is very difficult (if not impossible) to answer the question whether Jābir adopted the method from Ibrāhīm b. Sinān (either directly from the latter's two aforesaid works or through the medium of Bīrūnī) without mentioning its inventor

---

8 On his life and works, see Rashed 2012, p. 459f.

9 Ibrāhīm b. Sinān 1983, pp. 275–302, 287–289.

10 Bīrūnī, *Chronology*, p. 326, En. translation, p. 322; *Taḥdīd*, pp. 100–101, English translation, pp. 69–70, E. S. Kennedy's commentary, p. 43.

11 Jābir b. Aflaḥ, *al-Kitāb fī al-hay'a*, B: ff. 95v–97r, E1: ff. 96r–97r, E2: ff. 114r–115r, T: —.

or independently discovered it. On the one hand, the fact that Jābir's extensive and manifold criticisms of Ptolemy's mathematical procedures in the *Almagest* seems to have no precedents in Eastern (Middle Eastern) Islamic astronomy as well as the difference in the mathematical procedures adopted by the two astronomers for implementing the method may point to an independent achievement of the Andalusian astronomer. But on the other hand, since the original account of Ibrāhīm b. Sinān has not survived, we cannot be sure whether what Jābir penned originally belongs to Ibrāhīm or he adopted the primary rationale behind the method from the Middle Eastern sources but elaborated on its underlying computational procedure in a truly significant way. As shown in a detailed investigation by J. Samsó,[12] there are some traces of the arrival of mathematical and astronomical materials composed in the Middle East between the mid-tenth century and the first half of the eleventh century to al-Andalus, indicating, in part, the diffusion of al-Bīrūnī's works (including *al-Qānūn*) in some specific scientific circles in al-Andalus. Consequently, the possibility exists that Jābir learned the method somehow from the Middle Eastern sources, formulated it anew, and included it among his criticisms of Ptolemy.[13] If this were the case, then we may conclude that it is another aspect of the transmission of astronomical knowledge between the two separated domains of medieval Islamic astronomy (the Middle East and al-Andalus), and that one of the original astronomical methods invented in medieval Islamic astronomy was transferred into medieval Latin astronomy through the Western Islamic astronomical literature. However, these statements require further evidence in order to be established with certainty. Nevertheless, this provides further motivation to investigate whether Jābir's other alternatives proposed for Ptolemy's mathematical procedures of planetary astronomy may be rooted in the medieval Middle Eastern sources.

It is curious that the Eastern Islamic astronomers became familiar with Jābir's principal work from *circa* the middle of the thirteenth century. Quṭb al-Dīn al-Shīrāzī (d. 1311 CE) composed a brief summary of this work (completed at the middle of Rabīʿ I 663 H/the beginning days of January 1265 CE), including the four-point method.[14] Athīr al-Dīn al-Mufaḍḍal b. ʿUmar al-Abharī (d. between 660 H/1263 CE and 663/1265) wrote a treatise on the new cosmography of Jābir, according to which Mercury and Venus are placed above the Sun.[15]

6.5 Other References to the Method in Medieval Middle Eastern Astronomy

The author of the anonymous *Ṣināʿat al-majisṭī* (*Art/Industry of the Almagest*) (Athīr al-Dīn Abharī?), dedicated to Najm al-Dīn ʿAlī b. ʿUmar b. ʿAlī Dabīrān/

12 Samsó 1996.
13 See also Samsó 2020, p. 511.
14 Shīrāzī, *Fawāʾid*, ff. 90v–91r.
15 See Bellver 2020, p. 186.

al-Kātibī al-Qazwīnī (d. 1276 CE), a member of the main staff of the Marāgha observatory, briefly described the method in the case both of the Sun and of the outer planets.[16]

It is interesting that, in his *Taḥrīr al-majisṭī* (completed on February 13, 1247 CE), Naṣīr al-Dīn al-Ṭūsī (d. 1274 CE) refers to the method, after explaining the Ptolemaic iterative procedure in *Almagest* X.7 and XI.1 and 5, in these critical remarks:

> I say: Ptolemy's method involved some of the moderns, because it is not on the basis of a straightforward principle of proof (*aṣl mustaqīm min al-burhān*), but consists of an iterative procedure (*tikrār al-ʿamal*). It is said [NB. Ṭūsī does not identify the inventor of the method or his source] that if [a superior planet] is observed at four situations [viz., four oppositions to the mean sun] [under the conditions that] the apparent arc of the ecliptic intercepted by two of them and the time [taken for the planet to travel through it] are equal to the [apparent ecliptic] arc and the time interval between the remaining two, then the diameter passing through the apogee would be located in the middle of the extremities of the two symmetrical arcs. This is in accord with the rule of geometry [i.e., it is correct]. Nevertheless, if the observer prefers to make observations, for its ease, rather than do difficult calculations, then [he should bear in mind] that **[1]** the increase of the number of observations and the conditions [which must be fulfilled in them] make it extremely difficult; moreover, **[2]** [a method] in which more calculations are given [i.e., a computational method] is more reliable than [another method] in which more sense (*al-ḥiss*) [data] are given [i.e., an empirical method].[17]

Ṭūsī does not put forward a reasonable comparison between Ptolemy's iterative procedure and the four-point method, although all essential elements for making a sound comparison between the two are present in the quoted passage, as will be explained presently.

The two methods can be compared in two key aspects: on the empirical level, in the four-point method, it is required to conduct numerous observations of a superior planet near its oppositions to the mean sun over extended periods to carefully select four symmetric ones. However, this process is cumbersome, and achieving the ultimate goal of finding four perfectly symmetrical mean oppositions is practically impossible. In contrast, Ptolemy's three-point method requires observations of three arbitrary mean oppositions, with the only requirement being their distribution as far apart as possible across the ecliptic. On the mathematical (computational) level, as we have seen earlier, the iterative nature of

---

16 *Ṣināʿa* IV.6 & VI.1: P: ff. 48r, 65r, N: pp. 81, 113, I: ff. 40r, 55r.
17 al-Ṭūsī, *Taḥrīr*, P1: p. 345, P2: f. 100r, P3: f. 130r, B: f. 134r, AS: ff. 96v–97r.

the algorithmic procedure demands meticulous calculations and adjustments; this approach is time-consuming and prone to error. In contrast, the four-point method is much simpler and more straightforward than Ptolemy's *approximate* iterative procedure. Therefore, we can safely conclude that Ptolemy's method is preferrable, because it sacrifices a higher degree of the computational accuracy in favor of operational practicality.

Ṭūsī's first point is highly relevant if we assume that, in his general statement in [1], he intends to address the serious empirical difficulties in the four-point method. However, he does not thoroughly examine the matter to realize that fulfilling the symmetry condition is impossible. Nevertheless, Ṭūsī's second point does not involve a direct comparison of the mathematics of the two methods. Instead, in his general conclusion in [2], we encounter a philosophically minded theoretical astronomer's unexpected perspective favoring computational methods over empirical ones. He does not continue his line of argumentation from the opening sentence but mistakenly and misleadingly attributes the primary reason for preference to the iterative procedure in Ptolemy's method.

Ṭūsī's two later commentators, Niẓām al-Dīn Aʿraj al-Nīshābūrī (d. 1328/1329)[18] and ʿAbd al-ʿAlī al-Bīrjandī (d. 1525/1526),[19] endorsed his views.

References

Abū Muḥammad Jābir b. Aflaḥ, *al-Kitāb fī al-hay'a/Islāḥ al-majisṭī* (*Book on astronomy/ Correction of the* Almagest), MSS. E1: Biblioteca Real Monasterio de San Lorenzo de el Escorial, ár. 910, E2: Biblioteca Real Monasterio de San Lorenzo de el Escorial, ár. 930, B: Staatsbibliothek PreußischerKulturbesitz zu Berlin, Landberg, no. 132, P: Tehran, Parliament Library, no. 1440S.

Anonymous (Abharī, Athīr al-Dīn al-Mufaḍḍal b. ʿUmar?), *Kitāb fī ṣināʿat al-majisṭī* (*Book on the art/industry of the* Almagest), MSS. P: Iran, Parliament Library, no. 6195, N: Iran, National Library, no. 2607560, I: Istanbul, Süleymaniye, Ayasofia, no. 2583.

Bellver, J., 2006, "Jābir b. Aflaḥ on the four-eclipse method for finding the lunar period in anomaly", *Suhayl* **6**, pp. 159–248.

Bellver, J., 2008, "Jābir b. Aflaḥ on lunar eclipses", *Suhayl* **8**, pp. 47–91.

Bellver, J., 2020, "The Arabic versions of Jābir b. Aflaḥ's al-Kitāb fī l-Hay'a", in: Juste *et al.* 2020, pp. 181–200.

al-Bīrjandī, Niẓām al-Dīn ʿAbd al-ʿAlī b. Muḥammad b. al-Ḥusayn, *Sharḥ* Taḥrīr al-majisṭī (*Commentary upon* al-Ṭūsī's *Exposition of the* Almagest), MS. P: Iran, Parliament Library, no. 6167.

al-Bīrūnī, Abū al-Rayḥān, 1962, *Taḥdīd nihāyāt al-amākin li-taṣḥīḥ masāfāt al-masākin* (Determination of the Directions of Places for the Correction of the Distances of Localities), Bulgakov, P. and Ahmad, I. (eds.), Cairo, reprinted in *Islamic Geography*, Vol. 25, Frankfurt: Institute for History of Arabic–Islamic Science, 1992; English translation:

---

18 Nīshābūrī, *Sharḥ*, PN: 172r–v; P: ff. 182r–v.

19 His comments are found in the marginal glosses scattered throughout MS. P1 of Ṭūsī's *Taḥrīr*, as well as are gathered up in his *Sharḥ* (P: ff. 327r–v).

Ali, J., *The Determination of the Coordinates of Positions for the Correction of Distances between Cities*, Beirut: 1967; E. S. Kennedy's commentary: Kennedy, E. S., 1973, *A Commentary Upon Bīrūnī's Kitāb Taḥdīd al-Amākin: An 11th Century Treatise on Mathematical Geography*, Beirut: American University of Beirut.

al-Bīrūnī, Abū al-Rayḥān, 1954–1956, *al-Qānūn al-mas'ūdī* (*Mas'ūdic canons*), 3 Vols., Hyderabad: Osmania Bureau.

al-Bīrūnī, Abū al-Rayhān, 1878/1923, *Chronology: Athār al-bāqiyya 'an al-qurūn al-khāliyya* (*The Chronology of Ancient Nations*), Edition: Sachau, C. E. (ed.), Leipzig: Deutsche Morgenl. Gesellschaft and Otto Harrassowitz, English Translation: Sachau, C.E. (En. tr.), London: William H. Allen, 1879.

Duke, D., 2005, "Ptolemy's treatment of the outer planets", *Archives for History of Exact Sciences* **59**, pp. 169–187.

Hartner, W., and Schramm, M., 1961, "Al-Bīrūnī and the theory of the solar apogee: An example of originality in Arabic science", in Crombie, A. C. (ed.), *Scientific Change. Historical Studies in the Intellectual, Social and Technical Conditions for Scientific Discovery and Technical Invention, from Antiquity to the Present.* Symposium on the History of Science, University of Oxford 9–15 July, London: Heinemann, pp. 206–218.

Hill, G. W., 1900, "Ptolemy's problem", *The Astronomical Journal* **21** (issue 485), pp. 33–35.

Ibrāhīm b. Sinān b. Thābit b. Qurra, 1983, *Rasā'il ibn Sinān*, Sa'īdān, Aḥmad Salīm (ed.), Kuwait.

Juste, D., van Dalen, B., Nikolaus Hasse, D., Burnett, C., 2020, *Ptolemy's Science of the Stars in the Middle Ages*, Ptolemaeus Arabus et Latinus, Studies, Vol. 1, Turnhout, Belgium: Brepols.

Kennedy, E. S., 1971, "Al-Bīrūnī's Masudic Canon", *Al-Abhath* **24**, pp. 59–78; reprinted in Kennedy *et al.* 1983, pp. 573–592.

Kennedy, E. S., colleagues, and former students, 1983, *Studies in the Islamic Exact Sciences*, Beirut: American University of Beirut.

Kennedy, E. S., 1991–1992, "Transcription of Arabic letters in geometric figures", *Zeitschrift fur Geschichte der Arabisch-Islamischen Wissenschaften* 7, pp. 21–22.

King, D. A. and Saliba, G. (eds.), 1987, *From Deferent to Equant: A Volume of Studies on the History of Science of the Ancient and Medieval Near East in Honor of E. S. Kennedy*, New York: New York Academy of Sciences.

Lorch, R. P., 1995, *Arabic Mathematical Sciences: Instruments, Text, Transmission*, Aldershot: Ashgate-Variorum.

Neugebauer, O., 1975, *A History of Ancient Mathematical Astronomy*, 3 Vols., Berlin-Heidelberg-New York: Springer.

al-Nīshābūrī, Niẓām al-Dīn A'raj, *Sharḥ Taḥrīr al-majisṭī (Commentary upon al-Ṭūsī's Exposition of the Almagest)*, MSS. PN: USA, Rare Book & Manuscript Library of University of Pennsylvania, LJS 392 (dated 13 Dhi al-Qa'da 813/9 March 1411), P: Iran, Parliament Library, no. 6372 (dated 15 Ṣafar 817/6 May 1414).

Pedersen, O., 2010/1974, *A Survey of the Almagest*, Odense: Odense University Press, with annotation and new commentary by A. Jones, New York: Springer.

Rashed, R., 2012, *Founding Figures and Commentators in Arabic Mathematics: A History of Arabic Sciences and Mathematics*, Vol. 1, El-Bizri, N. (ed.), Wareham et al. (En. trs.), London-New York: Routledge.

Samsó, J., 1996, "Al-Bīrūnī in al-Andalus", in: Casulleras, J., and Samsó, J. (eds.), *From Baghdad to Barcelona*, Barcelona: Universitat de Barcelona; reprinted in Samsó 2007, Chapter VI.

Samsó, J., 2001, "Ibn al-Haytham and Jābir b. Aflaḥ's criticism of Ptolemy's determination of the parameters of Mercury", *Suhayl* **2**, pp. 199–225.

Samsó, J., 2007, *Astronomy and Astrology in al-Andalus and the Maghrib*, Aldershot: Ashgate-Variorum.

Samsó, J., 2020, *On Both Sides of the Strait of Gibraltar: Studies in the History of Medieval Astronomy in the Iberian Peninsula and the Maghrib*, Leiden-Boston: Brill.

Shīrāzī, Quṭb al-Dīn, *Fawā'id min al-Kitāb al-mawsūm bi-l-Majisṭī li-Ibn Aflaḥ al-Maghribī* (*Useful notes from the Book referred to as the Almagest by Ibn Aflaḥ al-Maghribī*), MS. Oxford, Bodleian, Thurston 3, ff. 75v–92v (copying finished on 7 Rajab 675/15 December 1276).

Swerdlow, N. M., 1987, "Jābir ibn Aflaḥ's interesting method for finding the eccentricities and direction of the apsidal line of a superior planet", in: King and Saliba 1987, pp. 501–512.

Toomer, G. J. (ed. & tr.), 1984/1998, *Ptolemy's Almagest*, Princeton: Princeton University Press.

al-Ṭūsī, Naṣīr al-Dīn Muḥammad, *Taḥrīr al-majisṭī* (*Exposition of the* Almagest), MSS. Iran, Parliament Library, P1: no. 3853 (commented by ʿAbd al-ʿAlī al-Bīrjandī, d. 1525/1526 CE), P2: no. 6357, P3: no. 6395, B: Staatsbibliothek zu Berlin, Sprenger 1838 = Ahlwardt 5655, ff. 1v–152r (copying finished at Marāgha on Wednesday, 22 Ramaḍān 655 H. = 3 October 1257 CE, for a certain Maḥmūd al-Khāṭīb: f. 152r), AS: Istanbul, Aya Sofia, no. 2583, ff. 1v–112v (copying completed on Thursday, 5 Shawwāl 686 H. = 13 November 1287 CE: f. 112v).

7

PLANETARY LATITUDES IN MEDIEVAL ISLAMIC ASTRONOMY

An Analysis of the Non-Ptolemaic Latitude Parameter Values in the Marāgha and Samarqand Astronomical Traditions[1]

7.1 The Ptolemaic Latitude Models and Their Reception in Medieval Islamic Astronomy

In the Ptolemaic latitude theory for the superior planets in the *Almagest* (Figure 7.1(a)), the eccentric is inclined to the ecliptic at a fixed angle i_0, which is equal to 2;30° for Saturn, 1;30° for Jupiter, and 1° for Mars. The center of the epicycle revolves on the eccentric in the direction of increasing longitude, reaches the northern and southern *limits*[2] at opposite locations on the ecliptic, and crosses it at the nodes. The epicycle itself is inclined to the eccentric with its perigee at the same direction as the eccentric, but at a *varying* angle i_1. i_1 is minimum and equal to i_0 when the center of the epicycle is at the nodes and reaches its maximum when the center of the epicycle is located at the *limits*: $i_{1\max}$ = 4;30° for Saturn, 2;30° for Jupiter, and 2;15° for Mars. The direction of these inclinations is fixed with respect to the planets' apsidal lines. $\angle ATN$ shows the longitudinal difference ω_A between the apogee A and the northern limit N, which is equal to +50° for Saturn, −20° for Jupiter, and 0° for Mars. The model in the *Handy Tables* (Figure 7.1(b)) is as same as in the *Almagest*, with the exception that i_1 is *fixed* at the same values in the *Almagest*, and that ω_A = +40° for Saturn, a value less accurate than that in the *Almagest*. In the *Planetary Hypotheses* (Figure 7.1(c)), i_1 is *fixed* but equal to i_0, that is, the epicycle always remains parallel to the ecliptic. The

<hr>

1 Original publication: S. M. Mozaffari, "Planetary latitudes in medieval Islamic astronomy: An analysis of the non-Ptolemaic latitude parameter values in the Maragha and Samarqand astronomical traditions," *Archive for History of Exact Sciences* **70** (2016), pp. 513–541. © 2016 Springer Nature, and republished by permission.

2 They are the locations in the ecliptic where a superior planet reaches its extremal northern and southern latitudes (*Almagest* XIII.1: Toomer 1998, p. 598).

DOI: 10.4324/9781003481966-11

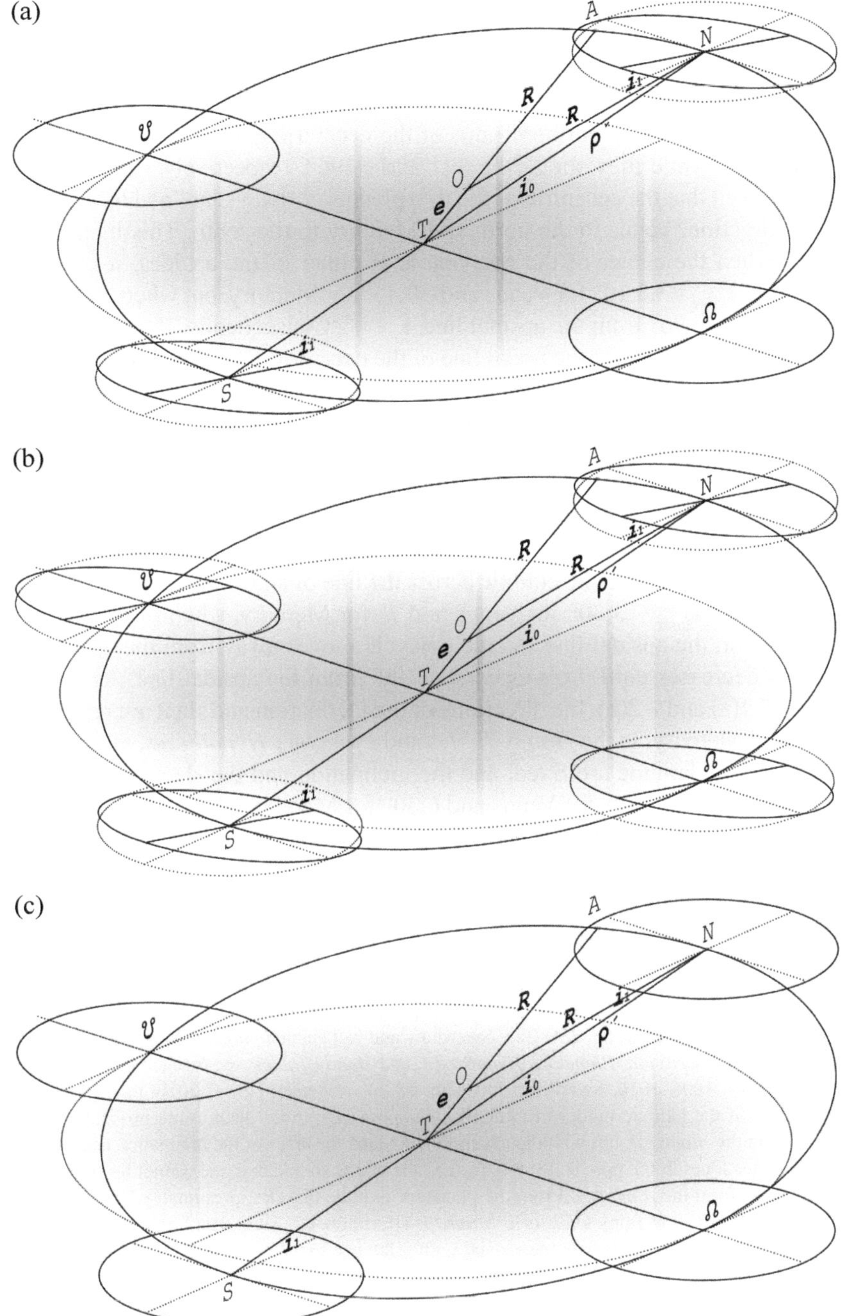

Figure 7.1 Ptolemaic latitude model for the superior planets in (a) the *Almagest*, (b) the *Handy Tables*, and (c) the *Planetary Hypotheses*.

inclinations are equal to i_0 in the *Almagest/Handy Tables*, with an improvement in the case of Mars, where $i_0 = i_1 = 1;50°$.

The Ptolemaic model for latitude of the inferior planets is shown in Figure 7.2(a) for Venus and Figure 7.2(b) for Mercury. Ptolemy observed that the distance ω_A from the apogee to the northern limit of the eccentric is 0° for Venus and $+180°$ for Mercury, while they should be $-1°$ and $+196°$, respectively, at his time. He also observed that the eccentric of the two planets is always inclined to the ecliptic in one direction, Venus to the north and Mercury to the south. This inclination is varied: when the center of the epicycle is at either of the apsides, it reaches its maximum, $i_{0max} = +0;10°$ for Venus and $-0;45°$ for Mercury, but when the center of the epicycle is $\pm90°$ from the apsidal line, $i_0 = 0°$. Consequently, Ptolemy defines a conventional position for the nodal line of the eccentric $\pm90°$ from the apsidal line. Next, when the center of the epicycle is $\pm90°$ from the apsidal line, the diameter passing through the epicyclic apogee and perigee is inclined to the ecliptic in the line of sight at an angle i_1. This inclination is also varied, $i_{1max} = 2;30°$ for Venus and $6;15°$ for Mercury. As the epicycle moves toward the apsidal line, it decreases until it vanishes at either of the apsides. Then, when the center of the epicycle is in the apsidal line, the orthogonal to the diameter passing through the epicyclic apogee and perigee is slanted to the ecliptic across the line of sight at an angle i_2. The slant is also varied, $i_{2max} = 3;30°$ for Venus and 7° for Mercury, when the center of the epicycle is in the apsidal line. As the epicycle moves away from the apsidal line, the slant decreases until it disappears at $\pm90°$ from the apsidal line. As shown in Figures 7.2(a) and 7.2(b), the directions of the inclination and slant are reversed for Venus and Mercury. In the *Handy Tables* and *Planetary Hypotheses*, all three components of the latitude are fixed, and the inclination and the slant of the epicycle are equal: $i_1 = i_2 = 3;30°$ for Venus and $6;30°$ for Mercury and $i_0 = 0;10°$ for both.[3]

Although Ptolemy's latitude models in the *Handy Tables* and *Planetary Hypotheses* are simpler and more coherent than those in the *Almagest*, they nevertheless appear to have been of little influence on medieval Islamic astronomers who adhered to the latitude models in the *Almagest*. They had, however, adopted a few

---

3 For the latitude models and parameters in the *Almagest*, see Pedersen [1974] 2010, Chapter 14, Neugebauer 1975, vol. 1, pp. 206–226, Swerdlow and Neugebauer 1984, Chapter 6, Riddell 1978. For *Canobic Inscription, Planetary Hypotheses*, and *Handy Tables*, see Neugebauer 1975, vol. 2, pp. 908–917, 1006–1016, Swerdlow 2005, pp. 58–68, Pedersen [1974] 2010, pp. 398–400. The structures of the latitude model in Ptolemy's *Canobis Inscription* seem quite probably to be the same as in the *Almagest*, but with the exception of Mars, the sizes of the inclination angles are different (after Jones 2005, pp. 70–73, 89). In fact, the text is so bad that one cannot be certain of any differences from the *Almagest*. From the planetary latitude theories, presumably, prior to Ptolemy that are referred to in Pliny's *Historia Naturalis* (II.xiii.66–67: 1938–1962, vol. 1, pp. 212–215), only the maximum latitudes are mentioned with reference to a 12° ecliptical/zodiacal belt: Moon, ±6°; Mercury, +5°/−3°; Venus, ±7°; Mars, ±2°; Jupiter, $\pm2\frac{1}{2}°$ or +3°/−1°; Sun and Saturn: ±1° (also, see Neugebauer 1975, vol. 2, p. 782, Eastwood and Grasshoff 2003, pp. 203–6, Eastwood 2007, pp. 119–126, 136–137). The pre-Ptolemaic solar model having the latitudinal component, in which the Sun reaches the maximum latitude of ±0.5°, is discussed in Jones 2000. For the planetary latitudes in Babylonian astronomy, see Steele 2003.

(a)

(b)

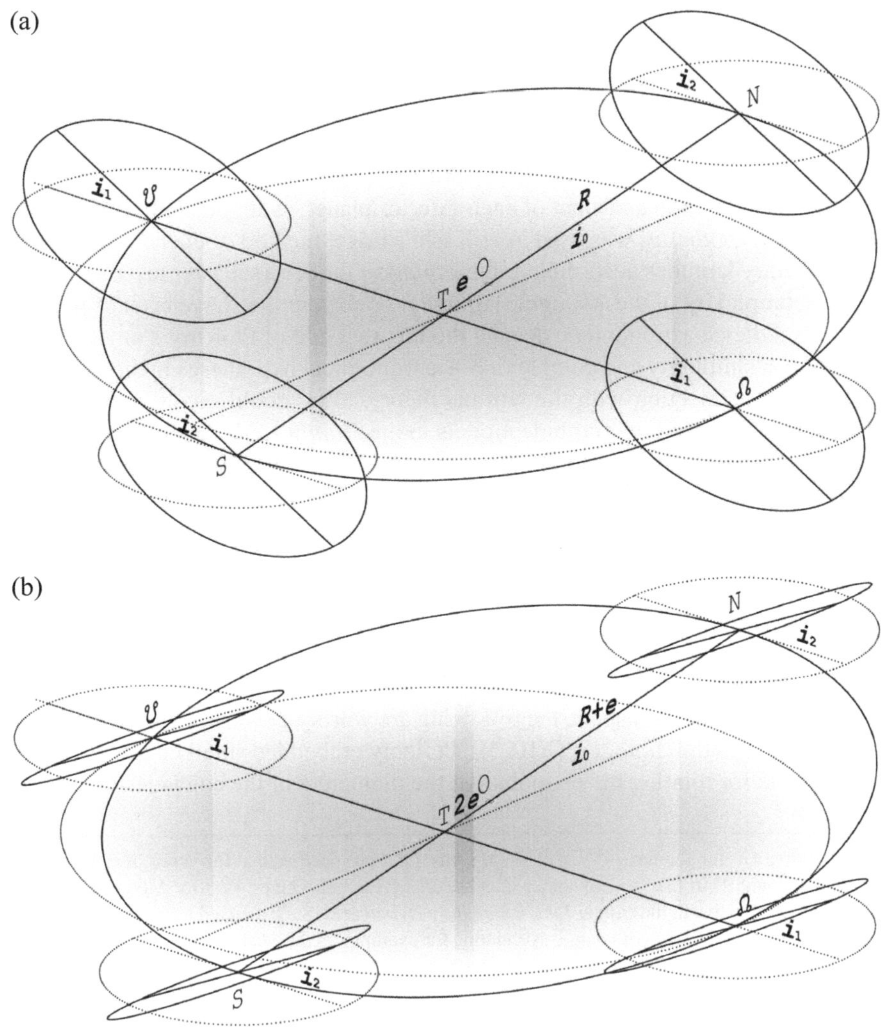

Figure 7.2 Ptolemaic latitude model for (a) Venus and (b) Mercury.

elements of the latitude models in the *Handy Tables* and *Planetary Hypotheses*. For instance, the extremal values for the latitude of all the planets in the *Handy Tables* have been applied to the *Mumtaḥan Zīj* (see note 19); the maximum values for the latitudes of the inferior planets from the *Handy Tables/Planetary Hypotheses* are also mentioned in al-Battānī's *Ṣābiʾ zīj* and al-Farghānī's *Book on astronomy* (although both authors do not identify their sources)[4]; the table for the latitude of

4 See Nallino [1899–1907] 1969, vol. 1, p. 116, vol. 3, p. 175; al-Farghānī XVIII, pp. 73–4, P: f. 67r: al-Farghānī gives the rounded values to the nearest $0;20° = {}^1\!/_3°$ for the maxima of the latitudes: ±3°

147

Venus from the *Handy Tables* is mentioned in Ibn Yūnus's *Ḥākimī zīj* (Cairo, d. 1009) (see following). Ibn al-Shāṭir of Damascus (1304–1375/6) discussed how to adapt the latitude theory in the *Almagest* and *Planetary Hypotheses* to his own planetary model, in which two additional small circles are incorporated into the epicycle–eccentric mechanism in order that the motions of heavenly objects be produced solely by uniform circular motions.[5] The equality of the maximum values of the inclination and slant of each inferior planet, as in the *Handy Tables* and *Planetary Hypotheses*, was used in al-Kāshī's novel method of the computation of the planetary latitudes according to the *Almagest* models (see later text).

The complexity of the *Almagest* latitude models seems to have been of no concern to medieval astronomers, despite the fact that one of Ptolemy's most notable remarks on simplicity and complexity in astronomical hypotheses in the *Almagest* is stated in connection with the latitude theory.[6] This could give the impression that, since the underlying latitude models in the *Handy Tables*/*Planetary Hypotheses* are simpler than those in the *Almagest*, they might be preferable. But instead, they followed the *Almagest* models with some attempt to remedy them in some aspects, as will be mentioned briefly later.

In the observational aspect, the Middle Eastern Islamic astronomers seem also to have possessed some persuasive observational evidence of problems with the planetary latitudes in some specific *zījes*; nevertheless, such criticisms had been directed to the adopted parameters, numerical tables, and so on, rather than to the models themselves.[7]

In the philosophical aspect, the following are worthy of consideration: in the latitude models in the *Almagest* (XIII.2),[8] Ptolemy embedded small circles that were responsible for rotating the endpoints of the diameters of the epicycles in order to

---

for Saturn, $\pm 2°$ for Jupiter, $+4^1/_3°$ and $-7°$ for Mars, but he specifies that for Venus, it is $6^1/_3°$ in the *Almagest* and 9° in the "sources other than the *Almagest*," and gives $4^1/_3°$ for Mercury, which is indeed equal to that in the *Handy Tables*/*Planetary Hypotheses*. Such a mixed tradition can be found in the Andalusian branch of Islamic astronomy, for example, in the *Muqtabis Zīj* of Ibn al-Kammād (twelfth century) (see Chabás and Goldstein 1994, p. 32).

5 See Roberts 1966.

6 *Almagest* XIII.2: Toomer 1998, p. 600. The two others are mentioned in III.1 and III.4 (Toomer 1998, pp. 136, 153). These passages have been rendered faithfully into Arabic (*cf. Arabic Almagest*, S: ff. 29r, 34r, 212r; PN: ff. 32v, 161v).

7 For example, ʿAlī b. Amājūr's criticisms of the planetary latitudes as computed from the *Mumtaḥan Zīj* (see note 19) or Wābkanwī's claim that the directions of the observed latitudes of the interior planets are opposite to what are derived from his contemporary *zījes* (Wābkanawī, T: f. 2v, Y: f. 3v, P: f. 3v; on Wābkanawī, see Chapter 3). Of the criticisms of other kinds, it is worth noting Ibn Yūnus's severe critical remarks on Ḥabash al-Ḥāsib's (d. after 869) knowledge of the *Almagest* latitude model for the inferior planets (see Ḥabash, ff. 63v–65v; Debarnot 1987, p. 54), as well as his berating al-Battānī's understanding of how the direction of latitude of Mercury should be determined in the *Almagest* model (Ibn Yūnus, L: p. 4; Caussin 1804, pp. 53, 55). Factually, in the rules set forth in Chapter 47 of the *Ṣābiʾ zīj*, the directions given for the slant component of latitude of the inferior planets are diametrically opposed to what should be taken into account according to *Almagest* XIII.6 (Toomer 1998, p. 635); corrected in Nallino [1899–1907] 1969, vol. 3, p. 175.

8 Toomer 1998, pp. 599–601.

account for their variable inclinations. Some spherical versions of the Ptolemaic models were proposed in Islamic astronomy in order to remove the small circles and give a philosophically justified *physical* model for the latitudinal tilting of the epicycles: Ibn al-Haytham (965–*ca.*1040) utilized two contacting concentric spheres with different poles whose distance is equal to the maximum inclination or slant of the epicycle and rotating in opposite directions with the same velocity as the center of the epicycle of the planet as seen from the Earth.[9] Naṣīr al-Dīn al-Ṭūsī (1201–1274) deployed three concentric interconnected spheres, the two of them producing the geometrical device known as the "Ṭūsī couple," to solve the difficulties arising from Ptolemy's small circles.[10]

In the mathematical aspect, it merits mentioning that a *geometrical* spherical version of the latitude models in the *Almagest* was proposed by Jamshīd Ghiyāth al-Dīn al-Kāshī (1380–1429); although apparently influenced by al-Ṭūsī, it was essentially intended to improve upon the *mathematics* of the models.[11] As mentioned earlier, it uses a distinct feature of the Ptolemaic latitude models in both the *Handy Tables* and *Planetary Hypotheses*, that the extremal values of the inclination and slant of each inferior planet are equal.[12] Al-Kāshī also presented two-argument tables for deriving longitude and latitude; dispensing with lengthy computational procedures, a practitioner only need enter them with the adjusted eccentric and epicyclic anomalies to derive the ecliptical coordinates.[13] Another example is this: in the *Almagest* models, the slant component of the latitude of Mercury from the tables should be corrected such that when the planet is in the apogeal half of the eccentric, one-tenth of the slant is subtracted, and inversely, when the planet is in the perigean part of the eccentric, the same amount is added. Muḥyī al-Dīn al-Maghribī rectified the procedure,[14] and Wābkanawī later commented upon al-Maghribī's method and improved it in the manner that the amount of the slant itself should be multiplied by the absolute value of the cosine of the planet's eccentric anomaly;[15] over one century later, Kāshī instead proposed a new interpolation function in order to calculate the slant component of the latitude of Mercury, by means of which there is no need to include this correction.[16]

---

9 See Mancha 1990, Ragep 1993, pp. 214–217, 2004.

10 See Ragep 1987, pp. 344–348, 1993, pp. 218–222, Saliba and Kennedy 1991.

11 See Brummelen 2006, pp. 357–358.

12 See Brummelen 2006, p. 360. Al-Kāshī uses the *Almagest* maximum value of the slant of each inferior planet for its maximum inclination as well, that is, $i_{1max} = i_{2max} = 3;30°$ for Venus, as is in the *Handy Tables* and *Planetary Hypotheses*, and $i_{1max} = i_{2max} = 7;0°$ for Mercury, while the *Handy Tables* and *Planetary Hypotheses* have 6;30°. Al-Kāshī states (IO: ff. 102v, 103v, 106r) that this is what the moderns have found through new observations, which seems to be a conclusion derived from examining the latitude tables in the *zījes* of the Marāgha tradition (see Section 7.2).

13 al-Kāshī, IO: ff. 142r–156r, P: pp. 136–152; in these two MSS, only the tables for the longitude of the Sun, Moon, Jupiter, and Venus, and those for the latitude of the inferior planets, are available.

14 al-Maghribī, *Adwār* II.5.3: M: ff. 18v–19r, CB: ff. 17v–18r.

15 Wābkanawī, T: f. 55r, Y: f. 100r, P: f. 83v.

16 al-Kāshī, IO: f. 104r; Qūshčī, pp. 340–341.

Despite all the aforementioned refinements, the most important alternatives to Ptolemy's planetary latitude models during the medieval period were Indian models.[17] They were adopted in some *zījes* of the early ninth century, such as al-Kh^{w}ārizmī's *Sindhind Zīj* (*ca.* 840)[18] and Yaḥyā b. Abī Manṣūr's *Mumtaḥan Zīj* (*ca.* 832),[19] and were later passed to the Western Islamic astronomical tradition (the Maghrib and Andalus) through al-Kh^{w}ārizmī's *zīj*, for example, in the *Toledan Tables*[20] and Ibn ʿAzzūz al-Qusanṭīnī's *Muwāfiq zīj* (d. 1354).[21] The only medieval astronomer who proposed latitude models different from Ptolemy's is apparently Levi ben Gerson (1288–1344), who supplied computational procedures and numerical tables as well.[22]

Among the equatoria invented in the late Islamic period for finding the ecliptical positions of the heavenly objects, al-Kāshī's instrument for computing the planetary latitudes may be worth mentioning.[23] However, because of the smallness of the inclinations, such a device could probably be in use only for demonstrational purposes.

7.2 The Values for the Inclinations and ω_A in Medieval Islamic Astronomy

In the late Islamic period, there was a great deal of systematic planetary observations to determine their underlying parameters, while in the early Islamic period, the investigation was of solar and lunar motions. The new values for planetary inclinations and longitudes of ascending nodes are found in the *zījes* written in

---

17 Although the Indian models are inferior to Ptolemy's, nevertheless, they have two distinct features: the eccentrics of the inferior planets coincide with the ecliptic, and the epicycles of the superior planets are parallel to the ecliptic (see Kennedy and Ukashah 1969).

18 See Neugebauer 1962, pp. 34–40, Kennedy and Ukashah 1969.

19 The *Mumtaḥan Zīj* was later revised by some astronomers working in Iraq and Damascus up to about the mid-ninth century, although the manuscripts preserved probably go back to a recension made in the tenth century (see van Dalen 2004a, esp. p. 11, Mozaffari 2016–2017). For the planetary latitudes, the following elements have been adopted in this *zīj*: the Indian values for the longitudes of the ascending nodes, Ptolemy's values for the extremal latitudes as tabulated in the *Handy Tables*, and a simple sinusoidal function, seemingly influenced by the Indian latitude models (see Viladrich 1988, esp. pp. 264–6, Kennedy 1990, pp. 173–177). As Ibn Yūnus (L: p. 100; Caussin 1804, pp. 111, 113) states, ʿAlī b. Amājūr found some errors in the latitudes of the planets and their directions with respect to the ecliptic as calculated from the *Mumtaḥan Zīj*. The earliest observation Ibn Yūnus reports from the Banū Amājūr family is the conjunction of Regulus with Venus made on September 10, 885, and the latest lunar eclipse on November 4/5, 933; see Ibn Yūnus, L: pp. 102, 109; Caussin 1804, pp. 123, 125, 157; concerning the solar and lunar eclipses observed by them, see Stephenson 1997, pp. 471–2, 479–482, Steele 2000, pp. 116–117.

20 Toomer 1968, pp. 8, 69–72, van Dalen 1999, pp. 323–324.

21 Samsó 1997, p. 92, 1999, pp. 16–17.

22 Its preliminary philosophical aspects are introduced in Goldstein 2002; see also Glasner 2003. The models and their parameters and technical aspects still await further research.

23 See Kennedy 1951, 1960, pp. 176–180, 198–214.

association with the well-known Islamic observatories established at Marāgha (*ca.* 1260–1320), Beijing (*ca.* the 1270s), and Samarqand (*ca.* 1430–1449).

7.2.1 Inclinations

In the Marāgha tradition, new values were observed for the inclinations of the epicycles of the inferior planets. These are mentioned nowhere and can only be derived from the latitude tables. Al-Maghribī explains his own observations and computations made at the Marāgha observatory for the purpose of determining the solar, lunar, and planetary parameters in a treatise titled *Talkhīṣ al-majisṭī* (*Compendium of the Almagest*). Nevertheless, since the last two books, IX and X, of this treatise are unfortunately missing from its only surviving manuscript, which is in the author's handwriting (Leiden: Universiteitsbibliotheek, no. Or. 110), we do not know whether and how he derived his values for the inclinations of the inferior planets and ω_A of the superior planets (see later) from his own observations. According to the list of its contents, Book IX is devoted to the planetary retrograde motions and latitudes, and Book X to the stereographic projection of the celestial sphere onto a plane surface tangent to the north celestial pole, which is used for the drawing of astrolabe plates.[24]

The maximum latitude of an inferior planet is due to the inclination of its epicycle and takes place when the planet is at inferior conjunction with the mean sun, that is, at the true epicyclic anomaly of 180°. We indicate the inclination component of the latitude as $\beta_{icl}(\alpha)$, where α is the true epicyclic anomaly; $\beta_{max} = \beta_{icl}(180°)$ in Table 7.1.

In the *Īlkhānī zīj*, the maximum tabular value for the latitude of Venus is 8;40°,[25] corresponding to $i_{1max} \approx 3;25°$. This is close to the extremal value 8;52° as derived from the *Handy Tables* and *Planetary Hypotheses*, in which $i_{1max} = 3;30°$.[26] Nevertheless, the table itself suffers from some difficulties, as will be explained later in Section 7.4, so that it is actually impossible to assign one single value to i_{1max} from which the table has been computed. It is noteworthy that Ibn Yūnus, in his *Ḥākimī zīj*, has an extra table for the latitude of Venus where $\beta_{icl}(0°) = 1;29°$ and

Table 7.1 New values for the inclination of the inferior planets in the Marāgha tradition

	Almagest		*Īlkhānī zīj*		al-Maghribī's *Adwār*	
	β_{max}	i_{1max}	β_{max}	i_{1max}	β_{max}	i_{1max}
Mercury	4; 4°	6;15°	[*Alm.*]	[*Alm.*]	4;35°	7; 2°
Venus	6;22	2;30	8;40°	3;25° (?)	6;40	2;37

24 al-Maghribī, *Talkhīṣ*, f. 2r; Mozaffari 2014, pp. 68–71.
25 al-Ṭūsī, C: p. 128, T: f. 76r, M: f. 77v, P: f. 44r. Also, see note 31.
26 Neugebauer 1975, vol. 2, pp. 1011–1016.

$\beta_{icl}(180°) = 8;52°.$[27] With Ibn Yūnus's non-Ptolemaic $e \approx 1;3$ for the eccentricity of Venus (corresponding to the maximum equation of center = $2;0,30°$)[28] and $r \approx 43;28$ for the radius of the epicycle (corresponding to the maximum epicyclic equation = $46;25°$ at mean distance),[29] it can be found that both latitude values result from the rounded $i_{1max} = 3;30°$. It seems that Ibn Yūnus obtained this table either directly or *via a medium* from the *Handy Tables*. Note that a mixed tradition of the *Almagest/Handy Tables* for the latitudes of the inferior planets already existed in earlier Islamic works (see note 4 earlier). The values $3;25°$ and $3;30°$ are in better agreement with the correct inclination of the planet than $2;30°$ in the *Almagest* and other Islamic *zījes*.

In al-Maghribī's *Adwār al-anwār*,[30] the maximum latitude of Venus is $6;40°$, which corresponds to $i_{1max} \approx 2;37°$. In his first *zīj*, the *Tāj al-azyāj*, written in Damascus in 1258 before his joining Marāgha, he has the value $8;30°$ for the maximum latitude of Venus,[31] which corresponds to $i_{1max} \approx 3;21°$. His value for the maximum latitude of Mercury in the *Adwār* is $4;35°$, corresponding to $i_{1max} \approx 7;2°$.[32] This is approximately equal to the Ptolemaic maximum value for the slant of Mercury, that is, $i_{2max} = 7;0°$. Since al-Maghribī's table of the slant is identical to the table in *Almagest* XIII.5,[33] it is straightforwardly deduced that al-Maghribī took the maximum inclination of Mercury equal to its maximum slant, as in the *Handy Tables/Planetary Hypotheses*, although in the *Handy Tables* and *Planetary Hypotheses*, Ptolemy takes in the inclination as $6;30°$, not $7;0°$. It should be noted that the values near $7;0°$ are in excellent agreement with the true value of the inclination of the planet.[34] These values are summarized in Table 7.1.

It appears to have been generally accepted after the Marāgha observatory that the value $3;30°$ for the inclination of Venus was a recent improvement upon the parameters of the *Almagest*. For instance, Ibn al-Shāṭir refers to it as "what the moderns have amended [in the *Almagest*]."[35] Also, one may conclude that in each of the two independent observational programs carried out at the Marāgha

27 Ibn Yūnus, O: ff. 101r–103v.

28 Ibn Yūnus maintained the Indian-originated idea adopted in early Islamic astronomy that Venus's eccentricity is equal to the Sun's one and their apsidal lines coincide (see Ibn Yūnus, L: p. 121; Caussin 1804, p. 221).

29 Ibn Yūnus, L: p. 121; Caussin 1804, p. 221. Also, see Section 7.4.

30 al-Maghribī, *Adwār*, CB: f. 87v, M: f. 89v; Wābkanawī, T: f. 163v.

31 Dorce 2003, p. 218. It is noteworthy that some close values $8;35°$ and $8;36°$ are found in the Western Islamic, Hebrew, and Spanish astronomical tables; for example, the *Muqtabis zīj* of Ibn al-Kammād (Chabás and Goldstein 1994, p. 32), the *Alfonsine Tables of Toledo*, and the canons to the tables of Judah ben Asher II of Burgos (d. 1391) (Chabás and Goldstein 2003, pp. 164–165). No relation between them and the *Tāj al-azyāj* or *Īlkhānī zīj* seems to exist.

32 A close value of $4;38°$ is found in the *Alfonsine Tables of Toledo* and in the canons to the tables of Judah ben Asher II (Chabás and Goldstein 2003, p. 164). No relation between it and al-Maghribī seems to exist, however.

33 Toomer 1998, p. 634.

34 Also, see Swerdlow 2005, pp. 63, 68.

35 Roberts 1966, p. 216; the addition in brackets is ours.

observatory, that is, by al-Maghribī and by the main staff of the observatory, who were engaged in preparing the *Īlkhānī zīj*, it was found that the maximum values of the inclination and slant of one of the inferior planets are equal, although neither maintained this for the other inferior planet. This can be considered a partial rediscovery of Ptolemy's latitude model in the *Handy Tables* and *Planetary Hypotheses*. Over a century after the Marāgha observatory, al-Kāshī appears to have achieved the same result, that in the new observations made by recent predecessors and modern scholars both the inclination and the slant of Venus and Mercury are, respectively, equal to 3;30° and 7;0°.[36] He was familiar with both the *zījes* of the Marāgha tradition so that he was involved in the task of revising the *Īlkhānī zīj* and hence named his own work *Zīj-i Khāqānī dar takmīl-i Zīj-i Īlkhānī* (*Khāqānī zīj*; *The improvement of the Īlkhānī zīj*), and in his discussion of the latitude theory, he explicitly refers to al-Maghribī and his *Zīj al-kabīr* (*Great zīj*),[37] an alternative name for al-Maghribī's *Adwār al-anwār*.[38]

The tables of the latitude of Venus in Ibn al-Shāṭir's *Jadīd zīj*,[39] Ulugh Beg's *Sulṭānī zīj*,[40] and al-Kāshī's *Khāqānī zīj*[41] are identical to the corresponding table in the *Īlkhānī zīj*, as are the tables of the latitude of Mercury aside from the correction for distance in computing the slant.[42] This illustrates the influence that the formal work of the Marāgha astronomy exerted on later Middle Eastern astronomers, and that the inclinations of the inferior planets from the Marāgha observatory appeared trustworthy to them. Perhaps for this reason, the astronomers at the Samarqand observatory turned to measuring the inclinations of the superior planets. The values derived for the inclinations of the superior planets from the critical entries in the northern and southern latitude tables in Ulugh Beg's *Sulṭānī zīj*[43] are as listed in Table 7.2. The critical entries are the latitudes for the true anomaly of

36 al-Kāshī, IO: ff. 102v, 103v, 106r.

37 al-Kāshī, IO: ff. 104r.

38 This title can be found in other sources as well; for example, al-Kamālī, ff. 230v and 231r.

39 Ibn al-Shāṭir, O: f. 56r, K: f. 71r; the table is *Almagest*-type, namely, the entries are for each 6° of the argument in the range from 0° to 90°/270° to 360°, and for each 3° of the argument in the interval from 90° to 270°. This table can also be found in Muḥammad al-Ṭabīb al-Muhtadī al-Mūṣilī's commentary on Aḥmad b. Ghulām Allāh's *al-Lum'a fī ḥall al-kawākib al-sab'a* (f. 58r), which is based upon Ibn al-Shāṭir's *Jadīd zīj*. (Muḥammad al-Ṭabīb appended this commentary to another treatise of his own on the sine quadrant, *Risāla fī al-rub' al-mujayyab*, and then called the two altogether as *al-Jam' al-mufīd*.)

40 Ulugh Beg, P1: f. 146v, P2: f. 163v.

41 al-Kāshī, IO: ff. 139v–140r, P: p. 124.

42 *Īlkhānī zīj*, C: p. 137, T: f. 83v, M: f. 83r, P: f. 47v; al-Kāshī, IO: ff. 140v–141r, P: pp. 125–126; Ulugh Beg, P1: f. 149v, P2: ff. 166v–167r. In the *Īlkhānī zīj*, the values of the slant of Mercury are tabulated as $^{11}/_{10}\beta_{sl}$ and $^{9}/_{10}\beta_{sl}$ in order to eliminate the additional step for correcting this latitude by adding/subtracting one-tenth of the value obtained from the table in the *Almagest*. In Ulugh Beg's and al-Kāshī's *zījes*, only the entries related to $^{11}/_{10}\beta_{sl}$ can be found, which is because of the new interpolation function proposed by al-Kāshī.

43 Ulugh Beg, P1: ff. 137v, 140v, 143v; P2: ff. 153v, 156v, 160r. The latitudes of the superior planets in al-Kāshī's *Khāqānī zīj*, as well as the values he mentions for their inclinations, are Ptolemaic (IO: ff. 100v, 139v, P: p. 131).

Table 7.2 New values obtained for the inclinations of the superior planets at the Samarqand observatory

	$\beta_N(90°)$	$\beta_S(90°)$	i_0	$\beta_N(180°)$	$\beta_S(180°)$	i_{1max}
Saturn	+2;29°	−2;29°	2;30° [*Alm.*]	+3;14°	−3;18°	6; 0°
Jupiter	+1;28	−1;29	1;30 [*Alm.*]	+2; 8	−2;16	2;45
Mars	+1; 9	−1; 4	1;20	+4;32	−7;15	2; 7

90° and of 180°; the first is used to determine the inclination i_0 of the eccentric, and the latter to derive the maximum inclination i_{1max} of the epicycle. The subscripts N and S denote, respectively, the northern and southern latitudes. Note that our derivation of the inclinations from the extremal latitudes, as explained in Section 7.4, is based on the new values measured for the eccentricities and the radii of the epicycles of the superior planets at the Samarqand observatory, which are summarized in Tables 7.8 and 7.9. This explains why the values of i_0 for Saturn and Jupiter are Ptolemy's, while the corresponding latitudes are non-Ptolemaic.

7.2.2 The Values of ω_A

In Ptolemaic astronomy, the apsidal and nodal lines of the planets are sidereally fixed. Indian models have completely different features: the apsidal lines of the planets move in the direction of increasing longitude, but their nodal lines in the retrograde direction, like that of the Moon, at unequal rates.[44] The majority of (especially early) Islamic *zījes* have Ptolemy's values for ω_A.[45] The non-Ptolemaic values for ω_A from the Marāgha and Samarqand traditions are listed in Table 7.3 (the true modern values at that time are indicated in the columns headed "Mod."). These values are either explicitly mentioned in the canons to the tables or can be extracted from the tables or columns for "the minutes of proportion." The minutes of proportion c_5 are a simple cosine function of the argument of latitude $\omega_p = \kappa + \omega_A$, where κ is the true eccentric anomaly of the planet. In the *Almagest*, the user has to add the values of ω_A to κ and then enter the table with the result. This is also the case with both al-Maghribī's and Wābkanawī's *zījes*,

---

44 For example, *Súrya Siddhánta* I.43–44: pp. 29–30.
45 For example, the *Zīj* of Ḥabash (f. 66r); al-Battānī's *Ṣābiʾ zīj* (Nallino, [1899–1907] 1969, vol. 2, pp. 140–141); Bīrūnī's *al-Qānūn al-masʿūdī* X.10 (vol. 3, p. 1323); Ibn Yūnus, O: ff. 78r–79r; al-Kāshī, IO: f. 101v (see Kennedy 1951, p. 19). The Ptolemaic values were also prevalent in the Latin sources, insofar as the materials reflected in the secondary literature indicate; for example, Copernicus's *De Revolutionibus* (Swerdlow and Neugebauer 1984, p. 498), Bianchini (mid-fifteenth century; see Goldstein and Chabás 2004, p. 460, Chabás and Goldstein 2009, p. 95). The values of ω_A are not properly defined in the *Alfonsine Tables of Toledo* (Chabás and Goldstein 2003, p. 163).

but in the *Sulṭānī zīj*, the entries are shifted by $-\omega_A$ with respect to the corresponding entries in the *Almagest* tables. As a result, the procedure is shortened by one step.

Table 7.3 Difference ω_A between the longitudes of the apogees and the northern limits of the superior planets in the *zījes* of the Marāgha and Samarqand observatories

	Ptolemy [Alm.][1]	Mod.	Al-Maghribī[2]	Wābkanawī[3]	Mod.	Ulugh Beg[4]	Mod.
Saturn	+53°/+50°	+49°	+60°	+60°	+61°	+60°	+63°
Jupiter	−19/ −20	−11	−10	−10	− 4	− 8	− 3
Mars	− 4.5/ 0	− 8	+ 6	−12	+ 4	+ 4	+ 6

Notes
1. *Almagest* XIII.1 (Toomer 1998, p. 598). For Saturn: +27° in the *Canobic Inscription* and +40° in the *Handy Tables* and *Planetary Hypotheses*; cf. Neugebauer 1975, vol. 2, pp. 910, 916, 1010; Goldstein 1967, pp. 20–25; Jones 2005, pp. 74–75, 90.
2. Al-Maghribī, *Adwār al-anwār* II.5.2: CB: f. 17v, M: f. 18v. Tables of the latitudes: CB: f. 87v, M: f. 89v (also, Wābkanawī, T: f. 163v). In his commentary on the *Īlkhānī zīj*, al-Nīshābūrī (P1: f. 117v, P2: ff. 141r–v) gives ω_A of each superior planet, according to "the new observations, as Muḥyī al-Dīn has mentioned in his *zīj*," as Saturn 150°, Jupiter 80°, and Mars 96°.
3. Wābkanawī, *Zīj* III.5.3: T: f. 54r, P: f. 82v, Y: ff. 98v–99r.
4. Ulugh Beg, *Sulṭānī zīj*, tables of the latitudes: P1: ff. 137v, 140v, 143v, P2: ff. 153v, 156v, 160r. The tables of the minutes of proportion give $\cos(\kappa + 60°)$ for Saturn, $\cos(\kappa - 8°)$ for Jupiter, and $\cos(\kappa + 4°)$ for Mars. No further corrections are mentioned in the related explanatory text in III.4 (P1: ff. 107v–108r, P2: ff. 119r–v). Consequently, the values of ω_A for Saturn, Jupiter, and Mars should be, respectively, +60°, −8°, and +4°.[46] These are in agreement with the values 150°, 82°, and 94° which ʿAlī b. Muḥamamd Qūshčī (*ca.* 1402–74), one of the Samarqand astronomers contributing to the preparation of the *Sulṭānī zīj*,[47] gives in his *Commentary on Zīj of Ulugh Beg* (pp. 332–333) for ω_A of these planets in the *Sulṭānī zīj*.

7.3 Discussion and Conclusion

Unlike the detailed examples in the *Almagest* for determining the parameters of the planets' motions in longitude, Ptolemy did not mention specific observations for planetary latitudes, only what appear to be approximate extreme values, which gave rise to some medieval objections.[48] Consequently, the medi-

46 It is also worthwhile that also in al-Ṭūsī's *Īlkhānī zīj*, the entries in the tables of the minutes of proportion for Saturn and Jupiter are displaced by the Ptolemaic values for $-\omega_A$ (C: pp. 101, 110, 119, P: ff. 35r–v, 37v, 39v, T: ff. 54v, 62r, M: ff. 62r, 66v, 72r). But we are told in the canons of this *zīj* (II.3: C: p. 42, P: f. 15r, T: f. 21v, M: f. 27r) to add the values 7° and 12°, respectively, to the adjusted eccentric anomaly of Saturn and Jupiter prior to entering to the tables. These are *only* due to the asymmetrical tables of the equations of the epicyclic anomaly of the two planets in this *zīj*, the entries of which are displaced and always additive by adding, respectively, 7° and 12° to those in the corresponding *Almagest* table, and thus, these amounts should have been subtracted from the mean eccentric anomalies before tabulating. Consequently, after *adjusting* the mean eccentric anomaly by the equation of center, the result is still smaller than the *true* eccentric anomaly, respectively, by 7° and 12°.
47 His name is explicitly mentioned in the prologue of the *Sulṭānī zīj* (P1: f. 1v, P2: f. 1r).
48 For example, Goldstein 2002, p. 27.

eval astronomers had no access to specialized methods and techniques required for making observations to determine the latitude parameters.[49] The required observations for the derivation of the inclinations and the values of ω_A should essentially be made at some specific conditions which are very rarely possible to satisfy: in order to measure the inclinations of the eccentric and the epicycle of a superior planet, it should be observed when it is *both* close to a limit and at opposition to (or as near as possible to conjunction with) the mean sun. This can take place once in about 30 years for Saturn, 12 years for Jupiter, and 15 or 17 years for Mars. The longitude of the ascending node can be determined when the planet crosses the ecliptic, in this case, when the center of the epicycle is at a node, whose longitude is thus equal to the mean longitude of the planet.[50] As mentioned briefly in Section 7.1, early Islamic astronomers employed Indian methods or the ungrounded (if not meaningless) mixed versions of the Hindu and Ptolemaic latitude theories, as well as having difficulties with understanding the latitude models in the *Almagest*. Among the surviving early Islamic *zījes*, presumably al-Battānī's is one of the first works in which the *Almagest* latitude theory appeared in an understandable and correct abridged form (though having a minor fault; see earlier, note 7), that is, at the turn of the tenth century. These perhaps explain the reason it took a relatively long period until the first attempts were made to measure the inclinations of the planets (at the Marāgha observatory in the mid-thirteenth century). Also, for the same reason, the derivation of the latitude parameters from the observational data can be considered the acme of medieval observational astronomy.

Concerning the Eastern branch of Islamic astronomy, a partial improvement upon Ptolemy's values for the inclinations of the inferior planets in the *Almagest* was made at the Marāgha observatory. This can, however, be considered a rediscovery of Ptolemy's latitude models of the inferior planets both in the *Handy Tables* and *Planetary Hypotheses*, of which the Marāgha astronomers were quite probably aware. Of course, al-Maghribī's value of $7;2°$ for the inclination of Mercury is in excellent agreement with modern data and thus clearly better than Ptolemy's values in the *Almagest* ($6;15°$) and in the *Handy Tables*/*Planetary Hypotheses* ($6;30°$). As the contents of al-Maghribī's *Talkhīṣ al-majisṭī* reveal, there is no doubt that he had made the actual observations at the Marāgha observatory, as well as having the technical and mathematical skills to derive the parameters from the observational data. In the beginning of 1262, the observations were begun in the Marāgha observatory (al-Maghribī's first documented observation in the *Talkhīṣ* is the lunar eclipse on March 7, 1262), and in April–May 1276,

---

49 We can address some alternative strategies developed in Islamic astronomy to facilitate obtaining specific parameters, to remove anticipated difficulties, or to secure desired results. But these alternative methods themselves are based upon already-existing ones, mostly proposed by Ptolemy himself (e.g., the three alternative methods for measuring the solar eccentricity; see Chapters 1 and 6).
50 For the difficulties with the latitude observations, see Swerdlow 2005, pp. 47, 49.

he completed the canons of his last *zīj*, *Adwār al-anwār*, and then engaged in preparing its tables.[51]

Ulugh Beg's planetary parameters (Table 7.4; also, see Section 7.4, Tables 7.8–7.9) can explicitly indicate systematic observations at Samarqand, although nothing about them appears to have been documented. Unlike the improvement made at the Marāgha observatory in the case of the inferior planets, the values the Samarqand observers determined for the inclinations of the superior planets are either equal to or larger than the corresponding values in the *Almagest*, while they should be very close (if not essentially equal) to each other. Accordingly, it may be that the attempts made in the Samarqand observatory resulted in a deterioration in the *Almagest*'s underlying values for the inclinations of the superior planets, while these had already been improved significantly in Ptolemy's other two works. We are not told how and when the Samarqand observers reached their own values for the extremal latitudes of the superior planets; in his *Commentary on Zīj of Ulugh Beg*, Qūshčī only mentions the maximum and minimum latitudes at the limits, expressing "we found" them, but without giving any more information.[52] The true values of the maximum latitudes of the three planets at oppositions, the zodiacal signs where they occurred in the period of 1430–1449 when the Samarqand observatory was active, and their differences from the extremal values in the *Sulṭānī zīj* (Table 7.2) are given in what follows. It is noteworthy that these errors are comparable with those found in the stellar latitudes in Ulugh Beg's star catalogue incorporated into his *zīj*.[53]

Mars	−6;38°	Aquarius	−37′
	+4;30	Cancer	+ 2
Jupiter	−1;40	Pisces	−36
	+1;36	Virgo	+32
Saturn	−2;49	Pisces	−29
	+2; 7	Leo	+67

On the values of ω_A, it should be noted that observing no new value in the case of the inferior planets cannot be so strange, because, unlike the superior planets, the directions of the nodal lines of the eccentrics of the inferior planets in Ptolemaic astronomy are an intrinsic convention of the model, that they are ±90° from the apsidal lines. It is not imaginable that an observer could recognize that the two are not perpendicular to each other when the center of the epicycle is at either of the apsides but make angles as small as about 1° and 16° to the perpendicular. Especially for Mercury, it should be considered that Ptolemy's longitude of the apogee of the planet, 191° in 135 CE, is over −30° from the true value of about

---

51 Mozaffari 2014, p. 71.

52 Qūshčī, p. 329. When mentioning the extremal latitudes of the inferior planets, his statement changes to "they determined" (pp. 336, 340).

53 See Shevchenko 1990, Krisciunas 1994, esp. p. 273, Verbunt and van Gent 2012, esp. pp. 7–8.

223°. The same error can also be found in the majority of the Islamic *zījes*, with the exception of those based upon Indian parameters (e.g., al-Kh^{w}ārizmī's).[54]

The two interrelated problems in the values of ω_A for the superior planets in Table 7.3 are as follows:

1. These might initially be interpreted as indicating that the medieval astronomers were aware of the change in the relative directions of the apsidal and nodal lines. Then, an inevitable consequence of the continuous increase in ω_A with respect to Ptolemy's values would be that *the apsidal and nodal lines of the superior planets do not share the same motions*, in contrast to the fundamental assumption in Ptolemaic astronomy that they are sidereally fixed. Nevertheless, such a conclusion was never posed in the Eastern (Middle Eastern) Islamic astronomy, where permitting any secular change in the Ptolemaic constants or fundamental assumptions and hypotheses was considered very unlikely. Accordingly, any new value, variation, or change found in a parameter was undoubtedly attributed to faulty observations, defects, and discrepancies in instruments and/or untrustworthy ancient data. However, a diametrically opposed point of view was adopted in the Western Islamic astronomy, where it was allowed to consider proper motions and/or models to account for the observationally found secular changes.[55] For instance, about 1075, Ibn al-Zarqālluh found that the solar apogee had the proper motion of 1° in 279 Julian years in the direction of increasing longitude. Several Andalusian astronomers applied this proper motion to the apogees of the planets as well.[56] Nevertheless, whether this proper motion was thought to be the case for the nodal lines of the planets, or whether other such proper motions were proposed for them, must await a thorough inspection of the materials related to the planetary latitudes in the Western Islamic *zījes*.

2. How were they obtained? These medieval values perhaps were the result of actual observations. Needless to say, the values of ω_A are dependent upon the values adopted for the longitudes of the apogees. By means of the two, one can also examine to what extent the values obtained for the longitudes of the nodes/limits were accurate.

The longitudes of the apogees of the superior planets in the Marāgha and Samarqand astronomical traditions are listed in Table 7.4.[57] The values in brackets are the true values computed for the times. We round both the historical and modern values to the nearest 0.5°, which is sufficient for our purpose.

---

54 This surprising subject will be discussed by the present author elsewhere.

55 See Chapter 8.

56 See Samsó and Millás 1998, p. 269.

57 Note that Wābkanawī converted al-Maghribī's values to the epoch of his *zīj*, March 13, 1266, by adopting the precessional rate of 1° in 66 Persian years (of 365 days unvarying).

Table 7.4 Longitude of the apogee of the superior planets in the *zījes* of the Marāgha and Samarqand traditions

	Ptolemy [1]	*al-Maghribī* [2]	*Ulugh Beg* [3]
Saturn	233.0 + 1° [236.5]	258.5° [258.5]	250.0 + 7° [261.5]
Jupiter	161.0 + 1° [160.5]	178.0° [179.0]	167.5 + 12° [181.5]
Mars	115.5 + 1° [116.5]	137.0° [137.5]	142.0° [140.5]

Notes:

1. For July 8, 136; October 7/8, 137; and May 27/28, 139, respectively, for the three planets.
2. For February 27, 1273; August 12, 1274; and February 26, 1271, respectively, for the three planets. These are the dates when Ptolemy (Toomer 1998, pp. 484–498, 507–519, 525–537) and al-Maghribī (*Talkhīṣ*, ff. 123r–123v, 128r–129r, 132v–133r) observed the last of their own trios of the oppositions of a superior planet to the mean sun required for determining its eccentricity and the direction of its apsidal line in the *Almagest*. The addition of 1° to Ptolemy's values is due to his well-known error in determining the moment of the vernal equinox by more than 1 day, making a displacement in the position he determined for the vernal equinox by a little over 1°. Consequently, it is required to take this error into account by adding 1° to Ptolemy's values in order to fix his longitudes in a correct tropical frame of reference.
3. The date considered for the *Sulṭānī zīj* is the epoch of this *zīj*, July 4, 1437. The additional values are due to the fact that the longitudes of the apogees of Saturn and Jupiter have decreased by the values k_2 in the tables of the equations of anomaly and then tabulated (see Section 7.4).

As seen in Table 7.4, al-Maghribī has more precise values for the longitudes of the apogees of the superior planets than those applied to the *Sulṭānī zīj*. As regards Tables 7.3 and 7.4, one can deduce that the values obtained by al-Maghribī for the longitudes of the nodes/limits of the three planets are about +1°, +5°, −2.5° in error, respectively. Wābkanawī's value for ω_A of Mars increases the error in the longitude of the nodes/limits of the planet to +15.5°. The corresponding errors in the *Sulṭānī zīj* are −1.5°, +3°, and +3.5°.

As we shall see later, the possibility exists that al-Maghribī derived his values of ω_A from actual observations as accurate as the others documented in his treatise *Talkhīṣ al-majisṭī*. Following him, Wābkanawī mentioned in his *zīj* some observations performed during his 40-year career, which were directed to reconciling theory and observation.[58] Let us first consider whether and how the astronomers at Marāgha and Samarqand were able to obtain their own values for ω_A from observations in the periods of the activities in the two observatories.

In the period of al-Maghribī's observations at Marāgha (1262–1276), he had one opportunity in the case of Jupiter and Saturn to find that Ptolemy's value of ω_A for each planet is invalid at that time and another to determine a correct value. In this period, the mean longitude of Saturn changed from about 21° to about 197°, the longitude of its perihelion was about 79°, and its ascending node was at the longitude of about 107°. From the three observations of Saturn made on October 25, 1263; December 9, 1266; and February 27, 1273, when the planet crossed

---

58 See Chapter 3; see also his remark concerning the observed latitudes of the inferior planets in note 7 earlier.

the meridian of Marāgha, by means of Ptolemy's iterative algorithmic procedure, al-Maghribī obtained the longitude of the eccentric apogee of the planet as 258.5° (Table 7.4), which, of course, could be taken for the whole period of his observations at Marāgha. Then, with Ptolemy's value $\omega_A = +50°$, the latitude of the planet should have been zero when its mean longitude was about $\lambda_\Omega = \lambda_A - \omega_A - 90° = 258.5° - 50° - 90° = 118.5°$; this took place, according to modern theories, about mid-December 1269, and according to al-Maghribī's *Adwār*, about mid-November 1269. About this time, the planet culminated at Marāgha during nights and its latitude was about $+0;43°$,[59] which is in the range of the accuracy of the instruments of the Marāgha observatory.[60] It is thus probable that if al-Maghribī had made systematic observations in this time interval, he could have achieved the result that the planet had a sizable latitude while, adopting Ptolemy's $\omega_A = +50°$, its epicycle center should have been located very near the ascending node; thus, its latitude should have been about zero.

About mid-May to mid-December 1268, for some period before the last visibility of Saturn on June 11 and after its first visibility on July 18, the longitude of the planet changed from about 99° to 113°, and its latitude from about $-0;10°$ to $+0;10°$ and reached zero on September 14. From observations of the planet near the horizon as accurate as those recorded in his *Talkhīṣ al-majisṭī*, al-Maghribī might then conclude that the latitude of the planet had reached zero somewhere in this period. His tables give about 100° to 107° for the mean longitude of Saturn in this time interval.[61] This could lead to the result that ω_A for Saturn should be about 61.5°–68.5°. Therefore, it is probable that al-Maghribī's denial of Ptolemy's $\omega_A = +50°$ and his own rounded $+60°$ were found by observation.

In the similar situation for Jupiter, with the Ptolemaic $\omega_A = -20°$, the planet should have crossed the ecliptic when its mean longitude was equal to $\lambda_\Omega = \lambda_A - \omega_A - 90° = 178° + 20° - 90° = 108°$. This took place according to both the modern ephemeris and al-Maghribī's *Adwār* at about April 9–21, 1267, while the latitude of the planet was about $+0;26°$. Unlike Saturn, the culmination of Jupiter in this interval occurred during daylight, but an observer had time enough to observe the

59 al-Maghribī's *Adwār* gives a value ~$+0;40°$.

60 As we will see in Chapter 9, al-Maghribī's mean errors in the measurements of meridian altitudes by the central quadrant of the Marāgha observatory are (regardless of sign) about 0;3° for the Sun, 0;6° for the eight bright stars, and 0;5° for the planets. The mean errors in the latitudes of the stars and planets he computed from the observed meridian altitudes by the trigonometrical rules of spherical astronomy are, respectively, about 0;6° and 0;10°. In particular for Saturn, in the three aforementioned observations, the errors in the meridian altitude of the planet are, respectively, $-0;9°$, $-0;7°$, and $+0;6°$, which resulted in the errors of $-0;15°$, $-0;8°$, and $+0;4°$ in the latitude of the planet as calculated by al-Maghribī. The mean error in the latitudes of the 16 stars recorded in the non-Ptolemaic star table of the *Īlkhānī Zīj* (perhaps measured by the armillary sphere of the observatory) is about 0;15° (see also Chapter 10).

61 al-Maghribī's *Adwār* gives the planet's longitude as ~99° to ~113°, and its latitude as $-0;13°$ to $+0;6°$.

planet after sunset and find that it had a slight latitude, while according to Ptolemy its latitude should be about zero.

In the period of the activities at the Samarqand observatory (1430–1449), the longitude of Saturn changed from about 277° to 174°, its ascending node was at about 109°, and its perihelion/perigee was at about 82°. The Samarqand observers could thus observe the planet when crossing its nodes, southern limit, and perigee, and this 20-year period was sufficient for Jupiter and Mars to be observed at all the critical points.

To sum up, this chapter has established that new values for the latitude parameters, inclinations, and directions of the lines of the limits/nodes were observed at the Marāgha and Samarqand observatories in late medieval Islamic astronomy, and has analyzed their accuracy. However, in the absence of further documented information, it has not settled the question of how these values were determined, although some ways for deriving them from observations were proposed.

7.4 Planetary Inclinations and Latitude Tables

In order to derive the inclinations from the planetary extremal latitudes and parameter values adopted in the *zījes* of the Marāgha and Samarqand traditions, we make use of both modern and Ptolemy's methods. The first consists in solving trigonometrical equations resulting from the configuration of the ecliptic, eccentric, and epicycle at the northern or southern limit (in the case of the superior planets) and at the nodes (for the inferior planets). The latter is to extract the inclinations from the maximum and minimum latitudes by interpolation in the epicyclic equation tables, an approximate method explained in *Almagest* XIII.3.[62]

7.4.1 Marāgha: Inferior Planets

Figure 7.3 shows the configuration of the eccentric and epicycle of Venus when the center of the epicycle is at either of the nodes of the eccentric, that is, ±90° from the apsidal line; in this position, the inclination of the epicycle is in the line of sight. The radius of the eccentric $OC = R = 60$, the eccentricity $TO = e$, the distance from the Earth to the center of the epicycle $TC = (R^2 - e^2)^{1/2}$, and the radius of the epicycle $CP = r$. We let the true epicyclic anomaly $\alpha = \angle ACP$ so that $PP' = QQ' = r \sin \alpha$, $PQ = P'Q' = r \cos \alpha \sin i_{1max}$, and $CQ' = r \cos \alpha \cos i_{1max}$. For Mercury, the direction of the inclination of the epicycle is reversed, as the epicyclic apogee A is inclined to the north; also, $TO = 2e$. The latitude $\beta = \angle PTQ$ is found from the following formula (negative sign for Venus and positive sign for Mercury):

$$\sin \beta_{icl}(\alpha) = \pm \frac{PQ}{\sqrt{(TC + CQ')^2 + QQ'^2 + PQ^2}} \tag{7.1}$$

---

62 Toomer 1998, pp. 601–605. Ptolemy's method is adequately explained in Pedersen [1974] 2010, pp. 361–365, Neugebauer 1975, vol. 1, pp. 209–216, Swerdlow 2005, pp. 43–46.

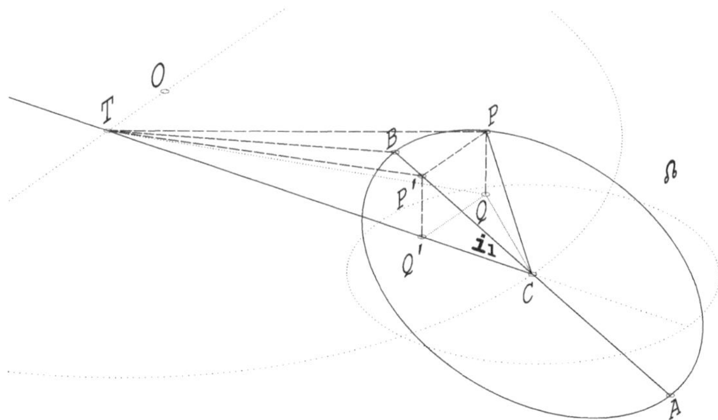

Figure 7.3 Configuration of the epicycle and the eccentric of Venus when the center of the epicycle is at the ascending node of the eccentric, ±90° from the apsidal line.

If $\alpha = 180°$, for which the latitude of the inferior planets reaches its maximum value, all that is necessary is to solve triangle TBC, in which $\angle BTC = \beta_{max}$:

$$\frac{TC}{\sin(\angle TBC)} = \frac{CB}{\sin(\angle CTB)} \quad \text{or} \quad \frac{TC}{\sin(\beta_{max} + i_{1max})} = \frac{r}{\sin(\beta_{max})}. \tag{7.2}$$

Thus, $i_{1max} = (\beta_{max} + i_{1max}) - \beta_{max}$. Therefore, i_{1max} can be conveniently found from TC, r, and β_{max}.

The underlying planetary parameters in the *zījes* of the Marāgha tradition are as follows: the eccentricity of Venus in the *Īlkhānī zīj* is $e \approx 1;2$, where $R = 60$, computed from the maximum tabular value $q_{max} = 1;59°$ for the equation of center of the planet,[63] in al-Maghribī's *Adwār al-anwār* $e \approx 1;3$, from $q_{max} = 2;0°$.[64] The value adopted for the eccentricity of Mercury, $e = 3;0$, and the radii of the epicycles of both the inferior planets in these two *zījes* are Ptolemaic: Venus, 43;10; Mercury, 22;30.

Due to Ptolemy's complicated model for Mercury's motion in longitude of a movable eccentric, in Figure 7.3, $TC \neq (R^2 - e^2)^{1/2}$, but with Ptolemy's value $e = 3;0$, $TC \approx 56;40$.[65] Then, with al-Maghribī's value $\beta_{max} = \beta_{icl}(180°) = 4;35°$ for the maximum latitude of Mercury, one can compute $i_{1max} \approx 7;2°$ from equation (7.2).

Our historical method, that is, Ptolemy's solution, of which al-Maghribī quite probably made use, is an analogy between the extremal latitudes and inclination and the epicyclic equations and true epicyclic anomalies, respectively, in the sense

63 al-Ṭūsī, C: pp. 125–6, T: ff. 73v–75r, P: ff. 42v–43r, M: ff. 76r–77r.
64 The equation tables of the inferior planets in al-Maghribī's *Adwār*, M: ff. 87v–89r, CB: ff. 85v–87r; also preserved in Wābkanawī, T: ff. 160v–163r and Kamālī, ff. 248v–251r.
65 Neugebauer 1975, vol. 1, p. 221, Toomer 1998, p. 609.

Table 7.5 Al-Maghribī's table of the inclination of Mercury $i_{1\max} = 7;2°$

	Tab.	*Re-c.*		*Tab.*	*Re-c.*
6°	2; 0°	1;59°	**96°**	0;17°	0;17°
12°	1;58°	1;58°	**102°**	0;35°	0;35°
18°	1;56°	1;55°	**108°**	0;54°	0;54°
24°	1;53°	1;51°	**114°**	1;14°	1;14°
30°	1;49°	1;47°	**120°**	1;35°	1;36°
36°	1;42°	1;41°	**126°**	1;57°	1;58°
42°	1;34°	1;34°	**132°**	2;21°	2;21°
48°	1;26°	1;26°	**138°**	2;44°	2;44°
54°	1;17°	1;17°	**144°**	3; 7°	3; 8°
60°	1; 7°	1; 7°	**150°**	3;29°	3;30°
66°	0;55°	0;56°	**156°**	3;50°	3;51°
72°	0;43°	0;44°	**162°**	4; 8°	4; 9°
78°	0;29°	0;30°	**168°**	4;20°	4;23°
84°	0;15°	0;16°	**174°**	4;30°	4;32°
90°	0°	0°	**180°**	4;35°	4;35°

Note: 144: MS. M: 3;6.

that the extremal latitudes $\beta_{icl}(0)$ (= $\angle ATC$ in Figure 7.3; line AT not drawn) and $\beta_{icl}(180°)$ (= $\angle BTC$) can be taken as the epicyclic equations in the correction tables for longitude corresponding, respectively, to the true epicyclic anomalies corresponding to the anomalies $\pm i_{1\max}$ and $180° \pm i_{1\max}$ (in Figure 7.3, assume that the epicycle is perpendicular to the ecliptic). When the center of the epicycle of an inferior planet is at either of the nodes, that is, $\pm 90°$ from the apsidal line, its true eccentric anomaly, $\angle OTC$ in Figure 7.3, is $\kappa = \pm 90°$, which corresponds to the mean eccentric anomaly $\bar{\kappa} \approx \pm 93°$. At this position, from al-Maghribī's table of the equations of Mercury, the following values for the epicyclic equation p can be derived:[66]

$$5;12° \text{ for } \alpha = 180° - i_{1\max} = 172°$$

and

$$4;34° \text{ for } \alpha = 180° - i_{1\max} = 173°$$

Al-Maghribī's $\beta_{\max} = 4;35°$ falls between these two values. So with taking it as the epicyclic equation, the linear interpolation between $p = 5;12°$ and $p = 4;34°$ results

66 The components of the epicyclic equation in the *Almagest* (about them, see Neugebauer 1975, vol. 1, pp. 183–186) with the tabular values in al-Maghribī's *Adwār* are given in the following:

$\bar{\kappa}$	$c_8(\bar{\kappa})$	$\alpha = 180° - i_{1\max}$	$c_6(\alpha)$	$c_7(\alpha)$	$p = c_6 + c_8 \cdot c_7$
93°	0;43°	172°	4;45°	0;38°	5;12°
		173	4;10	0;33	4;34

in $\alpha = 180° - i_{1max} \approx 172;58°$, and thus $i_{1max} \approx 7;2°$, which is in best agreement with our earlier derivation.

Table 7.5 shows that a table computed from this value is in good agreement with al-Maghribī's tabular entries.[67]

In the case of Venus, using formula (2) with al-Maghribī's value $\beta_{max} = \beta_{icl}(180°) = 6;40°$ results in $i_{1max} = 2;37°$ (note that because the eccentricity of Venus is small, $TC \approx 60$). But this value cannot be derived from al-Maghribī's table of the epicyclic equation of Venus according to Ptolemy's procedure. As mentioned earlier, both in the *Īlkhānī zīj* and in al-Maghribī's *Adwār*, the radius of the epicycle of Venus is Ptolemaic, and as shown in Table 7.6, the tables of the epicyclic equation of the planet at mean distance in both *zījes* are equivalent to the corresponding one in the *Almagest*. Nevertheless, al-Maghribī has different values for the epicyclic equation for arguments 171°–179° (enclosed by the dashed lines in Table 7.6). I cannot explain why these entries are different, larger than the *Almagest* entries. As mentioned earlier, Ibn Yūnus has a greater value for the radius of the epicycle of Venus than did Ptolemy ($r \approx 43;28$; see Table 7.6).[68] But al-Maghribī's equations for arguments 171°–179° can in no way be related to Ibn Yūnus's.

With Ptolemy's procedure, the maximum latitude $\beta_{max} = \beta_{icl}(180°) = 6;40°$ taken as the epicyclic equation at mean distance corresponds to an epicyclic anomaly $\alpha = 180° - i_{1max} = 177;28,38°$, and thus $i_{1max} = 2;31,22°$, which is nearly equal to the Ptolemaic 2;30°. It can be assumed that 6;40° is estimated from the observed latitudes of Venus near inferior conjunction with the Sun, although estimating latitude in this location is extremely difficult and al-Maghribī's error, like Ptolemy's, is close to −2°, but it cannot be determined which of the two values for i_{1max} al-Maghribī actually derived from it. However, since the tabular entries for the inclination of Venus in his *Adwār* are in better agreement with computation from $i_{1max} = 2;37°$ than from $i_{1max} = 2;30°$ (see later), it seems to be safe to conclude that it is the first value that, in reality, underlies al-Maghribī's table in the *Adwār*.

In Table 7.7, Cols. 2 and 3, respectively, indicate the *Almagest*[69] and al-Maghribī's table of the inclination of Venus in the *Adwār*. Cols. 4 and 5 give the recomputed values from the two values for i_{1max}. Neither of the two sets of the re-computed latitudes is as consistent with the tabular entries as we have seen earlier in the case of al-Maghribī's table of the inclination of Mercury (Table 7.5). The table was probably computed according to the following procedure: the entries for the arguments 0°–96° are nearly identical to the corresponding entries in the *Almagest*, except for the slightly improved increments of 0;1° for the first five entries; this seems reasonable, since a small change in i_{1max} from 2;30° to 2;37° causes no critical change in the latitudes in this range (at most, 0;3°). But most of the entries in the latter part of the table appear to have been derived from multiplying the corresponding entries in the *Almagest* table by 6;40/6;22 (see Col. 6). It is noteworthy

---

67 The agreement is maintained using the rounded value $i_{1max} \approx 7;0°$ as well.

68 Ibn Yūnus, L: pp. 188–190.

69 *Almagest* XIII.5: Toomer 1998, p. 633.

Table 7.6 Tables of the epicyclic equation of Venus in the *Īlkhānī zīj*, al-Maghribī's *Adwār*, and Ibn Yūnus's *Ḥākimī zīj*

1	2	3	4	5	6	7
	Almagest	*Īlkhānī zīj*	*al-Maghribī*	*Re-c.* $(r = 43;10)$	*Ibn Yūnus*	*Re-c.* $(r = 43;28)$
1°	—	0;26°	0;26°	0;25°	0;25°	0;25°
2°	—	0;51°	0;51°	0;50°	0;51°	0;50°
3°	—	1;16°	1;16°	1;15°	1;16°	1;16°
...	...	...	...	...	...	...
6°	2;31°	2;31°	2;31°	2;31°	2;32°	2;31°
...	...	...	...	...	...	...
12°	5; 1°	5; 1°	5; 1°	5; 1°	5; 4°	5; 2°
...	...	...	...	...	...	...
30°	12;30°	12;30°	12;31°	12;30°	12;37°	12;33°
...	...	...	...	...	...	...
135°	45;59°	45;59°	45;59°	46; 0°	46;25°	46;24°
136°	—	45;59°	46; 0°	46; 0°	46;25°	46;25°
...	...	...	...	...	...	...
170°	—	23;12°	23;12°	23;12°	23;25°	23;42°
171°	21;15°	21;15°	21;17°	21;15°	21;27°	21;43°
172°	—	19;11°	19;16°	19;12°	19;22°	19;38°
173°	—	17; 2°	17;10°	17; 3°	17;12°	17;27°
174°	14;47°	14;47°	14;59°	14;48°	14;54°	15; 9°
175°	—	12;27°	12;42°	12;29°	12;34°	12;47°
176°	—	10; 4°	10;20°	10; 5°	10;10°	10;20°
177°	7;38°	7;38°	7;53°	7;37°	7;42°	7;48°
178°	—	5; 9°	5;20° (1)	5; 6°	5;12°	5;14°
179°	—	2;35°	2;42°	2;34°	2;37°	2;38°
180°	0°	0°	0°	0°	0°	0°

Note: (1) MS. M: 5;22.

that applying similar procedures to computing the planetary equation tables can be found in Islamic astronomy, more remarkably, in al-Maghribī's *Tāj al-azyāj*[70] as well as in the *Īlkhānī zīj*.[71]

In the *Tāj al-azyāj*, al-Maghribī has for Venus $\beta_{max} = \beta_{icl}(180°) = 8;30°$. The table of the planet's epicyclic equation in this *zīj* does not have those strange entries found in the *Adwār* but is identical to the *Almagest/Handy Tables*, and so to the

70 See Dorce 2002–3, pp. 202–204, 2003, pp. 127–137.
71 For example, the table of the epicyclic equation of Mars in the *Īlkhānī zīj* (C: p. 116, P: ff. 38v–39r, M: ff. 70v–71v) has the maximum value 42;12°, corresponding to $r \approx 40;18$. Al-Kāshī (IO: f. 99r, 112r) notices that this value is different from Ptolemy's and that the entries of the table are derived from multiplying the entries of the *Almagest* table (with the maximum value 41;9°; see Toomer 1998, p. 551) by 42;12/41;9. Spot checks show that this is indeed the case.

Īlkhānī zīj (Table 7.6). Both formula (7.2) and the interpolation in the anomalistic equation table of Venus result in $i_{1max} \approx 3;21°$. Al-Maghribī's values for the inclination and the recomputed values are shown in Table 7.7 (Cols. 7 and 8).

The entries in the latitude tables of the inferior planets in the *Īlkhānī zīj* are given for each integer degree of the argument. Col. 9 in Table 7.7 shows the entries in the table of the inclination of Venus in the *Īlkhānī zīj*. Both solving the trigonometrical equation (A2) with $\beta_{max} = \beta_{icl}(180°) = 8;40°$ and the interpolation in the table of the epicyclic equation of Venus result in $i_{1max} \approx 3;25°$. But the first half of the table, in the region of the arguments 0–90°, is identical to the corresponding table in the *Almagest* computed from $i_{1max} = 2;30°$. Moreover, re-computation of the table from $i_{1max} = 3;25°$ (Col. 10) shows that the results are by no means in agreement with the entries in the latter part of the table. In MS. P (f. 44r) of the *Īlkhānī zīj*, an unknown commentator has restored the latter half of the table to the Ptolemaic numbers and thus reduced the maximum latitude of the planet to 6;22°. We are told that this modification was done on the basis of the corresponding table in al-Maghribī's *zīj*, while, as we have seen earlier, his *Tāj* and *Adwār* have, respectively, the values 8;30° and 6;40° for the maximum latitude of Venus. It is worth noting that in his commentary on the *Īlkhānī zīj*, al-Nīshābūrī briefly refers to this problem in the table of the latitude of Venus as a subtle point, but he offers no solution.[72] Since the table is intrinsically contradictory, I cannot speculate about the reasoning behind it. However, it was apparently reasonable enough for Ibn al-Shāṭir and the Samarqand astronomers to quote it in their own *zījes*.

7.4.2 Samarqand: Superior Planets

The new planetary parameters obtained at the Samarqand observatory, as deduced from Ulugh Beg's *Sulṭānī zīj*, are summarized in Table 7.8 for the equation of center, and Table 7.9 for the equation of anomaly. In both, Cols. 2 and 4 show, respectively, the maximum and minimum equations tabulated in this *zīj*, and Cols. 3 and 5 show the arguments, mean eccentric or true epicyclic anomalies, for which the extremal values are given.

The tables of the equation of center in the *Almagest* are symmetrical and additive for half the table and subtractive for the other half, but in the *Sulṭānī zīj*, they are displaced and always additive; the values $k_1 = \frac{1}{2}(\text{Max} + \text{Min})$ of the displacements and $q_{max} = \frac{1}{2}(\text{Max} - \text{Min})$ of the maximum equations of center (note that $k_1 > q_{max}$), and the corresponding eccentricities $e = R \cdot \tan \frac{1}{2} q_{max}$, where the radius of the eccentric $R = 60$, are given in the last three columns of Table 7.8.

The tables of the equation of the epicyclic anomaly are also always additive, but they use a different type of the displacement explained as follows: For all the planets, the first half of the table, for arguments 0°–180°, corresponds to the equation p_A of the epicyclic anomaly at greatest distance ($R + e$; for Mercury:

---

72 al-Nīshābūrī, P1: f. 119r, P2: f. 143v.

Table 7.7 Tables of the inclination β_{icl} of Venus in the *Almagest*, al-Maghribī's *Tāj*, and *Adwār* the *Īlkhānī zīj*

1	2	3	4	5	6	7	8	9	10
	Almagest	al-M. Adwār	Re-c. i_{1max}= 2;37°	2;30°	[Alm.] $\times\frac{6;40}{6;22}$	Al-M. Tāj	Re-c. i_{1max}= 3;21°	Īlkh. zīj	Re-c. i_{1max}= 3;25°
6°	1; 2°	1; 3°	1; 5°	1; 3°	1; 5°	1;24°	1;24°	1; 2°	1;26°
12°	1; 1°	1; 2°	1; 5°	1; 2°	1; 4°	1;24°	1;23°	1; 2°	1;25°
18°	1; 0°	1; 1°	1; 3°	1; 0°	1; 3°	1;22°	1;21°	1; 1°	1;24°
24°	0;59°	1; 0°	1; 1°	0;59°	1; 2°	1;20°	1;19°	0;59°	1;23°
30°	0;57°	0;58°	0;59°	0;56°	1; 0°	1;17°	1;15°	0;57°	1;20°
36°	0;55°	0;55°	0;56°	0;53°	0;58°	1;13°	1;11°	0;55°	1;17°
42°	0;51°	0;51°	0;52°	0;50°	0;53°	1; 9°	1; 7°	0;51°	1;13°
48°	0;46°	0;46°	0;48°	0;46°	0;48°	1; 5°	1; 1°	0;47° (1)	1; 8°
54°	0;41°	0;41°	0;43°	0;41°	0;43°	1; 0°	0;55°	0;41° (2)	0; 3°
60°	0;35°	0;35°	0;38°	0;36°	0;37°	0;52°	0;48°	0;35°	0;56°
66°	0;29°	0;29°	0;32°	0;30°	0;30°	0;43°	0;41°	0;29°	0;49°
72°	0;23°	0;23°	0;25°	0;24°	0;24°	0;32°	0;32°	0;23°	0;41°
78°	0;16°	0;16°	0;17°	0;17°	0;17°	0;21°	0;22°	0;16° (3)	0;33°
84°	0; 8°	0; 8°	0; 9°	0; 9°	0; 8°	0;11°	0;12°	0; 8° (4)	0;23°
90°	0°	0°	0°	0°	0°	0°	0°	0°	0;12°
93°	0; 5°	——		0; 5°		——		0; 7°	0; 6°
96°	0;10°	0;10°	0;10°	0;10°	0;10°	0;13°	0;13°	0;14°	0;13°
99°	0;15°	——		0;15°		——		0;22°	0;20°
102°	0;20°	0;21°	0;21°	0;20°	0;21°	0;25°	0;27°	0;29°	0;28°
105°	0;26°	——		0;26°		——		0;37°	0;36°
108°	0;32°	0;33°	0;34°	0;32°	0;34°	0;43°	0;43°	0;45°	0;44°
111°	0;38°	——		0;39°		——		0;54°	0;53°
114°	0;44°	0;46°	0;48°	0;45°	0;46°	0;58°	1; 1°	1; 2°	1; 2°
117°	0;50°	——		0;53°		——		1;11°	1;12°
120°	0;59°	1; 2°	1; 3°	1; 0°	1; 2°	1;19°	1;21°	1;20°	1;22°
123°	1; 8°	——		1; 9°		——		1;30°	1;34°
126°	1;18°	1;21°	1;21°	1;17°	1;22°	1;45°	1;44°	1;43°	1;46°
129°	1;28°	——		1;27°		——		1;57°	1;58°
132°	1;38°	1;48°	1;41°	1;37°	1;43°	2;11°	2;10°	2;12°	2;12°
135°	1;48°	——		1;48°		——		2;28°	2;27°
138°	1;59°	2; 8°	2; 5°	2; 0°	2; 5°	2;40°	2;40°	2;44° (5)	2;43°
141°	2;11°	——		2;13°		——		3; 6°	3; 1°
144°	2;23°	2;38°	2;33°	2;27°	2;30°	3;14°	3;16°	3;30°	3;20°
147°	2;43°	——		2;42°		——		3;58°	3;41°
150°	3; 3°	3;13°	3; 7°	2;59°	3;12°	4; 5°	4; 0°	4;27° (6)	4; 4°
153°	3;23°	——		3;18°		——		4;54°	4;30°
156°	3;44°	3;55°	3;48°	3;38°	3;55°	5; 0°	4;52°	5;22°	4;58°
159°	4; 5°	——		4; 1°		——		5;50°	5;28°
162°	4;26°	4;39°	4;37°	4;25°	4;39°	5;58°	5;54°	6;20°	6; 1°
165°	4;49°	——		4;51°		——		6;53°	6;36°
168°	5;13°	5;29°	5;31°	5;17°	5;28°	6;56°	7; 3°	7;27° (7)	7;11°
171°	5;36°	——		5;41°		——		7;57° (8)	7;44°
174°	5;52°	6; 9°	6;19°	6; 3°	6; 9°	7;44°	8; 4°	8;20°	8;13°
177°	6; 7°	——		6;17°		——		8;35°	8;32°
180°	6;22°	6;40°	6;40°	6;22°	6;40°	8;30°	8;30°	8;40°	8;40°

Notes (1) M: 0;46 | (2) M: 0;40 | (3) C: 0;17, M: 0;15 | (4) M: 0;7 | (5) C: 2;45 | (6) M, T: 4;26 | (7) M, T: 7;26 | (8) M, T: 7;56.

Table 7.8 Equations of center and eccentricities in Ulugh Beg's *Sulṭānī zīj*

1	2	3	4	5	6	7	8
	Max	*Arg(s)*	*Min*	*Arg(s)*	k_1	q_{max}	e
Saturn	13;37,45°	260°	0;22,15°	86°	7°	6;37,45°	3;28,30 ≈ 3;29
Jupiter	11;18,39°	261–262°	0;41,21°	86–87°	6°	5;18,39°	2;46,58 ≈ 2;47
Mars	23;50,48°	252°	0; 9,12°	84°	12°	11;50,48°	6;13,30 ≈ 6;14
Venus	3;39,19°	267°	0;20,41°	89°	2°	1;39,19°	0;52
Mercury	7; 2°	259–263°	0;58°	89–93°	4°	3; 2°	3; 0 [*Alm.*]

Note: Sulṭānī zīj, P1: ff. 135r, 138r, 141r, 144r, 147r; P2: ff. 152r, 155r, 158r, 161v, 165r.

Table 7.9 Equations of the epicyclic anomaly and radii of the epicycles in Ulugh Beg's *Sulṭānī zīj*

1	2	3	4	5	6	7	8	9
	Max	*Arg(s)*	*Min*	*Arg(s)*	k_2	$p_{A, max}$	$p_{\Pi, max}$	r
Saturn	13;11,42°	96°	0; 2,26°	263°	7°	6;11,42°	6;57,34°	6;51
Jupiter	22;49, 2°	101°	0; 7, 1°	258°	12°	10;49, 2°	11;52,59°	11;47
Mars	36;50,57°	127°	312;22,51°	101°	360°	36;50,57°	47;37,9°	39;43
Venus	45;10,10°	155°	313; 7,44°	223°	360°	45;10,10°	46;52,16°	43;10 [*Alm.*]
Mercury	19; 1°	108–110°	336; 8°	245–248°	360°	19; 1°	23;52°	22;30 [*Alm.*]

Note: Sulṭānī zīj, P1: ff. 135v, 138v, 141v, 144v, 147v; P2: ff. 152v, 155v, 158v–159r, 162r–v, 165v.

$R + 3e$), when the center of the epicycle is located at the apogee, but the latter half of the table, for arguments 180°–360°, to the equation p_Π of the epicyclic anomaly at the least distance ($R - e$, except for Mercury), when the center of the epicycle is located at the perigee. In the case of Saturn and Jupiter, all the entries are increased by an amount k_2. For Mars and the two inferior planets, the first half of the table gives p_A, while the entries in the latter part of the table were increased by 360°, that is, 360° − p_Π.[73] The values k_2 are shown in Col. 6. Cols. 7 and 8 display the maximum values of the epicyclic equation, respectively,

73 It is noteworthy that the method used in the *Sulṭānī zīj* for the displacement of the entries in the tables of the equation of center of all the planets and the displacement method for the tables of the equation of anomaly of Saturn and Jupiter are those invented by Kushyār b. Labbān in the latter part of the tenth century (see Brummelen 1998); they are also employed in the *Īlkhānī zīj* (see earlier, note 45). The method of 360° displacements appeared as early as the tables of the equation of center of all the planets in Ibn al-Fahhād's *'Alā 'ī zīj* (*ca.* 1172) (see Kamālī, ff. 48v–49r). A simple version of the 360° displacements in the tables of the equation of anomaly is adopted (to the best of my knowledge, for the first time) in the *Īlkhānī zīj*, of which, it can be said, the method employed in the *Sulṭānī zīj* is a more developed version. For a brief survey of the methods of the displacements, see van Dalen 2004b, pp. 841–843.

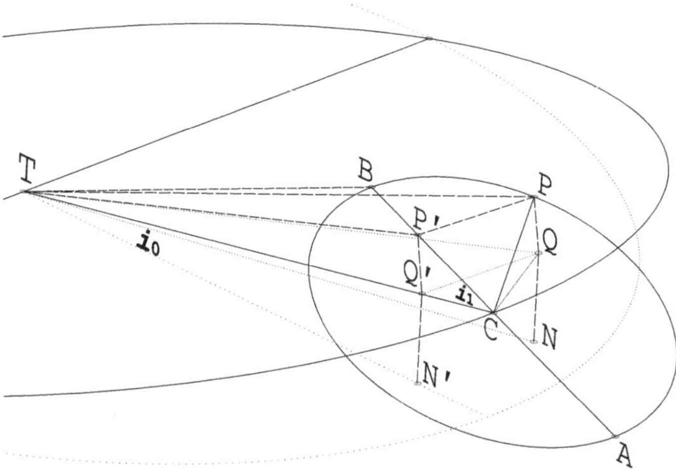

Figure 7.4 Configuration of the epicycle and the eccentric of the superior planets with respect to the ecliptic when the center of the epicycle is at the northern limit.

at apogee and perigee (for Saturn and Jupiter: $p_{A, max} = Max - k_2$ and $p_{\Pi, max} = k_2 - Min$). The last column gives the corresponding epicyclic radii (for all the planets, except Mercury, $r = \sin(p_{A, max}) \cdot (R + e) = \sin(p_{\Pi, max}) \cdot (R - e)$). In the *Sulṭānī zīj*, there is a table for the differences $p_{\Pi} - p_A$ and two interpolation tables for each planet, one for the epicyclic anomaly $0°–180°$ and another for the epicyclic anomaly $180°–360°$. The entries of the equation tables are carried to two sexagesimal places, except Mercury to one place, and are given for each integer degree of the argument (for Mars, from $150°$ to $210°$, the arguments are in steps of $0;20°$). Their reconstruction on the basis of the values derived earlier is in good agreement with the tabular entries.

It merits noting that in his *Commentary on Zīj of Ulugh Beg*, Qūshčī explains the layout of the equation and mean motion tables in this *zīj*. The values he mentions for the maximum equations agree with what can be extracted from the tables, with the exception of very slight differences in q_{max} of Saturn ($6;37,46°$) and Jupiter ($5;18,42°$). The parameters we deduced from the tables are in agreement with those he gives. As Qūshčī remarks, for all the planets, the mean eccentric anomalies are reduced and, inversely, the mean epicyclic anomalies are increased by the aforementioned values for k_1 and are then tabulated; and for Saturn and Jupiter, the longitudes of the apogees are decreased by the aforementioned values for k_2.[74]

Figure 7.4 shows the configuration of the epicycle and eccentric, inclined to the ecliptic, of a superior planet when the center of the epicycle is at the northern limit.

74 Qūshčī, pp. 273–274, 320–324, 371.

In order to derive the inclinations from the extremal latitudes in the *Sulṭānī zīj*, we should first know the distance $TC = \rho'$ between the Earth T and the center C of the epicycle of a superior planet at the northern limit, which can be computed from the eccentricity (Table 7.8) and the value adopted for ω_A, as mentioned earlier in Table 7.3. Where $R = 60$:

Planet	ρ' northern	ρ' southern
Saturn	61;39,43	58;11,13
Jupiter	62;45,18	57;14,33
Mars	66;12,30	53;47,19

In Figure 7.4, the distance $TC = \rho'$ and the radius of the epicycle $CP = r$. We again let the true epicyclic anomaly $\alpha = \angle ACP$, and $\angle CTN' = i_0$, so that $PP' = QQ' = r \sin \alpha$, $PQ = P'Q' = r \cos \alpha \sin i_1$, and $CQ' = r \cos \alpha \cos i_1$, and now $QN = Q'N' = (TC + CQ') \sin i_0$. The latitude $\beta = \angle PTN$ at either limit can be computed from

$$\sin|\beta(\alpha)| = \frac{(TC + CQ')\sin i_0 - PQ \cos i_0}{\sqrt{(TC + CQ')^2 + QQ'^2 + PQ^2}}$$

$$= \frac{r \cos \alpha \sin(i_0 - i_{1\max}) + \rho' \sin i_0}{\sqrt{(\rho' + CQ')^2 + QQ'^2 + PQ^2}} \qquad (7.3)$$

First, in order to derive i_0, we let $\alpha = 90°$ in formula (7.3); thus, we have

$$\sin i_0 = \sqrt{1 + \frac{r^2}{\rho'^2}} \cdot \sin\left(|\beta_{\mathrm{icl}}(90°)|\right) \qquad (7.4)$$

Then, in order to derive i_1, analogous to what we have done earlier in the case of the inferior planets, we solve triangle TCB, in which $\angle BTC = \beta_{\max} - i_0$:

$$\frac{TC}{\sin(\angle TBC)} = \frac{CB}{\sin(\angle CTB)} \text{ or } \frac{\rho'}{\sin(\beta_{\max} - i_0 + i_{1\max})} = \frac{r}{\sin(\beta_{\max} - i_0)} \qquad (7.5)$$

And thus, $i_{1\max} = (\beta_{\max} - i_0 + i_{1\max}) - (\beta_{\max} - i_0)$.

Our formulas (7.4) and (7.5), applying the planetary parameters and the tabular latitudes in Ulugh Beg's *zīj* (as mentioned in Tables 7.2, 7.8, and 7.9), result in the inclinations in Table 7.2. Using Ptolemy's solution, in which the extremal latitudes at the limits are taken as the epicyclic equations and the inclinations as the true

Table 7.10 Ulugh Beg's table of the latitude of Mars at the southern limit $i_0 = 1;20°$, $i_{1max} = 2;7°$

	Tab.	Re-c.		Tab.	Re-c.		Tab.	Re-c.
0°	0;26°	0;26°						
6°	0;26°	0;26°	**93°**	1; 7°	1; 8°	**138°**	2;38°	2;38°
12°	0;27°	0;27°	**96°**	1;11°	1;11°	**141°**	2;49°	2;50°
18°	0;28°	0;27°	**99°**	1;15°	1;15°	**144°**	3; 2°	3; 2°
24°	0;29°	0;28°	**102°**	1;19°	1;18°	**147°**	3;16°	3;17°
30°	0;30°	0;30°	**105°**	1;23°	1;23°	**150°**	3;32°	3;33°
36°	0;32°	0;31°	**108°**	1;27°	1;27°	**153°**	3;50°	3;51°
42°	0;34°	0;33°	**111°**	1;32°	1;32°	**156°**	4;11°	4;12°
48°	0;36°	0;36°	**114°**	1;37°	1;37°	**159°**	4;34°	4;35°
54°	0;38°	0;38°	**117°**	1;42°	1;42°	**162°**	5; 1°	5; 1°
60°	0;41°	0;41°	**120°**	1;48°	1;48°	**165°**	5;29°	5;29°
66°	0;44°	0;45°	**123°**	1;55°	1;55°	**168°**	5;58°	6;58°
72°	0;48°	0;49°	**126°**	2; 2°	2; 2°	**171°**	6;27°	6;27°
78°	0;53°	0;53°	**129°**	2;10°	2;10°	**174°**	6;50°	6;52°
84°	0;58°	0;59°	**132°**	2;19°	2;18°	**177°**	7; 6°	7;10°
90°	1; 4°	1; 4°	**135°**	2;28°	2;28°	**180°**	7;15°	7;16°

epicyclic anomalies, as we have explained for the derivation of i_{1max} of the inferior planets, results in the same inclinations.

The tables in the *Sulṭānī zīj* were generally calculated carefully, showing only minor deviations in the cases that have been examined,[75] which is also true of the latitude tables at the northern and southern limits. The inclinations are as follows:

Planet	i_0	i_{1max}
Saturn	2;30°	6; 0°
Jupiter	1;30°	2;45°
Mars	1;20°	2; 7°

Re-computation of the latitudes from the values derived for the inclinations are in good agreement with the tables in the *Sulṭānī zīj*, with no difference exceeding 0;1° in the case of Saturn, 0;3° for Jupiter, and 0;4° for Mars. Table 7.10 is a specimen for Mars at the southern limit, which shows the greatest differences.

References

Arabic Almagest, Ḥunayn b. Isḥāq and Thābit b. Qurra (trs.), MSS. S: Iran, Sipahsālār Library, no. 594 (copied in 480 H/1087–8 CE), PN: USA, Rare Book and Manuscript

---

75 For example, the table of the equation of center of the Moon; see Mozaffari 2014, p. 106.

Library of University of Pennsylvania, no. LJS 268 (written in an Arabic Maghribī/ Andalusian script at Spain in 783 H/1381 CE; some folios between 37v and 38r, corresponding to *Almagest* III.3–IV.9, are omitted).

al-Bīrūnī, Abū al-Rayḥān, 1954–6, *al-Qānūn al-mas'ūdī* (*Mas'ūdīc canons*), 3 Vols., Hyderabad: Osmania Bureau.

Brummelen, G. V., 1998, "Mathematical methods in the tables of planetary motion in Kūshyār ibn Labbān's *Jāmi' Zīj*", *Historia Mathematica* **25**, pp. 265–280.

Brummelen, G. V., 2006, "Taking Latitude with Ptolemy: Jamshīd al-Kāshī's novel geometric model of the motions of the inferior planets", *Archive for History of Exact Sciences* **60**, pp. 353–377.

Caussin de Perceval, J.-J.-A., 1804, "Le livre de la grande table hakémite, Observée par le Sheikh,. . ., ebn Iounis", *Notices et Extraits des Manuscrits de la Bibliothèque nationale* **7**, pp. 16–240.

Chabás, J. and Goldstein, B. R., 1994, "Andalusian astronomy: *al-Zīj al-Muqtabis* of Ibn al-Kammād", *Archive for History of Exact Sciences* **48**, pp. 1–41.

Chabás, J. and Goldstein, B. R., 2003, *The Alfonsine Tables of Toledo*, Archimedes: New Studies in the History and Philosophy of Science and Technology, Vol. 8, Dordrecht: Springer Science + Business Media.

Chabás, J. and Goldstein, B. R., 2009, *The Astronomical Tables of Giovanni Bianchini*, Leiden: Brill.

van Dalen, B., 1999, "Tables of planetary latitude in the *Huihui li* (II)", in: Kim, Y. S. and Bray, F. (eds.), *Current Perspectives in the History of Science in East Asia*, Seoul: Seoul National University Press, pp. 316–329.

van Dalen, B., 2004a, "A second manuscript of the *Mumtaḥan Zīj*", *Suhayl* **4**, pp. 9–44.

van Dalen, B., 2004b, "The *Zīj-i Nāṣirī* by Maḥmūd ibn 'Umar: the earliest Indian Zij and its relation to the *'Alā'ī Zīj*", in: Burnett, C. *et al.* (eds.), *Studies in the History of the Exact Sciences in Honour of David Pingree*, Leiden: Brill, pp. 825–862.

Debarnot, M.-T., 1987, "The Zīj of Ḥabash al-Ḥāsib: A survey of MS Istanbul Yeni Cami 784/2", in: Saliba, G. and King, D. A. (eds.), *From Deferent to Equant: A Volume of Studies on the History of Science of the Ancient and Medieval Near East in Honor of E. S. Kennedy*, Vol. 500, New York: Annals of the New York Academy of Sciences, pp. 35–69.

Dorce, C., 2002–3, "The *Tāj al-azyāj* of Muḥyī al-Dīn al-Maghribī (d. 1283): Methods of computation", *Suhayl* **3**, pp. 193–212.

Dorce, C., 2003, "El Tāŷ al-azyāŷ de Muḥyī al-Dīn al-Maghribī", in: *Anuari de Filologia*, Vol. 25, Secció B, Número 5, Barcelona: University of Barcelona.

Eastwood, B. S., 2007, *Ordering the Heavens: Roman Astronomy and Cosmology in the Carolingian Renaissance*, Leiden-Boston: Brill.

Eastwood, B. S. and Grasshoff, G., 2003, "Planetary diagrams-descriptions, models, theories; From carolingian deployments to copernican debates", in: Lefèvre, W., *et al.* (eds.), *The Power of Images in Early Modern Science*, New York: Springer, pp. 197–226.

al-Farghānī, Aḥmad b. Muḥammad b. Kathīr, 1669, *Kitab fī 'l-ḥarakāt al-samāwiyya wa jawāmi' al-'ilm al-nujūm*, Arabic edition and Latin translation by Jacobus Golius: *Muhammedis Fil. Ketiri Ferganensis, qui vulgo Alfraganvs dicitur, Elementa Astronomica Arabicè et Latinè. Cum Notis ad res exoticas sive Orientales, quae in iis occurunt, Opera Jacobi Golii. Amstelodami.* MS. P: Princeton University Library, Islamic Manuscripts, Garrett no. 135L, retrieved from http://arks.princeton.edu/ark:/88435/x920fw91b.

Glasner, R., 2003, "Gersonides unusual position on 'position'", *Centaurus* **45**, pp. 249–236.

Goldstein, B. R., 1967, "The Arabic version of Ptolemy's planetary hypotheses", *Transactions of the American Philosophical Society* **57**, pp. 3–55.

Goldstein, B. R., 2002, "Levi ben Gerson's preliminary remarks for a theory of planetary latitudes", *Aleph* **2**, pp. 15–30.

Goldstein, B. R. and Chabás, J., 2004, "Ptolemy, Bianchini, and Copernicus: Tables for planetary latitudes", *Archive for History of Exact Sciences* **58**, 453–473.

Ḥabash al-Ḥāsib, *The zīj of Ḥabash al-Ḥāsib*, MS. Berlin, Ahlwardt 5750 (formerly Wetzstein I 90).

Ibn al-Shāṭir, ʿAlāʾ al-Dīn Abu ʾl-Ḥasan ʿAlī b. Ibrāhīm b. Muḥammad al-Muṭaʿʿim al-Anṣārī, *al-Zīj al-Jadīd*, MSS. K: Istanbul, Kandilli Observatory, no. 238; O: Oxford, Bodleian Library, no. Seld. A inf 30.

Ibn Yūnus, ʿAlī b. ʿAbd al-Raḥmān b. Aḥmad, *Zīj al-kabīr al-Ḥākimī*, MSS. L: Leiden, no. Or. 143, O: Oxford Bodleian Library, no. Hunt 331.

Jones, A., 2000, "Studies in the astronomy of the Roman Period IV: Solar tables based on a non-hipparchian model", *Centaurus* **42**, pp. 77–88.

Jones, A., 2005, "Ptolemy's *Canobic Inscription* and Heliodorus' observation reports", *SCIAMVS* **6**, pp. 53–97.

al-Kamālī, Muḥammad b. Abī ʿAbd-Allāh Sanjar (*Sayf-i munajjim*), *Zīj-i Ashrafī*, MS. Paris, Bibliothèque Nationale, no. 1488.

al-Kāshī, Jamshīd Ghiyāth al-Dīn, *Khāqānī zīj*, MSS. IO: London: India Office, no. 430, P: Iran: Parliament Library, no. 6198.

Kennedy, E. S., 1951, "An Islamic computer for planetary altitudes", *Journal of the American Oriental Society* **71**, pp. 13–21; reprinted in Kennedy *et. al.* 1983, pp. 463–471.

Kennedy, E. S., 1960, *The Planetary Equatorium of Jamshīd Ghiyāth al-Dīn al-Kāshī*, Princeton: Princeton University Press.

Kennedy, E. S., 1990, "Two topics from an astrological manuscript: Sindhind and planetary latitudes", *Zeitschrift für Geschichte der Arabisch-Islamischen Wissenschaften*, pp. 167–178; reprinted in Kennedy 1998, Trace VI.

Kennedy, E. S., 1998, *Astronomy and Astrology in the Medieval Islamic World*, Aldershot: Ashgate-Variorum.

Kennedy, E. S. and Ukashah, W., 1969, "Al-Khwārizmī's planetary latitude tables", *Centaurus* **14**, pp. 86–96; reprinted in Kennedy *et. al.* 1983, pp. 125–135.

Kennedy, E. S., colleagues, and former students, 1983, *Studies in the Islamic Exact Sciences*, Beirut: American University of Beirut.

Krisciunas, K., 1994, "A more complete analysis of the errors in Ulugh Beg's star catalogue", *Journal for the History of Astronomy* **24**, pp. 269–280.

al-Maghribī, Muḥyī al-Dīn, *Adwār al-anwār*, MSS. M: Iran, Mashhad, Holy Shrine Library, no. 332; CB: Ireland, Dublin, Chester Beatty, no. 3665.

al-Maghribī, Muḥyī al-Dīn, *Talkhīṣ al-majisṭī*, MS. Leiden: Universiteitsbibliotheek, no. Or. 110.

Mancha, J. L., 1990, "Ibn al-Haytham's homocentric epicycles in Latin astronomical texts of the XIVth and XVth Centuries", *Centaurus* **33**, pp. 70–89.

Mozaffari, S. M., 2014, "Muḥyī al-Dīn al-Maghribī's lunar measurements at the Maragha observatory", *Archive for History of Exact Sciences* **68**, pp. 67–120.

Mozaffari, S. M., 2016–2017, "A revision of the star tables in the *Mumtaḥan zīj*", *Suhayl* **15**, pp. 67–100.

Muḥammad al-Ṭabīb al-Muhtadī al-Mūṣilī, *al-Jamʿ al-mufīd*, MS. Iran: Parliament Library, no. 16238/3.

Nallino, C. A. (ed.), 1899–1907/1969, *Al-Battani sive Albatenii Opus Astronomicum*. Publicazioni del Reale osservatorio di Brera in Milano. n. XL, pte. I–III, Milan: Mediolani Insubrum. The Reprint of Nallino's edition: Minerva, Frankfurt.

Neugebauer, O., 1962, *The Astronomical Tables of al-Khwārizmī*, in: *Hist. Filos. Skr. Dan. Vid. Selsk.* **4**, no. 2, Copenhagen: The Royal Danish Academy of Sciences and Letters.

Neugebauer, O., 1975, *A History of Ancient Mathematical Astronomy*, 3 Vols., Berlin-Heidelberg-New York: Springer.

al-Nīshābūrī, Niẓām al-Dīn Aʿraj, *Kashf al-ḥaqāʾiq-i Zīj-i Īlkhānī* (*Opening of Truths of the Īlkhānī zīj*, a Commentary on al-Ṭūsī's *Īlkhānī zīj*), MSS. P1: Iran, Parliament Library, no. 1210, P2: Iran, Parliament Library, no. 1426.

Pedersen, O., 1974, *A Survey of Almagest*, Odense: Odense University Press. With annotation and new commentary by A. Jones, New York: Springer, 2010.

Pliny, 1938–1962, *Natural History*, English translation by H. Rackham, 10 Vols., Cambridge, MA-London: Harvard University Press-William Heinemann.

Qūshčī, ʿAlī b. Muḥammad, *Sharḥ-i Zīj-i Ulugh Beg* (*Commentary on Zīj of Ulugh Beg*), MS. Iran: National Library, no. 20127-5.

Ragep, F. J., 1987, "The two versions of the Ṭūsī Couple", in: Saliba, G. and King, D. A. (eds.), *From Deferent to Equant: A Volume of Studies on the History of Science of the Ancient and Medieval Near East in Honor of ES Kennedy*, Vol. 500, New York: Annals of the New York Academy of Sciences, pp. 329–356.

Ragep, F. J., 1993, *Naṣīr al-Dīn al-Ṭūsī's* Memoir on Astronomy (*al-Tadhkira fī ʿilm al-hayʾa*), 2 Vols., New York: Springer.

Ragep, F. J., 2004, "Ibn al-Haytham and Eudoxus: The revival of homocentric modeling in Islam", in: Burnett, C., Hogendijk, J. P., Plofker, K., and Yano, M. (eds.), *Studies in the History of the Exact Sciences in Honour of David Pingree*, Leiden-Boston: Brill, pp. 786–809.

Riddell, R. C., 1978, "The latitudes of Venus and Mercury in the *Almagest*", *Archive for History of Exact Sciences* **19**, pp. 95–111.

Roberts, V., 1966, "The planetary theory of Ibn al-Shāṭir: Latitudes of the planets", *Isis* **57**, pp. 208–219; reprinted in Kennedy 1983, pp. 72–83.

Saliba, G. and Kennedy, E. S., 1991, "The spherical case of the Ṭūsī couple", *Arabic Science and Philosophy* **1**, pp. 285–291; reprinted in Kennedy 1998, Trace VI.

Samsó, J., 1997, "Andalusian astronomy in 14th century Fez: al-Zīj al-Muwāfiq of Ibn ʿAzzūz al-Qusanṭīnī", *Zeitschrift für Geschichte der Arabisch-Islamischen Wissenschaften* **11**, pp. 73–110; reprinted in Samsó 2007, Trace IX.

Samsó, J., 1999, "Horoscopes and history: Ibn ʿAzzūz and his retrospective horoscopes related to the battle of El Salado (1340)", in: Nauta, L. and Vanderjagt, A. (eds.), *Between Demonstration and Imagination: Essays in the History of Science and Philosophy Presented to John D. North*, Leiden: Brill, pp. 101–124; reprinted in Samsó 2007, Trace X.

Samsó, J., 2007, *Astronomy and Astrology in al-Andalus and the Maghrib*, Aldershot: Ashgate.

Samsó, J. and Millás, E., 1998, "The computation of planetary longitudes in the *zīj* of Ibn al-Bannāʾ", *Arabic Science and Philosophy* **8**, pp. 259–286; reprinted in Samsó 2007, Trace VIII.

Shevchenko, M., 1990, "An analysis of errors in the star catalogues of Ptolemy and Ulugh Beg", *Journal for the History of Astronomy* **21**, pp. 187–201.

Steele, J. M., 2000, *Observations and Predictions of Eclipse Times by Early Astronomers*, Dordrecht-Boston-London: Kluwer Academic Publishers, reprinted by Springer.

Steele, J. M., 2003, "Planetary latitudes in Babylonian mathematical astronomy", *Journal for the History of Astronomy* **34**, pp. 269–289.

Stephenson, F. R., 1997, *Historical Eclipses and Earth's Rotation*, Cambridge: Cambridge University Press.

Súrya Siddhánta, 1860/1997, *The Súrya Siddhánta: A Textbook of Hindu Astronomy*, Gangooly, P. (ed.) and Burgess, E. (tr.), Delhi: Motilal Banarsidass.

Swerdlow, N. M., 2005, "Ptolemy's theories of the latitude of the planets in the *Almagest, Handy Tables*, and *Planetary Hypotheses*", in: Buchwald, J. Z. and Franklin, A. (eds.), *Wrong for the Right Reasons*, Netherlands: Springer, pp. 41–71.

Swerdlow, N. M. and Neugebauer, O., 1984, *Mathematical Astronomy in Copernicus's De Revolutionibus*, New York-Berlin-Heidelberg-Tokyo: Springer-Verlag.

Toomer, G. J., 1968, "A survey of the Toledan tables", *Osiris*, pp. 5–174.

Toomer, G. J. (ed.), 1998, *Ptolemy's Almagest*, Princeton: Princeton University Press.

al-Ṭūsī, Naṣīr al-Dīn, *Īlkhānī Zīj*, MSS. C: University of California, Caro Minasian Collection, no. 1462; T: University of Tehran, Ḥikmat Collection, no. 165 + Suppl. P: Iran, Parliament 6517 (Remark: The latter is not actually a separate MS, but contains 31 folios missing from MS. T. The chapters and tables in MS. T are badly out of order, presumably owing to the folios having been bound in disorder), P: Iran, Parliament Library, no. 181, M: Iran, Mashhad, Holy Shrine Library, no. 5332a.

Ulugh Beg, *Sulṭānī Zīj*, MSS. P1: Iran, Parliament Library, no. 72; P2: Iran, Parliament Library, no. 6027.

Verbunt, F. and van Gent, R. H., 2012, "The star catalogues of Ptolemaios and Ulugh Beg; Machine-readable versions and comparison with the modern HIPPARCOS Catalogue", *Astronomy & Astrophysics* **544**, p. A31.

Viladrich, M., 1988, "The planetary latitude tables in the *Mumtaḥan Zīj*", *Journal for the History of Astronomy* **19**, pp. 257–268.

al-Wābkanawī al-Bukhārī, Shams al-Dīn Muḥammad, *Zīj-i muḥaqqaq-i sulṭānī*, MSS. T: Turkey, Aya Sophia Library, No. 2694; Y: Iran, Yazd, Library of 'Ulūmī, no. 546, its microfilm is available in the University of Tehran central library, no. 2546; P: Iran, Library of Parliament, no. 6435.

8

HOLDING OR BREAKING WITH PTOLEMY'S GENERALIZATION

Considerations About the Motion of the Planetary Apsidal Lines in Medieval Islamic Astronomy[1]

8.1 Introduction

This chapter delves into the origin and evolution of certain ideas aimed at enhancing a fundamental aspect of Ptolemy's planetary models in medieval Islamic astronomy, namely, the progressive motion of the apsidal lines of the five planets.

For readers unfamiliar with Ptolemaic planetary models, it is important to know that both models Ptolemy constructs for the planets – one for Venus and the superior planets, and a more complex one for Mercury – consist of two components. A planet revolves on a small circle, known as an "epicycle," to account for the synodic phenomena. The center of this circle moves along a larger circle, called an "eccentric deferent," whose center is offset from the Earth (Figure 8.1). The distance between the Earth and the center of the eccentric defines the "eccentricity," one of the oldest known fundamental orbital elements of the planets. This forms the core basis of the simple ancient solar eccentric model, creating a periodic variation in the distance between a planet and the Earth to account for the observed differences in its apparent motion, which can sometimes be faster and sometimes slower. Except for Mercury, the extremal distances of a planet from the Earth lie along the extension of the eccentricity in both directions, known as the two "apses": the point of the eccentric located at the greatest distance from the Earth, "apogee," and the point indicating the least distance, "perigee." The line connecting the two is called the "apsidal line." These terms are still used for the Earth and Moon, although they have been replaced by "aphelion" and "perihelion" for the planets, respectively. In Ptolemaic astronomy, celestial motions and positions are visualized and measured

1 Original publication: S. M. Mozaffari, "Holding or breaking with Ptolemy's generalization: Considerations about the motion of the planetary apsidal lines in medieval Islamic astronomy," *Science in Context* 30 (2017), pp. 1–32. © 2017 Cambridge University Press, and republished by permission.

DOI: 10.4324/9781003481966-12

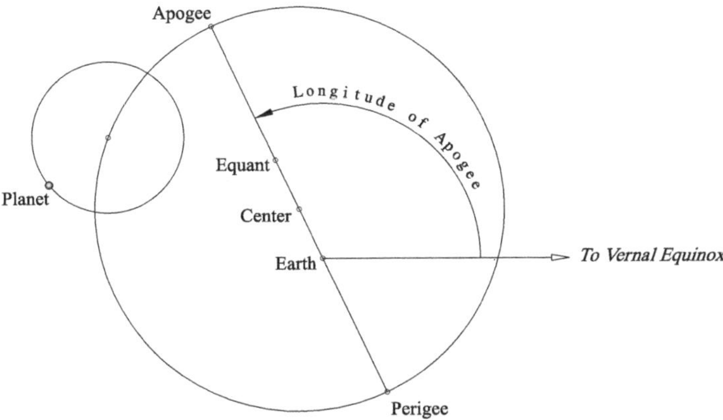

Figure 8.1 Ptolemy's model for Venus and the superior planets.

with reference to a tropical coordinate system, that is, a system referred to the ecliptic, or the Earth's orbital plane, or the path of the apparent annual motion of the Sun through the stars. Two coordinates are represented: longitude, showing the angular distance along the ecliptic, with the vernal equinox point serving as the zero point, and latitude, indicating the perpendicular angular distance from the ecliptic. The vernal equinox point itself has a regressive motion, known as precession, a consequence of which is a steady increase in the stellar longitudes. By definition, a *sidereal reference system* is one in which a fictitious star on the ecliptic having no proper motion serves as the zero point of longitudes.

This chapter is organized according to the following parts: in Section 8.2, we briefly review Ptolemy's demonstration of a progressive motion for the apsidal line of Mercury and his finding that the speed of its motion is equal to the rate of precession, which he generalizes to the apogees of the other planets. In Section 8.3, the modern values for the progressive motions of the planetary apogees in a tropical reference system are presented. In the next section, the values for the rate of precession measured during the medieval Islamic period are introduced. In Section 8.5, which constitutes the core of this chapter, we first discuss the discovery of the motion of the solar apogee in the ninth century, along with Bīrūnī and Ibn al-Zarqālluh's idea of assigning a proper motion to it. Then, focusing on the characteristic facts about planetary measurements both in early and late Islamic astronomy, we examine the values for the longitudes of the apogees of the Sun and planets in the most important works written in the Middle East during the medieval period. Next, we present Jamāl al-Dīn al-Zaydī's and Ibn al-Shāṭir's significant and fascinating contribution, which has not yet received proper attention in the modern literature: their idea that the motion of the apogees (for which both astronomers take the rate of $1°/60^y$) is greater than precession. Also, a plausible scenario is suggested for the reconstruction of the line of reasoning which may have led them to such an outcome. The last section coherently summarizes the findings set forth in this study and highlights its underlying purpose, that is, to show how medieval Islamic astronomers treated Ptolemy's generalization as a rational basis and a methodological axiom of astronomy in parallel with their noteworthy empirical discoveries that caused substantial changes in Ptolemy's proposition about the motion of the planetary apogees and its relation to precession.

8.2 Ptolemy's Considerations About the Spatial Motion of the Planetary Spheres

In *Almagest* VII.2, Ptolemy determines the rate of precession as 1° per century, based on observations of the star Regulus (α Leo) made by Hipparchus and himself. He tries to validate this result in the following chapter (VII.3) by meticulously examining the changes in the declination of some stars near the equinoctial points, using ancient Greek observations passed down to him.[2] Later, in *Almagest* IX.5 and IX.7, Ptolemy states that he finds it consistent with planetary phenomena that their apsidal lines shift slowly and uniformly around the center of the ecliptic in the direction of increasing longitude. As *a provisional estimation on the basis of evidence available to him*, namely, according to his derivation of the direction of the apsidal line of Mercury both for his own time and about 400 years earlier, he concludes that its motion equals the rate of precession of 1° in every 100 years.[3] He does not revisit the issue of the apsidal line motion for any of the other planets. Instead, he assumes that the apsidal lines of all five planets move collectively at a velocity equal to the rate of precession.[4] As a result, in Ptolemaic astronomy, which uses the ecliptic as a reference system, the spatial directions of the planetary orbs are considered to be fixed sidereally.

This approach forms a methodological cornerstone of Ptolemy's astronomy. For Ptolemy, it is logical to draw a conclusion based on the generalization of a fact that has been discovered and proven in a specific instance, particularly when the instance itself (i.e., Mercury) is one of only five members of the entire group (i.e., the five planets) and thus can reasonably represent the whole group. Medieval Middle Eastern astronomers strictly adhered to this methodological principle of the *Almagest*.[5] A notable example is, when they discovered the progressive motion

---

2 Toomer 1998, pp. 327–338, Pannekoek 1995, Duke 2006.

3 Toomer 1998, pp. 443, 453.

4 For instance, in order to determine the longitude of the apogee of a planet for the time of an ancient observation, Ptolemy only simply computes back in time from his own apogeal longitude values; for example, for the computation of the longitude of the apogee of Venus for one of Timocharis's observation in 272 BCE (*Almagest* X.4,6: Toomer 1998, pp. 477, 480), of Mars for an ancient observation dated 272 bce (*Almagest* X.9,10: Toomer 1998, pp. 502, 505), of Jupiter for an observation made in 241 BCE (*Almagest* XI.3,4: Toomer 1998, pp. 522, 525), and of Saturn for 229 BCE (*Almagest* XI.7,8: Toomer 1998, pp. 541, 544).

5 Another case is that medieval Middle Eastern astronomers attributed to observational errors the small but inexplicable or random deviations found between various observations, or between observation and theory. This treatment is also completely based upon Ptolemy's manner of the interpretation of observational data adduced and firmly established in the *Almagest* – for example, Ptolemy's statement in the case of the comparison of his own and Hipparchus's observations of the stellar latitudes that "if there is a discrepancy, it is of very small size, such as can be accounted

of the solar apogee about seven centuries after Ptolemy, they ultimately found it necessary to *assume* that its motion followed the precessional motion, just like the planetary apogees (see following).

For the purpose of comparison in the subsequent discussion, Ptolemy's values for the solar and planetary apogees are given in Table 8.1.[6] The numbers in bold indicate the true values at the times computed from the modern theories.[7]

---

for by small observational errors" (Almagest VII.3: Toomer 1998, p. 330). It should be mentioned, however, that the Middle Eastern Islamic astronomers attached less importance to the most crucial methodological insight Ptolemy set forth in the Almagest, that is, the idea or pre-condition of simplicity and complexity in astronomical hypotheses (Almagest III.1, III.4, and XIII.2: Toomer 1998, pp. 136, 153, 600). This insight was not taken into consideration in the theoretical non-Ptolemaic models constructed by them.

6 Almagest, III.4, IX.7,8, X.1,2, X.7, XI.1, XI.5: Toomer 1998, pp. 154–157, 449–450, 454, 470–471, 498, 519, 537.

7 Some notes occasionally appear to merit mentioning: (1) We add 1° to Ptolemy's longitudes, because of Ptolemy's well-known error in the determination of the moments of the vernal equinox by more than one day, which causes a displacement in the position he assigns to the vernal equinox point (the zero point of longitudes) by a bit over 1°, and consequently, the well-known systematic error of about −1° occurs in the longitudes of the mean sun and of the fixed stars (e.g., see Newton 1973, pp. 369–370). Thus, when comparing Ptolemy's longitudes of the planetary apogees with the modern values, it is required to take this error into account by adding 1° to Ptolemy's values given earlier in order to fix his longitudes in a correct tropical frame of reference. (2) The late Otto Neugebauer (1975, vol. 1, pp. 147, 208, 213) computes the following values for the geocentric longitude of the planetary apogees for 100 ce:

Mercury	Venus	Mars	Jupiter	Saturn
226;0°	47;0°	113;45°	143;0°	229;30°

The reason behind these erroneous figures is that in order to compute the geocentric from the heliocentric parameter values, he takes the heliocentric eccentricity of every planet as the "distance" between the center of the elliptical orbit of a planet and the Sun, while in the modern sense, it only denotes the "ratio" between the semi-minor and semi-major axes of the elliptical orbit of each planet; consequently, he did not take into account the semi-major axis in the computation of the distance of the center of the elliptical orbit of every planet from the Sun (in order to compute this distance, the eccentricity should be multiplied with the semi-major axis in terms of astronomical unit), as Figure 212 on p. 1,274 and Figure 217 on p. 1,277 in vol. 3 of his magnum opus conspicuously indicate. Note that since the speeds of the planetary apsidal lines are very low (their values are given in Section 8.4), the difference in time between 100 ce, which Neugebauer takes, and the times in the 130s adopted in our derivation causes no difference exceeding 1° in the longitudes of the apogees. (3) In the case of the Sun, the heliocentric perigee of the elliptical orbit of the Earth points to a longitude of about 71;7° at the beginning of 140 ce. We take into account the theoretical deviation between the Keplerian motion in an elliptical orbit and the Hipparchian eccentric model associated with the seasons method, of which both Hipparchus and Ptolemy made use in order to compute the true value for the longitude of the solar apogee at the times. (See Maeyama 1998; the results of Maeyama's thorough analysis were deployed in Chapter 1 for the examination of the accuracy of the values observed for the solar orbital elements in the medieval Islamic period before the sixteenth century.)

179

Table 8.1 Ptolemy's values for the longitude of the solar and planetary apogees

Sun/Earth	Saturn	Jupiter	Mars	Venus	Mercury
65;30° + *1*°	233; 0° + *1*°	161; 0° + *1*°	115;30° + *1*°	55; 0° + *1*°	190; 0° + *1*°
70;41	**236;18**	**160;27**	**116;44**	**~57;30**	**~222;30**
139–140 CE	July 8, 136	October 7/8, 137	May 27/28, 139	Mid-130s CE	Mid-130s CE
Tropically fixed			Sidereally fixed		

The dates given here for the superior planets are those of the last of Ptolemy's trios of observations of their oppositions to the mean sun as given in the *Almagest*, because he remarks that the values he derived for the longitudes of the apogees are for the times of those oppositions. In the case of the inferior planets, Ptolemy makes use of a variety of observations spreading from the late 120s to the early 140s.

As can be obviously seen, the aggregate error resulting from all the observational deficiencies and computational impreciseness committed by Ptolemy produces no error exceeding 3° in the case of the superior planets and Venus. The Sun and Mercury are indeed exceptional cases, each having its own historical and technical complications.

8.3 The Values for the Rate of Precession in Medieval Middle Eastern Astronomy

After Ptolemy, medieval Middle Eastern astronomers[8] measured non-Ptolemaic values for the rate of precession from the ninth through the fifteenth century, which are, without exception, better estimations than Ptolemy's. A concise summary follows:

1. 1°/76.4^y: in his *al-Qānūn al-mas ʿūdī* VI.8,[9] Bīrūnī compares his value of 189;24° for the longitude of Spica (α Vir), measured on July 2, 1009 CE (error ~−39′), with the value of 172;20° Ptolemy computes from his lunar theory for the date and time provided in Timocharis's observational record of the star's occultation by the Moon on March 9/10, 294 BCE (error ~+20′),[10] which results in the rate of precession of 16 cycles (of 360°) in 160,696,125 days (or about 1° in 76.4 years in the interval of time of 475,670 days between them).

---

8 About the life and works of the medieval astronomers referred to here, see DSB, NDSB, BEA, EI2, Sezgin 1978, Rosenfeld and İhsanoğlu 2003.

9 al-Bīrūnī 1954–1956, vol. 2, p. 676–678, F: ff. 100r–v, B: ff. 150r–v.

10 Almagest VII.3: Toomer 1998, 335–336, Pedersen 1974, 410, no. 13.

2. $1°/75^y$, Abu 'l-Wafā' al-Būzjānī's (June 10, 940–after May 25, 997) formal value, as mentioned in his *Majisṭī* VI.9.2.[11]

3. $1°/73^y$, used by Jamāl al-Dīn Muḥammad b. Ṭāhir b. Muḥammad al-Zaydī of Bukhārā (*fl.* the latter part of the thirteenth century). See later text.

4. $1°/72^{1/_2y}$, derived by the Banū Mūsā or Thābit b. Qurra (mentioned in the *On the solar year*)[12] from the comparison between a value for the longitude of Regulus (α Leo) observed in 830/831 CE and Hipparchus's corresponding value measured in 129/128 BCE.[13]

5. $1°/72^y$, adopted in the *Tāj al-Azyāj* (*Crown of the zījes*) by Muḥyī al-Dīn al-Maghribī (d. June 1283),[14] an excellent medieval approximation to the true value of $1°/71.6^y$.[15]

6. $1°/70^{1/_3y}$: Bīrūnī compares Abu 'l-Wafā''s value $\lambda = 135;30°$ for the longitude of Regulus, measured in 974/975 CE (error ∼−7′), with Hipparchus's corresponding value of 119;50°, determined in 129/128 BCE (error ∼−34′) and derives the value of 1° in 70 Persian/Egyptian years and 4 months.

7. $1°/70^{1/_4y}$, Ibn Yūnus's (d. 1009 CE) formal value for the rate of precession.[16]

8. $1°/70^y$, the formal value employed in late Middle Eastern Islamic astronomy. It comes back to the comparison between the longitudes of Regulus observed by Ibn al-Aʿlam (d. 985 CE) in 975/976 CE and the *Mumtaḥan* astronomers in 829/830 CE.[17] In his *Ḥākimī zīj*, Chapter 5,[18] Ibn Yūnus lists the 12 values for the longitude of Regulus measured by six individuals or groups of astronomers belonging to the early Islamic period (e.g. the *Mumtaḥan* observers and the Banū Mūsā and Banū Amājūr families) in the period from 826/827 through 975/976 CE and nicely tries to demonstrate that all of them are in accordance with the rate of precession of $1°/70^y$.

9. $1°/69^y$ (precisely speaking, "1° in 68 Persian/Egyptian years and $11^{1/_2}$ months"): Bīrūnī's formal value, which, quite probably, has its basis in the comparison between Ptolemy's (176;40°) and his values for the longitude of Spica, although he mistakenly (or because of an untraceable corruption in the manuscript tradition of his work) presents it as a result of (1).[19]

---

11 Abu 'l-Wafā', *Majisṭī*, f. 94r. See Mozaffari 2023a, pp. 481–484, for the reconstruction of a plausible way through which he might have reached that figure.

12 Neugebauer 1962, p. 279, Goldstein 1994, p. 190.

13 *Almagest* VII.2: Toomer 1998, p. 328, Pedersen 1974, p. 415, no. 46.

14 Dorce 2002–2003, p. 198, 2003, pp. 111, 180.

15 The value of $1°/72^y$ can be found in the *Barcelona Tables* (*ca.* 1381) and also was independently measured at Italy or France in 1306 (as documented in a codex preserved in Vienna, no. 5311, f. 137r; see Goldstein 1994, pp. 193, 196–197; for the star tables in this manuscript, see Kunitzsch 1986).

16 Ibn Yūnus, L: pp. 123, 125.

17 The star tables in the *Mumtaḥan zīj*: E: pp. 188–190 and L: ff. 31v, 153r. For a study of them and their relation to Ibn al-Aʿlam, see Mozaffari 2016–2017.

18 Ibn Yūnus, *Zīj*, L: pp. 106–108; Caussin 1804, pp. 143–155.

19 He also adopted the same value in order to construct his own table of the solar apogee's motion (al-Bīrūnī 1954–1956, vol. 2, p. 695).

10. $1°/66^y$, the frequently measured value in medieval Islamic astronomy that was usually derived whenever the Islamic astronomers compared their observed stellar longitudes with those in Ptolemy's star catalogue in *Almagest* VII.5–VIII.1. The examples are the *Mumtaḥan* astronomers,[20] the Banū Musā (using their value for the longitude of Regulus as compared to Ptolemy's corresponding value measured on February 23, 139),[21] al-Battānī (*Zīj al-Ṣābi'*, Chapter 51),[22] and Muḥyī al-Dīn al-Maghribī, adopted in his last *zīj*, the *Adwār al-anwār* (on the basis of his systematic stellar observations made in Marāgha from January 1267 to September 1268).[23]

Whenever the medieval Middle Eastern astronomers compared their own observed stellar longitudes with those registered in Ptolemy's catalogue, always with an extension of them to Menelaus's time, they obtained values close to $1°/66^y$ for the rate of precession; but when they compared their observed stellar longitudes either with those recorded by Ptolemy from his Greek predecessors or with those of their Islamic predecessors, they yielded smaller values, very often about $1°/70^y$. The latter value was dominant in the late Islamic period. It was adopted in the official works connected to the two observatories founded at Marāgha and Samarqand, specifically, the *Īlkhānī zīj*[24] and Ulugh Beg's *Sulṭānī zīj* III.13,[25] as well as in Ibn al-Shāṭir's *Jadīd zīj* (Chapter 19)[26] and al-Kāshī's *Khāqānī zīj*.[27]

8.4 The Modern Values for the Motion of the Planetary Apsidal Lines

The heliocentric apsidal lines of all the planets slowly rotate in the direction of increasing longitude at various rates. In order to derive the angular velocities of their geocentric apsidal lines, the motion of their heliocentric apsidal lines should be combined with that of the apsidal line of the Earth.[28] The annual motion of the

---

20 It should be noted that the rate of precession in the *Mumtaḥan zīj* (E: p. 187) is equal to 0;0,54,44,20° per *Arabic* year, which is approximately equal to 1° in 66 *Arabic* (*not* Egyptian/Persian) years. On the probable Indian origin of this value, see Pingree 1964, p. 138, 1972, p. 29, 1976, p. 113.

21 *Almagest* VII.2: Toomer 1998, p. 328, Pedersen 1974, p. 420, no. 83, Neugebauer 1962, pp. 281–282.

22 Nallino [1899–1907] 1969, vol. 3, pp. 187–190. See Chapter 10, pp. 283–284.

23 al-Maghribī, *Talkhīṣ* VII.1–2: L: ff. 111r–115r; *Adwār*, M: f. 82v, CB: f. 80v. See Chapter 9.

24 al-Ṭūsī, *Īlkhānī zīj*, P: f. 56v, T: f. 100r, M1: f. 100v, M2: f. 86v, C: p. 195.

25 Ulugh Beg, *Sulṭānī zīj*, P1: f. 116v, P2: f. 132v.

26 Ibn al-Shāṭir, *Zīj*, O: f. 61v, L1: ff. 23r–v, L2: ff. 28v–29r, PR: ff. 30v–31r.

27 Kāshī, *Zīj*, IO: f. 166r.

28 The explanation of the whole procedure is beyond the scope of this chapter. Nonetheless, it is worthwhile that since the eccentricities of the superior planets and Mercury are significantly larger than that of the Earth, the motion of their geocentric apsidal lines is greatly dependent upon those of their heliocentric apsidal lines, as can be seen in the preceding table; but the eccentricity of Venus is much less than that of the Earth, and as a result, its geocentric apsidal line always locates in the vicinity of the Earth's line of apsides.

geocentric and heliocentric apsidal lines is compared in Table 8.2 (both in degree per year and arc second per year).[29]

It is conspicuous that the motion of the planetary geocentric apsidal lines is not equal to the rate of precession, neither the modern value $1°/71.6^y$, nor any of the historical values mentioned earlier. Furthermore, the rates of the motion of the apsidal lines in an astronomical system based upon the tropical longitudes differ to a degree that will be evident from the comparison between the tolerably accurate observations made over a long interval of time, say, Ptolemy's observations and those made by any competent medieval observer. Accordingly, in a sufficiently long period after Ptolemy, the discovery that his assumption can no longer be held was probable.

8.5 New Findings for the Motion of the Apsidal Lines of the Sun and Planets in Islamic Astronomy

8.5.1 The Sun: A General Overview

Early Islamic astronomers in the ninth century faced a problem stemming from clear observational evidence: they discovered that the solar apogee, which was assumed to be tropically fixed in the *Almagest*, has a progressive motion. The Banū Mūsa extensively addressed this problem in their *On the solar year*. They drew an analogy between the assumed dependence of the motion of the planetary apogees on that of the fixed stars (i.e., precession) and the solar apogee motion. In fact, they generalized the Ptolemaic assumption of the equality of the motion of the apsidal lines of the planets with precession to include the observed motion of the solar apogee as well.[30] So by the start of the tenth century, it was accepted that the solar apogee is subject to precession, as seen in influential works like as al-Battānī's *Ṣābiʾ zīj* (chap. 33).[31]

Nevertheless, just over a century later, Bīrūnī found it necessary to dedicate a significant portion of his *magnum opus, al-Qānūn al-masʿūdī*, book VI, to a

Table 8.2 The modern values for the motion of the solar and planetary apogees

	Sun/Earth	*Saturn*	*Jupiter*	*Mars*	*Venus*	*Mercury*
Geocentric		$\sim1°/50.8^y$	$\sim1°/61.2^y$	$\sim1°/54.1^y$	$\sim1°/53.2^y$	$\sim1°/64.9^y$
	$\sim1°/58.2^y$	$70.9''/^y$	$58.9''/^y$	$66.5''/^y$	$67.6''/^y$	$55.5''/^y$
Heliocentric	$61.9''/^y$	$\sim1°/50.9^y$	$\sim1°/62.0^y$	$\sim1°/54.3^y$	$\sim1°/71.3^y$	$\sim1°/64.3^y$
		$70.7''/^y$	$58.1''/^y$	$66.3''/^y$	$50.5''/^y$	$56.0''/^y$

29 The heliocentric orbital elements of the planets and the rates of change in them can be computed, for example, by the formulas given in Simon *et al.* 1994, from which we derived the speed of rotations of their geocentric apsidal lines.

30 Neugebauer 1962, pp. 265–266.

31 Nallino [1899–1907] 1969, vol. 3, p. 108.

comprehensive analysis and interpretation of his Muslim predecessors' findings. Bīrūnī initially concluded that the motion of the solar apogee is faster than the rate of precession, and calculated the rates to be about $1°/46^y$ and $1°/60^y$ for its motion (note that the latter is close to the true rate of about $62''$ given earlier) by comparing his value about $85°$ for its longitude measured in 1016–1017 CE with the value $65.5°$ determined by Ptolemy in 140 CE and by Hipparchus around the middle of the second century BCE. But he later discarded this idea, for two clear reasons. The first was the sensitivity of the methods of deriving the solar orbital elements. The second was that Ptolemy's value for the longitude of the solar apogee was not trustworthy because, as Bīrūnī thought, it was expected that Ptolemy would have detected its motion in the long period that had passed after Hipparchus, as he had discovered the motion of the planetary apogees.

At the time of writing *al-Qānūn* in the 1030s, during the last years of Bīrūnī's career, he was still uncertain whether the center of the motion of the solar apogee coincides with the Earth or lies outside of it. He states that from what Ptolemy quotes from Hipparchus, one can infer that the latter believed that the apogee rotates about a point outside of the center of the universe, due to the various values he derived for its motion, sometimes speedy and sometimes retarded.[32] He concludes that the Sun has an annual equation (i.e., the so-called "equation of center"), and that the deviation or difference due to the assumption that the center of the apogee's motion does not coincide with the Earth makes a second equation for the Sun's motion, by which the first equation should be adjusted, and which would be perceptible in two years. He seems to have considered this idea, since he says next that "it circulates in my mind for my knowledge of such deviations" (he refers to his lengthy discussion of the determination of the solar parameters as regards a thorough analysis of his observations and those made by his Islamic predecessors). But he immediately changes his argumentation and concludes that Indian astronomical traditions and, afterward, his Islamic predecessors maintain that the motion of the solar apogee is around the center of the universe.[33] Thus, he decides to provisionally accept it, since nothing else appears,[34] but he passes

---

32 There is no doubt that the chapter in the *Almagest* giving such an impression is III.1, in which Ptolemy quotes a passage from Hipparchus, clearly presenting the latter's opinion of the "irregularity in the length of the year" (Toomer 1998, p. 133; also, for a detailed analysis of this chapter, see Jones 2005). Note that it is Bīrūnī who relates Hipparchus's idea to, and considers it as a consequence of, the assumption of a non-uniform motion for the solar apogee, due to the hypothesis that the center of its motion does not coincide with that of the Earth.

33 In Indian astronomy, the apogees move at unequal rates, but much less than the rate of precession; for example, see *Sūrya Siddhánta* [1860] 1997, pp. 29–32, [1861] 1974, p. 7, Kennedy and Pingree 1981, pp. 118–119, 282.

34 Apparently, during the medieval Islamic period, Indian astronomy serves as an alternative tradition to Ptolemaic astronomy for providing easy solutions and justifying tools for unprecedented problems. An amazing example is the problem of annular solar eclipses; about it, see Mozaffari 2013, 2015, and Chapters 2 and 3.

the problem to the next generations to find a solution for it in the future.[35] To the best of our knowledge, except for Jamāl al-Dīn al-Zaydī and Ibn al-Shāṭir of Damascus (*ca.* 1305–*ca.* 1375 CE), all the medieval Middle Eastern astronomers simply chose the latter option; no one dealt with this enigmatic problem again, and therefore, no one departed from the generalization of Ptolemy's assumption of the correlation between the motion of the planetary apogees and precession to the motion of the solar apogee.

However, Western Islamic astronomers, located in northwestern Africa and Andalus in Spain, pursued a different line of thought in their astronomical studies. Unlike their Middle Eastern counterparts, they seriously considered secular changes and periodic fluctuations in parameters that were assumed to be constant in Ptolemaic astronomy. Ibn al-Zarqālluh developed a solar model with a variable eccentricity and also adopted a proper motion of 1°/279 Julian years in the direction of increasing longitude for the solar apogee in a sidereal reference system of longitudes.[36] This value is about 1″ greater than the corresponding modern value of approximately 11.6″ (note that the latter value is the difference between the modern value 61.9″ for the annual motion of the solar apogee in a tropical reference system of longitudes, as given earlier, and the true rate of precession of 50.3″ per year).

Over five centuries later, Taqī al-Dīn Muḥammad b. Maʿrūf (1526–1585 CE), the director of the short-lived Istanbul observatory, significantly refined this value, following Bīrūnī's approach. He derived an annual motion of about 63″ from the value of about 95;33° for the longitude of the solar apogee he measured in 1579 CE and the Hipparchian value of 65;30°, using Hipparchus's observed time of the vernal equinox of March 24, 146 BCE, as the starting point. This is as accurate as Ibn al-Zarqālluh's value; both are indeed the best estimations of the motion of the solar apogee in medieval and early modern Islamic astronomy.[37] Of course, Taqī al-Dīn did not address the planets and fixed stars in his two *zījes*, as no material related to them can be found in either, perhaps because the Istanbul observatory

---

35 *al-Qānūn* VI.8: al-Bīrūnī 1954–1956, vol. 2, pp. 656, 661, 676, 680, 685, Hartner and Schramm 1961, p. 218.

36 Toomer 1969, pp. 316–317, 1987, p. 514, Samsó 1990, p. 10, Samsó and Millás 1989, p. 7.

37 The time of Hipparchus's observation is given as dawn, 27 Mechir (6th month) 178 death of Alexandria (= 602 Nabonassar) in *Almagest* III.1 (Toomer 1998, p. 138, Pedersen 1974, p. 414, no. 38). Taqī al-Dīn gives the difference in time between Hipparchus's and his observations as 2,53,42,16;21,13,28,30 days and then derives the daily motion of the solar apogee as

$$\frac{95;33,14,40,21,21° - 65;30°}{2,53,42,16;21,13,28,30\,\text{days}} = 0;0,0,10,22,51,5\,7,58°/\text{d} \qquad [\ldots,58,0].$$

Taqī al-Dīn, *Sidra*, K: f. 36v, N: f. 47v, V: f. 40r. His value for the interval of time is in error, but taking the correct value into account also results in the value 63″ for the annual motion of the Sun's apogee. For a detailed analysis of the accuracy of Taqī al-Dīn's observations in Istanbul, see Chapter 5.

was destroyed in the early 1580s,[38] before its observers had sufficient time to complete their ambitious observational program.[39] Consequently, it is unclear whether Taqī al-Dīn believed that the solar and planetary apogees are sidereally fixed, or whether his value of 63″ should be considered to include a proper motion for the apogee of the Sun. However, the former seems to be the case here, because Middle Eastern Islamic astronomers typically either measured the motion of the solar apogee and then applied that result to the motion of the planetary apogees and fixed stars (e.g., Ibn al-Fahhād; see following text) or determined the motion of some bright reference zodiacal stars and then used the value obtained for the rate of precession as the motion of the solar and planetary apogees (e.g., al-Battānī, Bīrūnī, and Muḥyī al-Dīn al-Maghribī) (the references have been mentioned earlier in Section 8.3, (8)–(10)).

8.5.2 The Planets

It is of great historical interest that if medieval Islamic astronomers also compared their values for the longitudes of the planetary apogees with Ptolemy's corresponding values, they would have inevitably had to confirm that the motions of the planetary apsidal lines are neither identical to each other nor equal to the rate of precession. Nevertheless, the majority of them strictly maintained the Ptolemaic assumption. However, some of the astronomers belonging to Western Islamic astronomical tradition applied Ibn al-Zarqālluh's value for the proper motion of the solar apsidal line to the planetary ones and, hence, maintained that the apogees of all planets are located at fixed distances from the solar apogee: Ibn al-Kammād (*fl.* 1116 CE), Ibn al-Hā'im (*fl.* 1205 CE), Ibn al-Raqqām (d. 1315 CE) in the *Shāmil zīj*, Ibn 'Azzuz al-Qusanṭīnī (*fl.* 1318–1354 CE) in the *Muwāfiq zīj*, and the authors of the *Tables of Barcelona*; but Ibn Isḥāq (*fl.* ca. 1193–1222 CE), Ibn al-Raqqām in the *Qawīm zīj* and *Mustawfī zīj*, and Ibn al-Bannā' (1256–1321 CE) in the *Minhāj al-ṭālib* restricted the idea to the apogees of the inferior planets.[40]

The principal problem a historian of astronomy comes across in the pursuit of this subject is that it is very often not known whether and how the unprecedented values for the planetary orbital elements adopted in the Islamic *zījes* were derived from actual observations. In fact, unlike in the case of the Sun and the Moon, the accounts of the observations and subsequent computations for the derivation of the planetary orbital elements from the observational data can be very seldom found in medieval Middle Eastern astronomy. The only surviving work on this topic is Muḥyī al-Dīn al-Maghribī's *Talkhīṣ al-majisṭī* (*Compendium of the* Almagest), which includes his extensive systematic observations carried out at the Marāgha observatory from 1262 to 1274 CE. Of course, scattered, random observations

---

38 Sayılı [1960] 1988, pp. 290–292.

39 In the account of his last documented observation dated to 11 December 1579 (Chapter 5, p. 123), Taqī al-Dīn states that he had not yet made any reliable stellar observation.

40 Samsó and Millás 1998, pp. 268–270, and later modifications in Samsó 2020, pp. 659–661.

can be found here and there, which have little or nothing to do with the measurement of the planetary orbital parameters.[41] Writing a while after the turn of the eleventh century, Bīrūnī complains that the Islamic astronomers from the day of al-Maʾmūn to al-Battānī and afterward did not mention their measurements in the same manner that Ptolemy mentioned his own ones and did not explain how they derived the planetary motions and positions through the continuous attempts those astronomers made to attain them.[42] Bīrūnī was very interested in collecting and precisely scrutinizing the findings of his predecessors. His complaint indicates that in the materials available to him (if not in the whole legacy of Islamic astronomy until his time), there was not a single account of the derivation of a specific planetary parameter from observations.

In fact, there is no evidence to believe that in a good number of the prominent works in Islamic astronomy, which exerted notable influence on later works, the longitudes of the apogees of the planets were derived from observational data. It should be noted that in order to measure the orbital elements of the superior planets, at least, the observations of three oppositions of a planet to the mean sun are required, from which, first, the eccentricity is computed according to Ptolemy's iterative procedure, and then the direction of the apsidal line is determined. The changes in the geocentric eccentricities of Saturn and Jupiter during one millennium after Ptolemy are noticeable to the extent that it is, by and large, improbable that they would remain obscure in a medieval systematic observational program.[43] Accordingly, in the case of a few Islamic astronomers for whom we can be certain that the new parameter values adopted in their works are observational achievements, both sets of the eccentricities and apogee longitudes have values independent from Ptolemy. Examples that provide undisputable pieces of evidence

41 For example, the observations of the planetary conjunctions or their near appulses to fixed stars that Ibn Yūnus collects from his Muslim predecessors in addition to his own, which is, indeed, the largest collection of such observations in Islamic astronomy, or those made by al-Kawāshī in Egypt in the 1270s and 1280s (see King and Gingerich 1982).

42 al-Bīrūnī 1954–1956, vol. 3, p. 1193, P: f. 169r; a similar remark can be found in *ibid.*, p. 1197.

43 The geocentric eccentricity of Jupiter increased from 2.70 to 2.91 (the radius of the geocentric orbit, the deferent, is taken as 60 units, according to Ptolemaic norm), and that of Saturn from 3.62 to 3.23, but that of Mars remained nearly constant, about 6, as Ptolemy derived, during the two past millennia. In late medieval Middle Eastern astronomy, there are 12 non-Ptolemaic values for the eccentricities of the superior planets (see Mozaffari 2014a), and 7 values for their epicycles radii. In the case of the inferior planets, the rates of changes in the geocentric eccentricities are too small that they can be neglected (~0.02 in every millennium for Venus, and ~0.04 per millennium for Mercury). The Islamic astronomers correctly found that the eccentricity of Venus is intrinsically correlated to that of the Sun/Earth, and so the majority of them made use of the values a bit less than those they measured for the solar eccentricity for that of Venus. There can be found only three non-Ptolemaic values for the eccentricity of Mercury in Islamic astronomy, measured by Ibn al-Aʿlam, Ibn Yūnus, and Ibn al-Zarqālluh (of course, Ibn Yūnus's value for the eccentricity of Mercury goes back to pre-Islamic Persian astronomy; already noted in King 1999, 502; see Ibn Yūnus, *Zīj*, L: pp. 121, 191–193, Caussin 1804, p. 221, Pingree 1968, p. 104, 1970, p. 113), and only one new value for the radius of the epicycle of either of the inferior planets, measured by Ibn Yūnus (see following, note 52).

for it are the non-Ptolemaic values derived from the observations carried out by Ibn al-Aʿlam at Baghdad in the latter part of the tenth century, by Muḥyī al-Dīn al-Maghribī in the Marāgha observatory, and contemporarily, by a group of Persian astronomers, under the supervision of the aforesaid Jamāl al-Dīn al-Zaydī, at the Islamic Astronomical Bureau established in Beijing by the Mongolian Yuan dynasty (1260–1368 CE) in the latter part of the thirteenth century,[44] as well as by the astronomers in the Samarqand observatory over one century later.[45]

As a first step, in Section 8.5.2.1, we shall select and discuss some examples from important *zījes* written in the medieval Middle East, where the values adopted for the longitudes of the apogees of the five planets either do not appear to have been based upon a firm observational ground or were, obviously, derived from the *Almagest*. Then, in Section 8.5.2.2, we will turn our attention to the works, mainly belonging to the late Islamic period, where the longitudes of the planetary apogees are, either certainly or quite probably, observational achievements.

8.5.2.1 Sets of Apogee Longitude Values Largely Based on Previous Data

EXAMPLE 1: THE *MUMTAḤAN* ASTRONOMERS AND IBN YŪNUS

First example is Ibn Yūnus, who gives, in his *Ḥākimī zīj*, Chapters 6,[46] 8, and 9[47] (in the beginning of the tables for the planetary mean motions in longitude), the values for the longitudes of the planetary apogees along with the corresponding values from the *Mumtaḥan zīj* (E: f. 86v, L: ff. 60v–61r), as summarized in Table 8.3.[48]

In Chapter 6, Ibn Yūnus's values, with the exception of that of Venus (whose apogee was assumed to be at the same longitude as the Sun), exceed the corresponding values of *Mumtaḥan* by 1;3°. This increase is less than half of the expected values for the apogeal motion resulting from any of the precessional rates of 1°/66^y, 1°/70^y, or 1°/70¼^y (which is Ibn Yūnus's own value) over a time span of 173 years following the *Mumtaḥan* epoch. However, the value provided in Chapter 8, as well as in the tables, shows an increase of 2;28° or 2;30°. This

---

44 About it, see Yabuuti 1997, van Dalen 2002a, 2002b, Yang 2017, Isahaya 2021.

45 The unprecedented planetary parameter values in Ulugh Beg's *Sulṭānī zīj*, the official fruit of the observational activities in Samarqand, have already been introduced thoroughly in Chapter 7.

46 Ibn Yūnus, *Zīj*, L: pp. 120–121; Caussin 1804, pp. 217–220.

47 Ibn Yūnus, *Zīj*, L: pp. 125, 145, 147, 149, 151, 153. King 1999, p. 502.

48 Early Islamic astronomers, like al-Battānī and Ibn Yūnus, adopted the wrong hypothesis that the orbital elements of the Sun and Venus are equal, so that the apsidal lines of the Sun and Venus coincide with each other. This hypothesis comes from the Indian midnight system (*Ārdharātrika*) developed by Āryabhaṭa (b. 476 CE), which is preserved in the *Pañcasiddhāntikā* of Varāha Mihira (505–587 CE) and the *Khaṇḍakhādyaka* of Brahmagupta (598–670 CE); see *Pañcasiddhāntikā* III.2–3, IX.7–8, XVI.12–14: Neugebauer and Pingree 1970–1971, vol. 1, pp. 39, 93, 149–151, vol. 2, pp. 24, 69–70, 101–102; *Khaṇḍakhādyaka* I.13 and II.6: Brahmagupta 1970, pp. 49–50, 54, 283, Kennedy 1958, pp. 256–257, Kennedy and Pingree 1981, p. 220, Pingree 1965, 1968, 1970, Van Der Waerden 1987a, esp. pp. 530–532, 1987b, Mozaffari 2019a.

Table 8.3 The *Mumtaḥan* and Ibn Yūnus's values for the longitude of the solar and planetary apogees

		Sun/Earth	Saturn	Jupiter	Mars	Venus	Mercury
The Mumtaḥan zīj 199 Y April 28, 830		82;39°	244;30°	172;32°	124;33°[49]	82;39° = Sun	201; 0°
		83;30	**249;48**	**171;33**	**129;27**	**70;21**	**233; 8**
Ibn Yūnus 372 Y March 16, 1003	Ch. 6 Ch. 8 and Tables	86;10	*245*;33[50] 24*7* (or 250?)[51]	173;35 175	125;36 130	86;10° = Sun	202; 3 203;30
		85;51	**253;11**	**174;20**	**132;38**	**73;34**	**235;46**

is consistent with the period between him and the *Mumtaḥan zīj* and his rate of precession $((372–199)/70¼ = 2;28°)$, with the exception of Mars (and probably, Saturn), which seem likely to be related to his own observations.

Indeed, we do not know how the *Mumtaḥan* observers arrived at their own values for the longitudes of the planetary apogees, except for Venus. For Venus, both the *Mumtaḥan zīj* and Ibn Yūnus simply used their values for the longitude of the solar apogee. All the astronomical observations we have from the astronomers of the *Mumtaḥan* scientific circle – who observed in Baghdad and Damascus in the first half of the ninth century under the orders of al-Maʾmūn, the seventh Abbāsid caliph, from 813 to 833 CE – are related to measurements of solar parameters, the obliquity of the ecliptic, and a number of stellar observations. As demonstrated elsewhere,[52] similar to the Almagest star catalogue, we find systematic errors ranging from $-0.5°$ to $-1°$ in the stellar longitudes in the Mumtaḥan star table. Their values for the longitudes of the planetary apogees might have been dependent upon the *Almagest* in some way, although the differences between them are not uniform, amounting to $11.5°$ in the case of Jupiter and Saturn, $11°$ for Mercury, and $9°$ for Mars.

We have found that Ibn Yūnus's values for the longitudes of the planetary apogees are majorly dependent upon the *Mumtaḥan zīj*. However, he criticizes the *Mumtaḥan* astronomical tradition for the deviations observed between the quantities computed on the basis of his *Mumtaḥan zīj* for some astronomical phenomena such as the planetary conjunctions and near approaches to the stars and the data that he and his Muslim predecessors obtained from making direct observations of them during the two centuries intervening between the *Mumtaḥan* astronomers and him. A worthwhile example is Ibn Yūnus's estimation of the time of the conjunction between Jupiter and Saturn in 1007 CE. From an observation made from Cairo early in the morning on November 7, 1007, he estimated that

49 In the text: *93*;33, obviously a scribal error in the *abjad* numerals (جـ جـ لـ → د د لـ).
50 A senseless figure in the manuscript, *30;20*,33, is clearly a scribal error in the *abjad* numerals, which seems to have arisen from the confusion with the Mumtaḥan epoch value of the mean longitude of Jupiter written down in the next line.
51 In Chapter 8: نحلج may be read as 250 (ن → حـ?); in the Tables: 246 (ن → و).
52 Mozaffari 2016–2017.

the conjunction would occur about noon on the given date, but according to the *Mumtaḥan zīj*, at that time Jupiter should be 22′ in longitude ahead of Saturn, and so this conjunction should have occurred about 6 equinoctial hours after noon on October 31, 1007, that is, about one week earlier.[53] A further example is Ḥabash's critical remarks about the *Mumtaḥan zīj*, as recorded in Ibn Yūnus's *Ḥākimī zīj*.[54]

It is important to note that the comprehensive planetary observations, either conducted by Ibn Yūnus himself or those he documented from his Muslim predecessors, only encompass the planetary conjunctions and the near approaches of the planets to stars (typically Regulus). At most, these observations can be used to assess the accuracy of the ephemerides calculated based on a specific set of astronomical tables, as well as to measure the sizes of the planets' epicycles. Ibn Yūnus seems to have used them in exactly these ways; in addition to his comments about the correction of the earlier *zījes*, he is the only medieval Middle Eastern astronomer who has non-Ptolemaic values for the radii of the inferior planets' epicycles. These values might have been derived from his observational data, marking his unique contribution to planetary astronomy.[55]

EXAMPLE 2: AL-BATTĀNĪ

The following observations can be made about al-Battānī's epoch values for the longitudes of the planetary apogees in Table 8.4.[56]

He obviously follows the Indian Midnight system, dominant in early medieval Middle Eastern astronomy, according to which the apogee of Venus shares the same longitude as that of the Sun; it results in a significantly large error of about $+11°$ for the longitude of the apogee of Venus, which is a common characteristic of the early Islamic values for the longitude of the apogee of the planet, as we have seen in the case of the *Mumtaḥan zīj* and Ibn Yūnus too.[57]

In the case of Saturn, Mars, and Mercury, the difference between al-Battānī's and Ptolemy's values amounts to 11;28°, which is very close to the apogeal/

53 Ibn Yūnus, *Zīj*, L: pp. 119–120; Caussin 1804, pp. 210–215. In fact, this conjunction took place on November 8, at 2:50 MLT, and thus, Ibn Yūnus's estimation is about -15 hours in error.

54 See Chapter 5, pp. 110–111.

55 For Venus, Ibn Yūnus's maximum value for the epicyclic anomaly is equal to 46;25° (corresponding to the radius of the epicycle $r \approx 43;28$), and for Mercury, 22;24° ($r \approx 22;52$); see Ibn Yūnus, L: pp. 121, 190, 192; Caussin 1804, p. 221.

56 al-Battānī, *Ṣābiʾ zīj*, Chapter 45: E: f. 117v; Nallino [1899–1907] 1969, vol. 3, p. 173; also, in the tables of planetary equations: E: ff. 208v, 211v, 214v, 217v, 220v; Nallino [1899–1907] 1969, vol. 2, pp. 108, 114, 120, 126, 132. MS. E was written in the Western Arabic script from the late eleventh or early twelfth century, and the alphanumerics in it are in the Maghribī *abjad* sequence. The following scribal errors in the *abjad* numerals can be found both in the end of Chapter 45 and in the planetary equation tables in it: Jupiter, فصد كح for قصد كح; and Mars, فكو يح for قكو نح. These radix values were missing from the end of Chapter 45 in the Latin translation published in 1537, f. 73r, and in 1645, p. 185. About the surviving materials of al-Battānī's *zīj* in Arabic, Castilian, and Latin, as well as an assessment of the tables of this work, see van Dalen and Pedersen 2008 and the references mentioned therein.

57 For more details in this regard, see Mozaffari 2019a.

Table 8.4 al-Battānī's values for the longitude of the solar and planetary apogees

	Sun/Earth	Saturn	Jupiter	Mars	Venus	Mercury
al-Battānī	82;14°	244;28°	1_64;28°	1_26;_58°	82;14° = Sun	201;28°
1191 Alexander September 1, 879	**83;44**	**250;46**	**172;21**	**130;22**	**71;16**	**233;53**

precessional motion of 11;25° which al-Battānī measured between Ptolemy's and his own times on the basis of a comparison between Ptolemy's and his own values for the longitudes of the three stars (β Sco, α Leo, and α CMa).[58]

The only surviving complete manuscript of al-Battānī's *zīj* in Arabic (MS. E) has a value of 164;28° for Jupiter, while it should be 172;28° in order to be in accordance with those given for the other three planets. Nevertheless, a look at the *zījes* in which the longitudes of the planetary apogees are taken from al-Battānī's radix values, indeed, confirms that the later astronomers, both in the Eastern and Western medieval Islamic domains, also read the value 164;28° for his epoch value for the longitude of the apogee of Jupiter. For example, in the table of the radix values for the longitudes of the solar and planetary apogees in Ṭabarī's *Mufrad zīj*,[59] al-Battānī's values have been updated for the beginning of 431 Y (1 Ādhār 1373 Alexander/March 1, 1062) by adding an increment of 2;45°, which is in agreement with the rate of precession of 1°/66^y and the interval of time of about 182 years between al-Battānī's and Ṭabarī's epochs, and the latter has 167;13° for Jupiter. Another example: in his *Ashrafī zīj*, Kamālī has added a value of 6;27,11° to al-Battānī's values for the longitudes of the apogees in order to update them for 1_3 *Adhar* [6] 161_4 Alexander/23 *Rajab* 702/13 *Khurdād* 672 (= March 13, 1303),[60] and the value for Jupiter is 170;55,11°. Moreover, Ibn al-Bannā''s epoch values for the longitudes of the apogees of the superior planets are less than al-Battānī's by 4;45,15°, which, in all likelihood, indicates al-Battānī's *zīj* as an actual source of Ibn al-Ishāq and Ibn al-Bannā', and the latter has 159;42,45° for Jupiter.[61] Therefore, Ṭabarī's, Kamālī's, and Ibn al-Bannā''s values evidently indicate the use of al-Battānī's radix value of 164;28°. We will show elsewhere that

58 See Section 10.4. Battānī mentions, in Chapter 51 of his *Sābi' zīj* (E: f. 127r; Nallino [1899–1907] 1969, vol. 3, pp. 187–188), that he carried out the stellar observations in 1627 Nabonassar/1191 Alexander (879–880 CE), and in Chapter 45 (E: f. 117v; Nallino [1899–1907] 1969, vol. 3, p. 172), the longitudes of the apogees are given for 1191 Alexander too. An explanation for the difference of 0;3° between al-Battānī's values for the increments of the apogeal and precessional motions in the period from Ptolemy to him may be that he took a time of about 3 years before the epoch of Ptolemy's star catalogue (the beginning of 885 Nabonassar/July 20, 137 CE) as that of Ptolemy's values for the longitudes of the planetary apogees, that is, about the mid-135s CE, which does not seem to be an unreasonable assumption with regard to the times of Ptolemy's derivations, as given earlier.

59 Ṭabarī, f. 175v.

60 Kamālī, F: f. 232v, G: f. 249r. In MS G, the Alexandrian date is wrongly given as 1_4 *Adhar* 161_2.

61 Samsó and Millás 1998, pp. 265–268. Note that Ibn al-Bannā''s values are for a date in which the precession according to the Western Islamic trepidation models is zero.

Battānī's modification of Ptolemy's value of 161;0° for the longitude of Jupiter's apogee as updated for his time by the precessional increment measured through his stellar observations (11;28°) by −8° might have had its basis in his observations performed in the periods of 888–892 and/or 900–904 CE, when the planet was very close to its apogee.

After all, for our purposes, it is sufficient to know that al-Battānī's epoch values for the longitudes of the apogees, at least for the three planets, were indeed derived from the *Almagest*.

EXAMPLE 3: BĪRŪNĪ

In his *al-Qānūn al-mas ʿūdī* X.4, Bīrūnī corrects Ptolemy's values for the planetary mean daily motions in longitude in a strange way. He first finds that the longitude of the mean sun as extracted from the *Almagest* for his epoch, that is, the noon of Tuesday, the first day of the year 400 Yazdigird (henceforth denoted as "Y")/ March 9, 1031, for the longitude of Ghazna (Modern Ghazni: latitude 33.55°, and longitude 68.43°) is less than the value he himself measured for the given time by 5;0,21,36,24,10,1°.[62] Next, he adds this amount to the increments of the mean longitudes of the superior planets in the time interval between Ptolemy's observations of the third oppositions of these planets to the mean sun (which Ptolemy used in the derivation of the orbital elements of these planets) and his epoch, computed from Ptolemy's values for the mean daily motions. Then, the resulting increments are divided by those time intervals, which results in a new set of values for the planetary mean daily motions, which, inevitably, turn out to be larger than Ptolemy's by about 0;0,0,3,19°.[63] This method cannot be employed for the correction of the mean daily motions in anomaly of the inferior planets as well as for the longitudes of the planetary apogees, as Bīrūnī himself states. Thus, he simply converts Ptolemy's values for the longitudes of the planetary apogees to his epoch on the basis of the value 1°/69^y he had already measured for the rate of precession (Section 8.3, (9)). Bīrūnī takes the mid-time of Ptolemy's observations of the first and third mean oppositions of the superior planets and the mid-time of Ptolemy's pairs of observations of the inferior planets, which he utilized in order to derive the longitudes of their apogees, as the origins (note that in the case of the superior planets, Ptolemy takes his measured values for the longitudes of the apogees of the superior planets for the times of his third mean oppositions). The incremental values in the longitudes of the apogees from Ptolemy's times in the

---

62 Actually, there is a difference of about −5° in the longitude of the mean sun between the values derived from the *Almagest* and computed according to the modern theories for the noon of March 9, 1031/1 Choiach 1779 Nabonassar in Alexandria; a modern computation gives 352;4°, but the *Almagest* gives 347;11°.

63 The periods between Ptolemy's third mean oppositions of the superior planets and Bīrūnī's epoch run from about 892 to about 895 Egyptian/Persian years, and so dividing 5;0,21,36,24,10,1° into them results in the values nearly equal to 0;0,0,3,19°.

192

130s to Bīrūnī's epoch amount to about 13°, which Bīrūnī adds to Ptolemy's values. In the end, he appears to have found it necessary to emphasize again that "we have said that the moderns (*al-muḥaddithīn*) did not mention how they did their works, as Ptolemy mentions [his measurements]. Thus, to us, they became as the puzzles and mysteries."[64]

EXAMPLE 4: IBN AL-FAHHĀD

A further example is Ibn al-Fahhād, who compiled six *zījes*. The first five of them are now lost, and only the last, titled the *'Alā'ī zīj*, survives in a single manuscript. The latter is a highly influential work in medieval Middle Eastern astronomy. For instance, Athīr al-Dīn al-Abharī's *Mulakhkhaṣ zīj* (d. between 660 H/1263 CE and 663/1265), the *Shāmil zīj*, and an anonymous treatise named *Kitāb fī Ṣinā'at al-majisṭī* (*Book on the art/industry of the* Almagest) dedicated to Najm al-Dīn Alī b. 'Umar Dabīrān al-Qazwīnī (d. 1276 CE), a member of the main staff of the Marāgha observatory, are entirely dependent upon Ibn al-Fahhād's *'Alā'ī zīj*.[65] Also, the parameter values adopted in this work were employed in Muḥyī al-Dīn al-Maghribī's *'Umdat al-ḥāsib*, the first *zīj* he compiled after joining the Marāgha team and before he started to make his own extensive observations in the Marāgha observatory, and in the *Īlkhānī zīj*.[66] It is interesting that for the derivation of the orbital elements of the Moon and Mars, Muḥyī al-Dīn provisionally made use of Ibn al-Fahhād's mean daily motions in order to compute the differences in the mean longitude of these celestial objects between his triple observations made in the Marāgha observatory.[67] The *'Alā'ī zīj* was translated into Greek by Gregory Chioniades under the supervision and instructions of Shams al-Dīn Muḥammad

---

64 al-Bīrūnī 1954–1956, vol. 3, pp. 1193–1197, P: f. 169r.

65 This is clearly mentioned in the prologue of the *Mulakhkhaṣ zīj* (K: f. 1v). Note that it is probable that both *Shāmil zīj* and *Ṣinā'a* were written by Athīr al-Dīn al-Abharī himself. The epoch of all these works is the year 600 Y, whose beginning is January 18, 1231. The values for the longitudes of the planetary apogees in them exceed those in the *'Alā'ī zīj* by 0;53,38°, which is in agreement with a precessional/apogeal motion of 1° in every 66 years and the interval of 59 Persian years between their epoch and that of the *'Alā'ī zīj*, 541 Y (Abharī, *Mulakhkhaṣ*, K: ff. 66r–v, 67v, 70v, 72v, 74v, 76v, U: f. 33v, F: f. 82r; *Shāmil*, P1: ff. 23r–v, F2: ff. 44r–v, D: f. 17r, Pa1: f. 22v; *Ṣinā'a*, P: ff. 48v, 65v–66v, 74v–75r, N: pp. 83, 114–116, 136–137, I: ff. 40v, 55v–56v, 64v–65r).

66 For example, the epoch longitude values of the planetary apogees in the *'Umda* (al-Maghribī, *'Umda*, M: ff. 23v, 24v, 25v, 26v, 27r) are more than Ibn al-Fahhād's values, given presently in this chapter, by 1;21,17°, so that all the values are ended with zero in the seconds; by a precessional rate of 1°/66^y and 90 years intervening the epoch of the *'Umda*, that is, the end of 600 Y, and that of the *'Alā'ī zīj*, that is, 541 Y, it should be about 1;21,49°. It seems that either the rounded increment of 1;22° was added to Ibn al-Fahhād's truncated epoch longitudes, or inversely, a truncated increment of 1;21° was added to the rounded Ibn al-Fahhād's longitude values. In the *Īlkhānī zīj*, Ibn al-Fahhād's value for the daily mean anomalistic motion of Mercury, 3;6,24,22,7,59°, was used (*Īlkhānī zīj*, C: p. 132, P: f. 45v, M1: f. 80v).

67 Muḥyī al-Dīn gives the differences in the lunar mean longitude between his trio lunar eclipses (March 7, 1262, April 7, 1270, and January 24, 1274) as 30;57,18° and 277;44,27°, respectively (see Mozaffari 2014b), both of which are in agreement with a daily mean motion of 13;10,35,1,55,32°

193

al-Wābkanawī, the outstanding figure of the second period of the astronomical activities in the Marāgha observatory, in the 1290s.[68]

It is curious to see how Ibn al-Fahhād derives his own values for the longitudes of the planetary apogees. No explanation or dated observation is given in the instructional chapters of the *'Alā'ī zīj*. The key evidence to help clarify this issue can be found in a treatise in Persian partially preserved in Iran, Parliament Library, MS. no. 184, ff. 243v–244r (only its first two pages are preserved in this MS), titled *On the proof of the solar equation (of center), its (longitude of) apogee, its mean motion, and its eccentricity by the observations*. The author gives the lengths of the solar year, spring, and summer from his observations performed in 539 Y/1170–1171 CE, from which he derives the same values 2;4,35 and 1°/66^y, respectively, for the solar eccentricity and the rate of precession as measured by the *Mumtaḥan* group about three centuries and a half before him. The length of the solar year is given as 365;14,30^d, as well as an *adjusted* value of 365^{d}5^{h}46^{m}52^{s}24$'''$ (= 365;14,27,11^d), corresponding to a mean daily motion of 0;59,8,20,35,25°. There exist strong reasons to believe that the author of this treatise is 'Alī b. 'Abd al-Karīm al-Fahhād. He was no doubt the most high-ranking astronomer at that time. The epoch of his *'Alā'ī zīj* is the beginning of the year 541 Y/February 2, 1172. In the prologue of *'Alā'ī zīj* (p. 5), Ibn al-Fahhād states that the data he obtained from his solar and lunar observations were close to those computed on the basis of the parameter values derived from the observations of Ḥabash, Yaḥyā b. Abī Manṣūr, Khālid b. 'Abd al-Malik, 'Abbās b. Sa'īd al-Jawharī (the so-called *Mumtaḥan* group). The most convincing reason is that the aforementioned unprecedented values for the length of the solar year and its correspondent solar mean daily motion were, indeed, adopted in the *'Alā'ī zīj* (the solar mean motion tables in the *'Alā'ī zīj* can be found on pp. 70–72). More support also comes from the values our author gives for the longitudes of the apogees of the planets for the year 539 Y (on f. 244r; see Table 8.5). In the *'Alā'ī zīj*, the same longitudes have been adjusted for the two Persian years later, as explained in the following:[69] Given the value 1°/66^y for the rate of precession, the increment in the longitudes of the apogees in two years amounts to 0;1,49°. The differences between the two sets of the values for the Sun, Venus, and Jupiter exactly match this increment, and the deviations found in the case of the other three planets can be plausibly explained

---

which is equal to Ibn al-Fahhad's value (see van Dalen 2004, p. 832). For Mars, see Mozaffari 2018–2019.

68 On Ibn al-Fahhād, see Kennedy 1956, p. 135, no. 8, Pingree 1985–1986, van Dalen 2004, Mozaffari 2019b, 2023b. For the computation of the solar and lunar eclipses in the *'Alā'ī zīj*, see Chapter 5, p. 114, note 18. On Wābkanawī, see Mozaffari and Zotti 2013, and Chapter 3.

69 The longitudes of the apogees can be extracted from a list given in the *'Alā'ī zīj* on p. 73. Owing to the especial method Ibn al-Fahhād used in the construction of his equation tables, the longitude of the solar apogee can be found in the table of its equation of center (opposite to the argument 0°), as well as the rounded values (to the first sexagesimal place) for the longitudes of the planetary apogees can be derived from the tables for their anomalistic equations (again opposite to the argument 0°); Ibn al-Fahhād, *'Alā'ī zīj*, pp. 73–76, 104, 106, 120, 122, 138, 140, 156, 158, 174, and 176.

Table 8.5 Ibn al-Fahhād's values for the longitude of the solar and planetary apogees

	Sun/Earth	Saturn	Jupiter	Mars	Venus	Mercury
On the proof...	87;48,54°	24*7*;*5*9,54°	177;41,54°	134;42,*5*4°	75;52,54°	*20*9;1*7*,54°
+ 0;1,49° =						
'Alā'ī zīj	87;50,43°	248; 1,4*3*	177;43,43	134;44,*4*3	75;54,43	209;19,43
	88;42	**256;28,58**	**177; 3,59**	**135;45,23**	**76;43,49**	**238;21,37**

by taking into account the likely scribal errors in writing down the *abjad* numerals with similar forms as follows. It can also be clearly recognized that all the longitudes in the treatise end with 54, while those in the *zīj* end with 43.[70]

Our main concern here is how Ibn al-Fahhād derived his values for the longitudes of the planetary apogees. In the *On the Proof. . .*, these values are listed immediately after our author explains how he measured the solar parameters, but without providing any details about how they are related to the solar parameters. That all numbers end with the same sexagesimal digits leaves no doubt that all of them should have been adjusted from an unknown earlier source by a procedure unknown to us (maybe analogous to what Bīrūnī did within two centuries before Ibn al-Fahhād). In any case, whatever the employed method was, and regardless of how precise his values are (notably, there are small errors of about ±1° for Jupiter, Mars, and Venus), it is less probable that he actually took the heavy burden of measuring the planetary orbital elements consisting in a long-term process of making observations and doing Ptolemy's iterative computational method for the superior planets, like what, for example, Muḥyī al-Dīn did in the Marāgha observatory about one century after him. However, it merits mentioning that Ibn al-Fahhād made some significant contributions to planetary astronomy in the late Islamic period, the most important of which is perhaps his rejecting the false hypothesis of the equality of the orbital elements of the Sun and Venus, which was dominant in early Islamic astronomy.[71] His contributions to observational astronomy were investigated in-depth in Mozaffari 2019b and 2023b.

The four examples briefly discussed earlier illustrate how uncertain and dubious it may be that the values for the longitudes of the planetary apogees in some highly influential works of medieval Middle Eastern astronomy actually stemmed

70 Saturn: in *On the Proof. . .*: 24*6*;*1*9,54 (a scribal error due to confusion between the similar *abjad* numerals in the form of ح وبط ← ح زنط), in the *'Alā'ī zīj*: 248; 1,4*8* (مح. . .← مح. . .); Mars: in *On the Proof. . .*: 134;42,*4*4 (ند. . . ← مد. . .), in the *'Alā'ī zīj*: 134;44,*5*3 (مح. . . ← نج. . .); Mercury: in *On the Proof. . .*: 1*9*9;1*6*,54 (. . .كط يز. . . ← . . .يو يط. . .).

71 See note 48; van Dalen 2004, p. 836. Of course, from Ibn al-Fahhād's statements in the prologue of the *'Alā'ī zīj* (p. 4), it can be deduced that Ibn al-A'lam was the first astronomer who opposed this hypothesis, which can be verified by the other sources, like Kamālī's *Ashrafī zīj* (F: f. 232v, G: f. 249r), written in Shiraz (central Iran) in the early fourteenth century.

from observational data, followed by the essential procedures in Ptolemaic context. Until now, curious readers have probably recognized an amazing feature of the earlier-mentioned sets of the Islamic values for the longitudes of the apogees: Ptolemy's value for the longitude of the Mercury's apogee is about $-32.5°$ off, an error which can also be conspicuously found in the corresponding values by the *Mumtaḥan* team, Ibn Yūnus, and Ibn al-Faḥḥād. It does not leave any room for doubt that the latter values are dependent upon Ptolemy's. In fact, this error is present in all the values adopted for the longitude of the apogee of Mercury in the Islamic *zījes*, except for the early works, which were based upon the Indian astronomical traditions (e.g., al-Khwārizmī's *zīj*).[72]

8.5.2.2 The Apogee Longitude Values Derived from Observational Data

As noted earlier, despite what we have seen in the four cases mentioned in the preceding sections, there are some works in Islamic astronomy where we can be certain that the values adopted for the longitudes of the apogees of the planets in them have to be considered as observational achievements in a true sense. Among them are Muḥyī al-Dīn al-Maghribī's *Talkhīṣ al-majisṭī*, which is, in reality, a unique treatise containing his dated observations and detailed accounts of the computational procedures for the derivation of the Ptolemaic parameters from the observational data,[73] and Ulugh Beg's *Sulṭānī zīj*, in which a good number of the unprecedented values for the planetary orbital elements are adopted.[74]

In his *Talkhīṣ*, Muḥyī al-Dīn presents his trio of observations of the oppositions of each superior planet to the mean sun and then explains his procedure of the re-quantification of Ptolemaic planetary models. The single surviving manuscript of this work is incomplete, so that all the parts pertained to the inferior planets and planetary latitudes are dropped from it. In Table 8.6, his values for the longitudes of the apogees are summarized.

Table 8.6 Muḥyī al-Dīn al-Maghribī's values for the longitude of the planetary apogees measured in the Marāgha observatory

Sun/Earth	Saturn	Jupiter	Mars	Venus	Mercury
88;51°	258;37,37°	177;56,32°	137; 7,6°	81;35°	210;20°
89;43	**258;27,41**	**178;43,37**	**137;35,4**	**77;51**	**239;17**
1264–1265 CE	February 26/27, 1273	August 11/12, 1274	February 25/26, 1271	January 17, 1232	

72 I will discuss this problematic situation at length elsewhere.

73 On this work, see Saliba 1983, 1985, 1986, Mozaffari 2014b, 2018–2019; and Chapter 9.

74 This is, perhaps, the only work in Islamic astronomy in which a set of non-Ptolemaic values for the inclinations of the epicycles and deferents of the superior planets is used; see Chapter 7.

In the case of the Sun and superior planets, the dates are for the last of his triple observations, from which he counts back in time in order to compute his own epoch positions for the last day of the year 600 Y/January 17, 1232, which are used in his third, and last, *zīj*, named the *Adwār al-anwār*; for the inferior planets, we quote the epoch values from the latter work.[75]

The values in Ulugh Beg's *Sulṭānī zīj*,[76] given for the epoch of its tables, the beginning of the year 841 H/July 4, 1437, are given in Table 8.7.

Muḥyī al-Dīn has excellent values for the longitudes of the apogees of the superior planets (with errors of only about +10′ for Saturn, within 1° for Jupiter, and less than $-^1/_2$° for Mars). Ulugh Beg has a more accurate value for the longitude of the apogee of Venus.[77] Nevertheless, both of them show an error of about −30° in the longitude of the apogee of Mercury, as in the other sets of values already put forward in this chapter, which gives rise to doubt that any Islamic observer in reality attempted to determine anew the orientation of the geocentric orbit of Mercury.

Considering the motion of the apogees, if Muḥyī al-Dīn, for example, compared his own measured values with Ptolemy's, he could readily find that the planetary apogees do not share the same rate in their progressive motion (the aggregate errors in Ptolemy's and Muḥyī al-Dīn's values result in the rates of about 1°/44^y for Saturn, 1°/67^y for Jupiter, and 1°/52^y for Mars). Nevertheless, dependent upon the tradition he belonged to, he found it necessary *to adhere to the authorities who maintained that the motion of the fixed stars equals that of the solar and planetary apogees.*[78] Consequently, he did not proceed to state any comparison between the values he measured for the longitudes of the solar and planetary apogees and Ptolemy's or those adopted by his Islamic predecessors. Perhaps, a main reason for this omission was to avoid explaining why the resulting rates are not equal to, but, rather, are significantly larger than, precession. Such a situation is also the case with Ulugh Beg and his team of astronomers in Samarqand.

8.5.2.3 Bisection of the Apogeal and Precessional Motions

In the late thirteenth century, Jamāl al-Dīn al-Zaydī and, about a century later, Ibn al-Shāṭir of Damascus significantly altered the Ptolemaic generalization by

Table 8.7 The values for the longitude of the planetary apogees measured in the Samarqand observatory

Sun/Earth	Saturn	Jupiter	Mars	Venus	Mercury
92;26°	256;55,31°	179;31,31°	141;56,48°	82;25,25°	214;28,28°
93;17	261;40,56	181;22,12	140;39,16	81;41,46	242;25,59

75 al-Maghribī, *Talkhīṣ*, ff. 126v, 131v, 135v; *Adwār*, CB: f. 80v, M: f. 82v.
76 Ulugh Beg, *Zīj*, P1: ff. 134r, 137r, 140r, 143r, 146r, P2: ff. 151r, 154r, 157r, 160v, 164r.
77 It merits noting that the *Sulṭānī zīj* has a significantly accurate value of 0;52 for the eccentricity of Venus.
78 al-Maghribī, *Talkhīṣ*, L: ff. 62r, 128r.

rejecting the notion that the motion of the solar and planetary apogees is equivalent to precession. (There is no evidence to show that the latter borrowed the idea from the former.)

The observational work carried out at the Islamic Astronomical Bureau in Beijing under Jamāl al-Dīn's supervision led to a new set of values for the planetary parameters. While the original work based on these parameter values seems to have been lost, some of its content, including a non-Ptolemaic star catalogue,[79] is preserved in the *Huihui lifa* (a Chinese translation of a Persian *zīj* from the Bureau, compiled in Nanjing in 1382–1383), the *Sanjufīnī zīj* (written in Arabic by Abū Muḥammad ʿAṭāʾ b. Aḥmad b. Muḥammad b. Khʷāja Ghāzī al-Samarqandī al-Sanjufīnī in northwestern Tibet in 1366), and a manuscript housed in the St. Petersburg branch of the Oriental Institute of the Russian Academy of Sciences (C 2460; dated 1383).

From the values tabulated for the apogeal and precessional motions in two separate columns in the table for the solar and the planetary mean motions from 764 to 895 H in the *Sanjufīnī zīj*,[80] as well as from the columns dedicated to the annual, monthly, and daily motions of the apogees and stars in the St. Petersburg manuscript (pp. 2, 4, 6, 8),[81] it can be clearly inferred that the apogeal motion at a rate of $1°/60^y$ is distinctly different from precession at a rate of $1°/73^y$.

In the prologue to his *Jadīd zīj*,[82] Ibn al-Shāṭir states:

> Understand that in my study of the observations, I followed the simplest methods and took the longest periods of time into account. In doing so, I relied on the ancient observations that all researchers agree upon. These observations come from Hipparchus the virtuous and those who came before him, *not* from Ptolemy.

> (emphasis added)

The most obvious outcome of such a decision was to validate the accomplishments of all his Islamic predecessors since the late tenth century by accepting a

---

79 See van Dalen 2000; an unusual value for the magnitude of Algol as recorded in this catalogue, together with other pieces of evidence, gave rise to conducting a detailed study on a probable secular variation in the brightness of the star: Mozaffari and Drake 2021.

80 Sanjufīnī, *Zīj*, ff. 44v–46r. An excerpt of the tables is given here:

Date	JDN	Apogeal motion	Precessional motion
24 Jumādā I 764	2218962	0; 0,0°	0; 0, 0°
27 Rabīʿ I 826	2240876	1; 0,0	0;49,15
21 Shaʿbān 839	2245625	1;13,5	0;59,55
17 Rabīʿ I 895	2265318	2; 7,8	1;44,18

81 For example, the apogeal motion of 0;29,7° and the precessional increment of 0;23,56° in 30 Arabic years.

82 Ibn al-Shāṭir, *Zīj*, K: f. 2v, O: f. 2v, L1: f. 2r, L2: f. 2r.

precessional rate of 1°/70^y. This was because "it aligned with Hipparchus's observations," as noted in the apparatus to the table of the precessional motion in his *Jadīd zīj*.[83]

He also mentions in the same place, "[T]he motion of precession is not equal to the motion of the apogees, which is 1° in 60 Persian years [of 365 days unvarying]. We provided proof of this in our book named the *Ta'līq al-arṣād* (*Accounting for observations*)." Unfortunately, this work is now lost, so we do not know how Ibn al-Shāṭir arrived at this conclusion. It is clear that he did not completely sever ties with the Ptolemaic generalization but only excluded the motion of the encompassing orb of the fixed stars from it. This exclusion could have been achieved even earlier, but Jamāl al-Dīn and Ibn al-Shāṭir are the only Islamic astronomers who discovered a part of the truth about the motion of the planetary apsidal lines. In an astronomical system using tropical longitudes, they found that these are not equal to precession but significantly larger than it.[84] Although this discovery could be verified and even improved in later observations, such as those at the Samarqand observatory, no later Muslim astronomers followed them in this regard.

Ibn al-Shāṭir's value for the motion of the solar and planetary apogees is very close to the true motions of the apsidal lines of the Sun/Earth and Jupiter. In his *Jadīd zīj*, he provides a table for the longitudes of the solar and planetary apogees from the beginning of the Hijra era through the year 900 H in intervals of 30 years. Assuming his discovery was based on actual observations, let us see how he could deduce it. We choose the values given for the beginning of the year 750 H/March 21, 1349 CE (shown in Table 8.8), which falls within his lifetime. He should have measured his own values for the longitudes of the apogees around this time and then computed those for other times in that 900-year interval, both ahead and back in time.

---

83 Ibn al-Shāṭir, *Zīj*, K: f. 51v, 52r, O: ff. 28r, 30v, 31r; L1: ff. 49v–50v; L2: ff. 62v, 65r, 65v; PR: ff. 61v, 99v, 100r.

84 It merits mentioning that Levi b. Gerson (South of France, 1288–1344 CE) was the first astronomer who pointed out that the planetary apogees are not subject to precession. Entirely rejecting the Ptolemaic generalization, he asserted that the comparison of his and Ptolemy's values for the longitude of the apogee of Saturn shows that the rate of the motion of the apsidal line of this planet is equal to 1°/44 Julian years; for Jupiter, 1°/60^y (equal to Ibn al-Shāṭir's value as well as close to the true rate of ~59$'''^y$); and for Mars, 1°/66^y (see Goldstein 1985, p. 113, 1988, pp. 382–383; and some notes and corrections in Mancha 1998, pp. 28, n. 30, and 31, n. 34). Levi measured no new value for the eccentricity of the superior planets, although he asserted that the eccentricity of Mars should be corrected (see Goldstein 1988, pp. 372, n. 8, 395). Thus, he transformed Ptolemy's planetary parameter values to the parameters of his own models (see Goldstein 1988, p. 372; for example, in the case of Saturn, see Mancha 1998, pp. 22–27). Copernicus (1473–1543 CE), in his *De Revolutionibus* V.7, 12, 16–17 (Copernicus 1543: English translation, [1939] 1990, pp. 755, 765, 773–774; also, see Swerdlow and Neugebauer 1984, pp. 333–334, 346–347, 362–363), has the following values for the motion of the apsidal lines of the superior planets: 1°/100^y for Saturn, 1°/300^y for Jupiter, and 1°/130^y for Mars, which are for a sidereal reference system of longitudes; the modern values are about 1°/175^y, 1°/421^y, and 1°/221^y, respectively, for the three planets. Note that the modern values for the motion of the solar and planetary apogees mentioned earlier in Section 8.3 are for a tropical reference system; consequently, if the true rate of precession is subtracted from them, the results will be equal to the true motions in a sidereal reference system.

Table 8.8 Ibn al-Shāṭir's values for the longitude of the planetary apogees

Sun/Earth	Saturn	Jupiter	Mars	Venus	Mercury
90;10,10°	255;10,10°	181;10,10°	138;10,10°	78;10,10°	213;10,10°
91;46, 1	**259;57, 5**	**179;56,12**	**139; 1,30**	**80; 2,36**	**241; 4,39**

We need to consider Ibn al-Shāṭir's fragment quoted earlier. What did he mean by the "ancient observations" he referred to? Ibn al-Shāṭir did not have any ancient values for the longitudes of the apogees of the superior planets and Venus to compare with his own values. It is only in the cases of the Sun and Mercury that the *Almagest* provides the ancient values for the longitudes of the apogees: Hipparchus's measurement of a longitude of 65;30° for the solar apogee for about 146 BCE and Ptolemy's derivation of a longitude of about 186° for the apogee of Mercury for some 400 years before him.[85] If Ibn al-Shāṭir's values are compared with them, the motions of about 1° in 60.5 years and 1° in 59.5 years result for, respectively, the apogees of the Sun and Mercury, which are nearly equal to his figure of $1°/60^y$. For the longitudes of the apogees of the superior planets and Venus, there can be found some values for the third century BCE in the *Almagest*, which, flawed by misinterpretation, can be counted as the *ancient* values, as shall be explained in what follows. In order to correct the planetary mean motions, Ptolemy makes use of the ancient observations carried out in the third century BCE, and in order to derive the longitudes of the apogees at these times, he only computes back in time from his values for the longitudes of the apogees measured in the 130s with the rate of precession/apogee motion of $1°/100^y$:[86]

Saturn	Jupiter	Mars	Venus
229;20°	157;13°	111;25°	50;55°
for 229 BCE	for 241 BCE	for 272 BCE	for 272 BCE

If we compare this set of Ptolemy's values with those given by Ibn al-Shāṭir, the following values for the motion for the planetary apogees result (rounded to one half-year):

Saturn	Jupiter	Mars	Venus
$1°/61.0^y$	$1°/6\underline{6}.5^y$	$1°/60.5^y$	$1°/59.5^y$

It can be readily seen that all the derived motions are about Ibn al-Shāṭir's figure of $1°/60^y$, except for Jupiter, in which case Ibn al-Shāṭir's value for the motion

85 *Almagest* IX.7: Toomer 1998, pp. 450–453.
86 See note 4.

of the apogees can be directly deduced from a comparison between his and Ptolemy's values for the longitude of its apogee: $((181;10,10° - 161°)/(1349 - 135) \approx 0;1^{°/y})$. Note that it is only in the case of Jupiter that the difference between Ibn al-Shāṭir's and Ptolemy's values for the longitude of its apogee is in good agreement with an apogee motion of $1°/60^y$, while in the case of the other planets, no agreement can be found between them.

The considerations we have discussed so far can generally illustrate a likely path through which Ibn al-Shāṭir could have arrived at his specific value for the motion of the solar and planetary apsidal lines. As we have already noted, Bīrūnī initially found a value of $1°/60^y$ for the motion of the solar apogee based on a comparison of Hipparchus's and his own values for its longitude, which is faster than any of the precessional rates available in his time. It is plausible to speculate that Bīrūnī's value might have served as a starting point for both Jamāl al-Dīn and Ibn al-Shāṭir to hypothesize that the longitudes of the solar and planetary apogees increase at a faster speed than the rate of precession. They might have later verified this result, presumably, by comparing their measurement of the longitude of the solar apogee with Hipparchus's corresponding value, dated to the middle of the second century BCE, or by comparing their own derivation of the longitudes of the planetary apogees (with the exception of Mercury; more likely the superior planets) with the supposed values attributed to the ancient observations recorded in the *Almagest* or Ptolemy's ones. Of course, the apogee longitude values Ptolemy gives for the times of the ancient observations (except for the Sun and Mercury) are actually his own values measured in the 130s that have been easily converted to the times in the four centuries before him by the rate of precession of $1°/100^y$. So if Jamāl al-Dīn and Ibn al-Shāṭir actually followed such a line of reasoning, their value for the motion of the planetary apsidal lines is partly dependent upon Ptolemy's longitude values. Note that the reason we exclude Mercury is the existence of an error of about $-30°$ in Ibn al-Shāṭir's value for the longitude of its apogee (indeed, the most prevalent error in Islamic planetary astronomy), which indicates that his value stems from Ptolemy's derivation. Similarly, the errors in both sets of Ptolemy's and Ibn al-Shāṭir's values for the longitudes of the apogees of the other planets are nearly of the same sign and order (Saturn, $\sim-3°/-5°$; Jupiter, $\sim+1°$; Mars, $\sim-1°$; and Venus, $\sim-2.5°/-2°$).

8.6 Conclusion

In summary, Ptolemy, based on his determinations of the longitude of Mercury's apogee in his time and observations recorded about 400 years prior, found that, with respect to a tropical reference system of longitudes, the motion of Mercury's apogee over this time interval was equal to the precessional motion calculated using his precession rate of $1°/100^y$. He then generalized this finding to cover the motions of the apogees of the other planets. He also found that the solar apogee is fixed and the lunar apogee has a regressive motion.

Shortly after the emergence of astronomy in medieval Islam, the motion of the solar apogee was discovered in the ninth century. This was undoubtedly an observational achievement, although it seems to have involved some theoretical considerations.[87] The Ptolemaic generalization of equating the motion of Mercury's apogee with precession to the motion of the apsidal lines of the other planets served as a useful methodological axiom. This showed medieval Middle Eastern astronomers a potential way to both justify and provisionally quantify the motion of the solar apogee. Thus, the Banū Mūsā argued that the motion of the solar apsidal line is equivalent to that of the apogees of the planets, an idea that appears to have been firmly established around the turn of the tenth century. However, through a detailed, precise analysis of the quantitative observational data achieved by his Muslim predecessors, Bīrūnī initially hesitated to accept this hypothetical equivalence and even expressed doubt about the uniformity of the motion of the solar apogee. Based on a comparison of his value for the longitude of the solar apogee measured in 1016–1017 CE with Ptolemy's and Hipparchus's value of 65.5° dated, respectively, to the middle of the second century CE and of the second century BCE, he found the angular velocity of the solar apogee to be equal to, respectively, $1°/46^y$ and $1°/60^y$. Treating this issue as an open question to be solved by future generations of astronomers, he provisionally accepted the extension of Ptolemy's generalization to cover the motion of the solar apogee. Nevertheless, despite the fact that planetary astronomy received considerable attention in late Islamic astronomy (especially at the Marāgha and Samarqand observatories) and a good number of non-Ptolemaic values were determined for the planetary orbital elements and inclinations, medieval Islamic astronomers, except for Jamāl al-Dīn and Ibn al-Shāṭir, neither revisited the motion of the solar apogee nor re-evaluated the Ptolemaic generalization in the case of the equality of the motions of the apogees.

The main reasons behind this, which form the fundamental characteristics of the medieval Middle Eastern astronomical thought system, can be summarized as follows. In the highly tradition-based scientific knowledge of the medieval period, methodological axioms like the Ptolemaic generalization were largely (if not entirely) necessary to address unprecedented fundamental questions, understand newly observed phenomena, and overcome the lack of information available about a specific subject at a given time. This applies specifically to the data available about the motion of the solar and planetary apogees, making it very challenging to ascertain their motions. If an underlying tradition could not provide the medieval astronomers with a suitable methodological axiom, a viable hypothesis, a widely accepted dogma, and so on, they had to turn to other astronomical

---

87 Apparently, the most important theoretical consideration contributing to the acceptance of the solar apogee motion was the Indian hypothesis that the orbital elements of the Sun are identical to those of Venus (see earlier, note 48), and consequently, if the apogee of Venus is subject to a progressive motion, then the solar apogee should have a progressive motion.

traditions, incorporating them into the formal tradition.[88] As we have already seen, despite his repeated references to the indisputable observational data indicating the motion of the solar apogee, Bīrūnī still found it necessary to briefly highlight that this motion existed in Indian astronomy. Consequently, such a system of astronomical thought rarely set aside a general hypothesis in favor of new empirical evidence that not only contradicted it but also did not agree with each other. Furthermore, at a practical level, another contributing factor was Islamic astronomers' awareness of the empirical limitations, primarily the difficulties with making astronomical observations. As a result, observed inexplicable deviations were typically attributed to the imperfection of observations or the deficiency of the instruments used.[89] Therefore, generally, any secular change in the constants of Ptolemaic astronomy was not allowed, and it was believed that the fixed stars and apogees share a common progressive motion. To Islamic astronomers, the

88 A very curious example of such situations in Islamic astronomy is the use of the Indian hypotheses for the estimation of the size of the apparent angular diameter of the Sun and the Moon within the framework of Ptolemy's solar and lunar models in order to justify and predict an annular solar eclipse, which is, theoretically, classified as an impossible and undefined phenomenon in Ptolemaic astronomy (see earlier, note 34). Bīrūnī made a significant contribution in this respect. Other examples are the colors of lunar eclipses and the optical limitation of the visibility of solar eclipses, which are discussed in Indian astronomy but are very superficially treated in Ptolemy's works, if at all (see Mozaffari 2015, p. 138).

89 Through Ptolemy's various references and suggestions about method sensitivity (for instance, his practical demonstration of the sensitivity of the method using observations of triple lunar eclipses to measure the size of the Moon's epicycle in *Almagest* IV.11: Toomer 1998, pp. 215–216), instrument accuracy (such as his statement about the equinoctial ring installed in Alexandria in *Almagest* III.1: Toomer 1998, p. 134), and his judgments regarding the precision achieved in the observations made before him (like his opinion about the stellar observations made by Timocharis in *Almagest* VIII.1: Toomer 1998, p. 321, as well as the solar observations made by Meton and Euktemon in *Almagest* III.1: Toomer 1998, p. 137), Islamic astronomers generally gained a good overview of the problem of errors in astronomy. This was enriched with numerous valuable insights about the critical methodological and empirical considerations, derived from their own experimental practices and their familiarity with the complexities of subtle observational matters, that should be considered both in the interpretation of observational data and in the derivation of parameters. Examples include: (1) ʿAbd al-Raḥmān al-Khāzinī, in his *Experimental Astronomy*, provides a detailed explanation on how to detect an error in a fundamental structural or motional parameter in a theory created within the framework of Ptolemy's models for the Sun, Moon, and planets, by evaluating the patterns of longitudinal errors (this implies a meticulous scrutiny of the behavior of the Ptolemaic models); he also introduces the concept of *inevitable/unavoidable errors*. (2) The recognition of the fundamental difference between irregular and systematic patterns of residuals by the Banū Amājūr and Ibn Yūnus (Mozaffari 2024) and proposals for some way to avoid the latter type as caused by any deficiency in instruments by Taqī al-Dīn Muḥammad b. Maʿrūf (Mozaffari 2016, p. 270). (3) The adoption of the average of the extreme values determined for a single parameter, established by Bīrūnī (al-Bīrūnī 1967, p. 59, Kennedy 1973, p. 32). (4) The adoption of the same value for a specific parameter determined by the use of two different methods, employed by Muḥyī al-Dīn al-Maghribī in his measurement of the solar eccentricity (*Talkhīṣ* III.5, f. 61r). (5) The disregard of small deviations or outlier values by Muḥyī al-Dīn in his stellar observations (*Talkhīṣ* VII.2, f. 115r; also, see Chapter 10, p. 288). For more on this fascinating topic, see Sheynin 1973, 1992, 1993, and further examples in Mozaffari 2016.

values for the rate of precession could be improved as new and presumably gradually better observations were made, but the equality of the spatial motions of the encompassing nesting celestial spheres of the stars, Sun, and planets could never be questioned.

Jamāl al-Dīn and Ibn al-Shāṭir made significant strides by differentiating between the motion of the fixed stars and that of the solar and planetary apogees, recognizing that the latter moves faster than the former by a motion of 1° in either 337 or 420 years. However, they upheld the primary aspect of the Ptolemaic generalization that the apogees of the Sun and all the planets move with the same motion. This can be considered a significant observational discovery that both Jamāl al-Dīn and Ibn al-Shāṭir made during their productive careers. Perhaps this is a more important achievement than the latter's other observational findings, such as his values for the range of variation of the apparent angular diameters of the Sun and Moon,[90] or his planetary models which were constructed solely to address the physical difficulties in Ptolemy's planetary models.[91]

Ibn al-Shāṭir claimed that in his derivation of celestial motions, he specifically used ancient observations that predated Ptolemy, particularly those conducted by Hipparchus. As we have already noted, Bīrūnī initially found a value of $1°/60^y$ for the motion of the solar apogee based on the comparison of his and Hipparchus's values for its longitude, which is faster than any of the precessional rates available in his time. It is plausible to speculate that Bīrūnī's value might have served as a starting point for both Jamāl al-Dīn and Ibn al-Shāṭir to hypothesize that the longitudes of the solar and planetary apogees increase at a faster speed than the rate of precession. They might have later verified this result, presumably, by comparing their measurements of the longitude of the solar apogee with Hipparchus's corresponding value, dated to the middle of the second century BCE, or by comparing their own derivation of the longitudes of the planetary apogees with the supposed values attributed to the ancient observations recorded in the *Almagest*.

In the astronomical system developed in Western Islamic lands, a shift or deviation from Ptolemy's generalization occurred with Ibn al-Zarqālluh's opinion that the solar apogee has a proper motion in longitude of $1°/279^y$ in a sidereal reference system. However, as mentioned earlier, some of the later astronomers belonging to that tradition generalized this proper motion to the apsidal lines of the planets. Thus, it can be said that the Ptolemaic generalization was revived in a new form and was utilized in the construction of a new hypothesis: the motion of the apogees is not related to precession, but greater than it, and so they are neither tropically nor sidereally fixed. This suggestion is similar to Jamāl al-Dīn's and Ibn al-Shāṭir's finding, but it is a purely theoretical achievement in essence, apparently resulting from following the main line of reasoning of Ptolemy in the *Almagest*, and does not have any basis in observational astronomy.

---

90 Saliba 1987, Mozaffari 2015, pp. 133–134.
91 On Ibn al-Shāṭir's planetary models, see Roberts 1957, 1966, Kennedy and Roberts 1959, Abbud 1962, Kennedy and Ghanem 1976; and a concise summary in Saliba 1996.

In summary, the ninth-century Islamic astronomers, followed by Bīrūnī, Ibn al-Zarqālluh, Jamāl al-Dīn, and Ibn al-Shāṭir, as well as, to some extent, later Western Islamic astronomers, each contributed to the substantial changes in a particular proposition of Ptolemy's astronomy, that is, the motion of the planetary apogees and the correlation between it and precession. These contributions took form, first, by including the motion of the solar apogee in Ptolemy's proposition and, later, by making a distinction between the motion of the apogees and precession but did not alter a fundamental axiom of Ptolemy's astronomical methodology, that is, the generalization in the case of the equality of the motions of all planets' apogees.

References

Unpublished Primary Sources

Abharī, Athīr al-Dīn al-Mufaḍḍal b. ʿUmar, *Al-Zīj al-mulakhkhaṣ ʿalā al-raṣad al-ʿAlāʾī* (*The abridged zīj based on the ʿAlāʾī observations*), MSS. K: Kolkata (Calcutta), National Library of India, including the Būhār collection, Arabic 347, U: Utrecht, Universiteitsbibliotheek, no. 1442, F: Florence, Biblioteca Medicea Laurenziana, Or. 106/2, ff. 72v–170r.

Abu ʾl-Wafāʾ al-Būzjānī, *Majisṭī*, MS. Paris: Bibliothèque Nationale de France, Arabe 2494; accessible through, retrieved from https://gallica.bnf.fr/ark:/12148/btv1b100374763.

Anonymous (Abharī, Athīr al-Dīn al-Mufaḍḍal b. ʿUmar?), *al-Zīj al-shāmil* (*The comprehensive zīj*), MSS. P1: Iran, Parliament Library, no. 6422, P2: Iran, Parliament Library, no. 6445, F1: Florence, Biblioteca Medicea Laurenziana, Or. 106/1, ff. 1r–71r, F2: Florence, Biblioteca Medicea Laurenziana, Or. 95/2 (116 folios, separately numbered from Or. 95/1), D: Dublin, Chester Beatty Library, Arabic 4076, Pa1: Paris, Bibliothèque nationale de France, arabe 2528, Pa2: Paris, Bibliothèque nationale de France, arabe 2529, Pa3: Paris, Bibliothèque nationale de France, arabe 2540, C: Cairo, Dār a l-Kutub, riyāḍī Taymūr 296/1, ff. 1r–80v. (For a comprehensive review of these MSS, see van Dalen 2021, pp. 517–522.)

Anonymous (Abharī, Athīr al-Dīn al-Mufaḍḍal b. ʿUmar?) (dedicated to Najm al-Dīn ʿAlī b. ʿUmar b. ʿAlī Dabīrān/al-Kātibī al-Qazwīnī), *Kitāb fī ṣināʿat al-majisṭī* (*Book on the art/industry of the* Almagest), MSS. P: Iran, Parliament Library, no. 6195, N: Iran, National Library, no. 2607560, I: Istanbul, Süleymaniye, Ayasofia, no. 2583.

al-Battānī, Abū ʿAbd-Allāh Muḥammad b. Jābir b. Sinān al-Ḥarrānī, *Zīj al-Ṣābiʾ* (*The Sabean Zīj*), MS. E: Biblioteca Real Monasterio de San Lorenzo de el Escorial, no. árabe 908 (edited in Nallino [1899–1907] 1969). Translated into Latin by Plato of Tivoli in Barcelona in the 12th century, printed twice, in Nuremberg in 1537 (with Farghānī's work) and in Bologna in 1645.

Ibn al-Fahhād: Farīd al-Dīn Abu al-Ḥasan ʿAlī b. ʿAbd al-Karīm al-Fahhād al-Shirwānī or al-Bākūʾī. *Zīj al-ʿAlāʾī*. MS. India, Salar Jung, no. H17.

Ibn al-Shāṭir, ʿAlāʾal-Dīn Abu ʾl-Ḥasan ʿAlī b. Ibrāhīm b. Muḥammad al-Muṭaʿʿim al-Anṣārī, *Al-Zīj al-Jadīd*. MSS. K: Istanbul, Kandilli Observatory, no. 238, O: Oxford, Bodleian Library, no. Seld. A inf 30, D: Damascus, Asad National library, no. 3093; L1: Leiden, Universiteitsbibliotheek, no. Or. 65; L2: Leiden, Universiteitsbibliotheek, Or. 530; PR: Princeton, Princeton University Library, no. Yahuda 145.

Ibn Yūnus, ʿAlī b. ʿAbd al-Raḥmān b. Aḥmad, *Zīj al-Kabīr al-Ḥākimī*, MS. L: Leiden, Universiteitsbibliotheek, no. Or. 143, MS. O: Oxford, Bodleian Library, no. Hunt 331.

al-Kamālī, Muḥammad b. Abī ʿAbd-Allāh Sanjar (Sayf-i munajjim), *Ashrafī zīj*, MSS. F: Paris: Bibliothèque Nationale, no. 1488, G: Iran–Qum: Gulpāyigānī, no. 64731.

al-Kāshī, Jamshīd Ghiyāth al-Dīn. *Khāqānī zīj*. MS. IO: London, India Office, no. 430; P: Iran: Parliament Library, no. 6198.

al-Maghribī, Muḥyī al-Dīn. *Adwār al-anwār*. MSS. M: Iran, Mashhad, Holy Shrine Library, no. 332; CB: Ireland, Dublin, Chester Beatty, no. 3665.

al-Maghribī, Muḥyī al-Dīn, *Kitāb al-Zīj ʿUmdat al-ḥāsib wa ghunyat al-ṭālib*, MS. M: Cairo: Egyptian National Library, no. MM 188.

al-Maghribī, Muḥyī al-Dīn, *Talkhīṣ al-majisṭī*, MS. Leiden: Universiteitsbibliotheek, Or. 110.

Sanjufīnī: Abū Muḥammad ʿAṭāʾ b. Aḥmad b. Muḥammad Khᵂāja Ghāzī al-Samarqandī al-Sanjufīnī, *Sanjufīnī zīj*, MS. Paris: Bibliothèque Nationale, Arabe 6040.

Ṭabarī, Abū Jaʿfar Muḥammad b. Ayyūb al-Ḥāsib, *Zīj-i Mufrad* (*The Unique Zīj*), MS. Cambridge, Browne Collection College, O.1.

Taqī al-Dīn Muḥammad b. Maʿrūf, *Sidrat muntaha ʾl-afkār fī malakūt al-falak al-dawwār* (*The Lotus tree in the seventh heaven of reflection*) or *Shāhanshāhiyya Zīj*, MSS. K: Istanbul, Kandilli Observatory, no. 208/1 (up to f. 48v; autograph), N: Istanbul, Süleymaniye Library, Nuruosmaniye Collection, no. 2930, V: Istanbul, Süleymaniye Library, Veliyüddin Collection, no. 2308/2 (from f. 10v).

al-Ṭūsī, Naṣīr al-Dīn, *Īlkhānī zīj*. MSS. C: University of California, Caro Minasian Collection, no. 1462; T: University of Tehran, Ḥikmat Collection, no. 165 + Suppl. P: Iran, Parliament Library, no. 6517 (Remark: The latter is not actually a separate manuscript, but contains 31 folios missing from MS T. The chapters and tables in MS. T are badly out of order, presumably owing to the folios having been bound in disorder), P: Iran, Parliament Library, no. 181, M1: Iran, Mashhad, Holy Shrine Library, no. 5332a; M2: Iran, Qum, Marʿashī Library, no. 13230.

Ulugh Beg. *Sulṭānī Zīj*. MS. P1: Iran, Parliament Library, no. 72; MS. P2: Iran, Parliament Library, no. 6027.

Yaḥyā b. Abī Manṣūr, *Zīj al-mumtaḥan*, MS. E: Madrid, Library of Escorial, no. árabe 927 (published in *The verified astronomical tables for the caliph al-Maʾmūn*, edited by Fuad Sezgin, with an introduction by Edward S. Kennedy. Frankfurt am Main: Institut für Geschichte der Arabisch-Islamischen Wissenschaften, 1986), MS. L: Leipzig, Universitätsbibliothek, no. Vollers 821.

Published References

Abbud, F. 1962. "The planetary theory of Ibn al-Shāṭir: Reduction of the geometric models to numerical tables", *Isis* **53**, pp. 492–499.

[*EI₂*:] Bearman, P., Bianquis, T., Bosworth, C. E., van Donzel, E., and Heinrichs, W. P., 1960–2005, *Encyclopaedia of Islam*, 2nd edn., 12 Vols., Leiden: Brill.

al-Bīrūnī, Abū al-Rayḥān, 1954–1956, *Al-Qānūn al-masʿūdī* (*Masʿūdīc canons*), 3 Vols. Hyderabad: Osmania Bureau; MSS. F: Paris, Bibliothèque Nationale de France, no. Ar. 6840, B: Berlin, Staatsbibliothek zu Berlin, no. Or. Oct. 275 = Ahlwardt 5667.

al-Bīrūnī, Abū al-Rayḥān, 1967, *Taḥdīd nahayāt al-amākin li-taṣḥīḥ masāfāt al-masākin* (*Determination of the Coordinates of Positions for the Correction of Distances between Cities*), Ali, J. (En. tr.), Beirut: American University of Beirut.

Brahmagupta, 1970, *The Khaṇḍakhādyaka of Brahmagupta*, Translated into English and edited by Bina Chatterjee, New Delhi: Bina Chatterjee.

Caussin de Perceval, J.-J.-A., 1804, "Le livre de la grande table hakémite, Observée par le Sheikh,. . ., ebn Iounis", *Notices et Extraits des Manuscrits de la Bibliothèque nationale* **7**, pp. 16–240.

Copernicus, N., 1543/1939, *De revolutionibus orbium coelestium*, Nuremberg: *On the Revolutions of the Heavenly Spheres*. Translated into English by Charles Glenn Wallis, Annapolis: St John's College Bookstore. Reprinted in: *The Great Books of the Western World*, 2nd edn, Vol. 15. Chicago: Encyclopædia Britannica, 1990.

Dorce, C., 2002–2003, "The *Tāj al-azyāj* of Muḥyī al-Dīn al-Maghribī (d. 1283): Methods of computation", *Suhayl* **3**, pp.193–212.

van Dalen, B., 2000, "A non-Ptolemaic Islamic star table in Chinese", in: Folkerts, M. and Lorch, R. (eds.), *Sic Itur Ad Astra: Studien zur Geschichte der Mathematik und Naturwissenschaften*, Wiesbaden: Harrassowitz, pp. 147–176.

van Dalen, B., 2002a, "Islamic and Chinese astronomy under the Mongols: A little-known case of transmission", in: Dold-Samplonius, Y., *et al.* (eds.), *From China to Paris: 2000 Years Transmission of Mathematical Ideas*, Stuttgart: Franz Steiner, pp. 327–356.

van Dalen, B., 2002b, "Islamic astronomical tables in China: The sources for the Huihui li", in: Ansari, R. (ed.), *History of Oriental Astronomy* (Proceedings of the Joint Discussion 17 at the 23rd General Assembly of the International Astronomical Union, Organised by the Commission 41 (History of Astronomy), Held in Kyoto, August 25–26, 1997), Dordrecht: Kluwer, pp. 19–30.

van Dalen, B., 2004, "The *Zīj-i Naṣirī* by Maḥmūd ibn Umar: The earliest Indian Zij and its relation to the *'Alā 'ī Zīj*", in: Burnett, C. (ed.), *Studies in the History of the Exact Sciences in Honour of David Pingree*, Leiden: Brill, pp. 825–862.

van Dalen, B., 2021, "The geographical table in the *Shāmil Zīj*: Tackling a thirteenth-century Arabic source with the aid of a computer database", in: Husson, *et al.* 2021, pp. 511–566.

van Dalen, B. and Pedersen, F. S., 2008, "Re-editing the tables in the *Ṣābi 'Zīj* by al-Battānī, *Mathematics Celestial and Terrestrial – Festschrift für Menso Folkerts zum 65. Geburtstag*", *Acta Historica Leopoldina* **54**, pp. 405–428.

Dorce, C., 2003, *El Tāŷ al-azyāŷ de Muḥyī al-Dīn al-Maghribī. Anuari de Filologia*, Vol. 25, Secció B, Número 5, Barcelona: University of Barcelona.

Duke, D., 2006, "Ancient declinations and precession", *DIO* **13**, pp. 26–34.

[*DSB*:] Gillipsie, C. C. (ed.), 1970–1980, *Dictionary of Scientific Biography*, 16 Vols., New York: Charles Scribner's Sons.

Goldstein, B. R., 1985, *The Astronomy of Levi ben Gerson (1288–1344), a Critical Edition of Chapters 1–20 with Translation and Commentary*, New York: Springer-Verlag.

Goldstein, B. R., 1988, "A new set of fourteenth century planetary observations", *Proceedings of the American Philosophical Society* **132**, pp. 371–399.

Goldstein, B. R., 1994, "Historical perspectives on Copernicus's account of precession", *Journal for the History of Astronomy* **25**, pp. 189–197.

Hartner, W. and Schramm, M. 1961, "Al-Bīrūnī and the theory of the solar apogee: An example of originality in Arabic science", in: Crombie, A. C. (ed.), *Scientific Change. Historical Studies in the Intellectual, Social and Technical Conditions for Scientific Discovery and Technical Invention, from Antiquity to the Present.* Symposium on the History of Science, University of Oxford 9–15 July, London: Heinemann, pp. 206–218.

[*BEA*:] Hockey, T. (ed.), 2014, *The Biographical Encyclopedia of Astronomers*, 2nd edn., New York [etc.]: Springer.

Husson, M., Montelle, C., and van Dalen, B., 2021, *Editing and Analysing Numerical Tables: Towards a Digital Information System for the History of Astral Sciences*, Turnhout: Brepols.

Isahaya, Y., 2021, "From Alamut to Dadu: Jamāl al-Dīn's armillary sphere on the Mongol silk roads", *Acta Orientalia Academiae Scientiarum Hungaricae* **74**, pp. 65–78.

Jones, A., 2005, "In order that we should not ourselves appear to be adjusting our estimates . . . to make them fit some predetermined amount", in: Buchwald, J. Z. and Franklin, A. (eds.), *Wrong for the Right Reasons*, The Netherlands: Springer, pp. 17–39.

Kennedy, E. S., 1956. "A survey of Islamic astronomical tables", *Transactions of the American Philosophical Society*, New Series **46**, pp. 123–177.

Kennedy, E. S., 1958, "The Sasanian astronomical handbook *Zīj-i Shāh* and the astrological doctrine of 'Transit' (*Mamarr*)", *Journal of the American Oriental Society* **78**, pp. 246–262.

Kennedy, E. S., 1973, *A Commentary Upon Bīrūnī's* Kitāb Taḥdīd al-Amākin, Beirut: American University of Beirut.

Kennedy, E. S. and Ghanem, I. (eds.), 1976, *The Life and Work of Ibn al-Shāṭir: An Arab Astronomer of the Fourteenth Century*, Aleppo: Institute for History of Arabic Science.

Kennedy, E. S. and Pingree, D. (eds.), 1981, *The Book of the Reasons behind Astronomical Tables*, New York: Scholars' Facsimiles & Reprints.

Kennedy, E. S. and Victor, R., 1959, "The planetary theory of Ibn al-Shāṭir", *Isis* **50**, pp. 227–235.

King, D. A., 1999, "Aspects of Fatimid astronomy: From hard-core mathematical astronomy to architectural orientations in Cairo", in: Barrucand, M. (ed.), *L'Égypte Fatimide: son art et son histoire – Actes du colloqie organisé à Paris les 28, 29 et 30 mai 1998*, Paris: Presses de l'Université de Paris-Sorbonne, pp. 497–517.

King, D. A. and Gingerich, O., 1982. "Some astronomical observations from thirteenth-century Egypt", *Journal for the History of Astronomy* **13**, pp. 121–128.

King, D. A. and Saliba, G. (eds.), 1987, *From Deferent to Equant: A Volume of Studies on the History of Science of the Ancient and Medieval Near East in Honor of E. S. Kennedy*, Vol. 500, New York: Annals of the New York Academy of Sciences.

[*NDSB*:] Koertge, N. (ed.), 2008, *New Dictionary of Scientific Biography*, 8 Vols., Detroit: Charles Scribner's Sons.

Kunitzsch, P., 1986, "The star catalogue commonly appended to the *Alfonsine Tables*", *Journal for the History of Astronomy* **17**, pp. 89–98; reprinted in Idem, 1989, *The Arabs and the Stars*, Northampton: Variorum, Trace XXII.

Maeyama, Y., 1998, "Determination of the Sun's orbit. Hipparchus, Ptolemy, al-Battânî, Copernicus, Tycho Brahe", *Archive for History of Exact Sciences* **53**, pp. 1–49.

Mancha, J. L., 1998, "Heuristic reasoning: Approximation procedures in Levi ben Gerson's astronomy", *Archives for History of Exact Sciences* **52**, pp. 13–50.

Mozaffari, S. M., 2013, "Historical annular solar eclipses", *Journal of British Astronomical Association* **123**, pp. 33–36.

Mozaffari, S. M., 2014a, "Ptolemaic eccentricity of the superior planets in the medieval Islamic period", in: Katsiampoura, G. (ed.), *Scientific Cosmopolitanism and Local Cultures: Religions, Ideologies, Societies; Proceedings of 5th International Conference of the European Society for the History of Science* (Athens, 1–3 November 2012), Athens: National Hellenic Research Foundation, pp. 23–30.

Mozaffari, S. M., 2014b, "Muḥyī al-Dīn al-Maghribī's lunar measurements at the Maragha observatory", *Archive for History of Exact Sciences* **68**, pp. 67–120.

Mozaffari, S. M., 2015, "Annular eclipses and considerations about solar and lunar angular diameters in medieval astronomy", in: Orchiston, W., *et al.* (eds.), *New Insights from Recent Studies in Historical Astronomy: Following in the Footsteps of F. Richard Stephenson*, New York: Springer, pp. 119–142.

Mozaffari, S. M., 2016, "A forgotten solar model", *Archive for History of Exact Sciences* **70**, pp. 267–291.

Mozaffari, S. M., 2016–2017, "A revision of the star tables in the *Mumtaḥan zīj*", *Suhayl* **15**, pp. 67–100.

Mozaffari, S. M., 2018–2019, "Muḥyī al-Dīn al-Maghribī's measurements of Mars at the Maragha observatory", *Suhayl*, **16**, pp. 149–249.

Mozaffari, S. M., 2019a, "The orbital elements of Venus in medieval Islamic astronomy: Interaction between traditions and the accuracy of observations", *Journal for the History of Astronomy* **50**, pp. 46–81.

Mozaffari, S. M., 2019b, "Ibn al-Fahhād and the Great Conjunction of 1166 AD", *Archive for History of Exact Sciences* **73**, pp. 517–549.

Mozaffari, S. M., 2023a, "A survey of Abu 'l-Wafā''s solar and stellar observations", *Journal of Astronomical History and Heritage* **26**, pp. 469–488.

Mozaffari, S. M., 2023b, "Sources of the planetary theories in Fahhād's *'Alā 'ī zīj*: Solving a medieval case of intellectual fraud", *Suhayl* **20**, pp. 143–221.

Mozaffari, S. M., 2024, "Ibn al-Zarqālluh's discovery of the annual equation of the Moon", *Archive for History of Exact Sciences* **78**, pp. x–x.

Mozaffari, S. M., and Drake, J. J., 2021, "Algol anomaly or careful observations of its brightness? The values recorded for the magnitude of Algol in the medieval astronomical corpus", *Journal for the History of Astronomy* **52**, pp. 77–103.

Mozaffari, S. M., and Zotti, G., 2013, "The observational instruments at the Maragha observatory after AD 1300", *Suhayl* **12**, pp. 45–179.

Nallino, C. A. (ed.), 1899–1907/1969, *Al-Battani sive Albatenii Opus Astronomicum*. Publicazioni del Reale osservatorio di Brera in Milano, n. XL, pte. I–III, Milan: Mediolani Insubrum, The, Reprint of Nallino's edition: Frankfurt: Minerva.

Neugebauer, O., 1962, "Thabit ben Qurra 'On the Solar Year' and 'On the Motion of the Eighth Sphere'", *Proceedings of the American Philosophical Society* **106**, pp. 264–299.

Neugebauer, O., 1975, *A History of Ancient Mathematical Astronomy*, Berlin-Heidelberg-New York: Springer.

Neugebauer, O., and Pingree, D., 1970–1971, *The Pañcasiddhāntikā of Varāhamihira*, 2 Vols., Copenhagen: Det Kongelige Danske Videnskabernes Selskab.

Newton, R. R., 1973, "The authenticity of Ptolemy's parallax data: Part 1", *Quarterly Journal of the Royal Astronomical Society* **14**, pp. 367–388.

Pannekoek, A., 1995, "Ptolemy's precession", *Vistas in Astronomy* **1**, pp. 60–66.

Pedersen, O., 2010/1974, *A Survey of Almagest*, Odense: Odense University Press, with annotation and new commentary by Alexander Jones, New York: Springer.

Pingree, D., 1964, "Gregory chioniades and palaeologan astronomy", *Dumbarton Oaks Papers* **18**, pp. 133–160.

Pingree, D., 1965, "The Persian 'Observation' of the solar apogee in ca. A.D. 450", *Journal of Near Eastern Studies* **24**, pp. 334–336.

Pingree, D., 1968, "The fragments of the works of Ya'qūb Ibn Ṭāriq", *Journal of Near Eastern Studies* **27**, pp. 97–125.

Pingree, D., 1970, "The fragments of the works of Al-Fazārī", *Journal of Near Eastern Studies* **29**, pp. 103–123.

Pingree, D., 1972, "Precession and trepidation in Indian astronomy before A.D. 1200", *Journal for the History of Astronomy* 3:27–35.

Pingree, D., 1976, "The recovery of early Greek astronomy from India", *Journal for the History of Astronomy* **7**, pp. 109–123.

Pingree, D. (ed.), 1985–1986. *Astronomical Works of Gregory Chioniades, Part 1: Zīj al-ʿAlāʾī*, 2 Vols. Amsterdam: Gieben.

Roberts, V., 1957, "The solar and lunar theory of Ibn ash-Shāṭir, a pre-Copernican Copernican model", *Isis* **48**, pp. 428–432.

Roberts, V., 1966, "The planetary theory of Ibn al-Shāṭir: latitudes of the planets", *Isis* **57**, pp. 208–219.

Rosenfeld, B. A. and İhsanoğlu, E., 2003. *Mathematicians, Astronomers, and Other Scholars of Islamic Civilization and Their Works (7th–19th c.)*, Istanbul: IRCICA.

Saliba, G., 1983. "An observational notebook of a thirteenth-century astronomer", *Isis* **74**, pp. 388–401; reprinted in Saliba 1994, pp. 163–176.

Saliba, G., 1985. "Solar observations at Maragha observatory", *Journal for the History of Astronomy* **16**, pp. 113–122; reprinted in Saliba 1994, pp. 177–186.

Saliba, G., 1986, "The determination of new planetary parameters at the Maragha observatory", *Centaurus* **29**, pp. 249–271; reprinted in Saliba 1994, pp. 208–230.

Saliba, G., 1987, "Theory and observation in Islamic astronomy: The work of Ibn al-Shāṭir of Damascus", *Journal for the History of Astronomy* **18**, pp. 35–43; reprinted in Saliba 1994, pp. 233–241.

Saliba, G., 1994, *A History of Arabic Astronomy: Planetary Theories during the Golden Age of Islam*, New York: New York University.

Saliba, G., 1996, "Arabic planetary theories after the eleventh century AD", in: Rashed, R. (ed.), *Encyclopedia of the History of Arabic Science*, Vol. 1, London-New York: Routledge, pp. 59–128.

Samsó, J., 1987, "Al-Zarqāl, Alfonso X and Peter of Aragon on the solar equation", in: King and Saliba 1987, pp. 467–476.

Samsó, J., 1990, "Trepidation in al-Andalus in the 11th century", in: Samsó 1994, Trace VIII.

Samsó, J., 1994, *Islamic Astronomy and Medieval Spain*, Ashgate: Variorum.

Samsó, J., 2007, *Astronomy and Astrology in al-Andalus and the Maghrib*, Aldershot: Ashgate-Variorum.

Samsó, J., 2020, *On Both Sides of the Strait of Gibraltar: Studies in the History of Medieval Astronomy in the Iberian Peninsula and the Maghrib*, Leiden-Boston: Brill.

Samsó, J. and Millás, E., 1989, "Ibn al-Bannāʾ, Ibn Isḥāq, and Ibn al-Zarqāllu's Solar Theory," in: Samsó 1994, Trace X.

Samsó, J. and Millás, E., 1998, "The computation of planetary longitudes in the zīj of Ibn al-Banna", *Arabic Science and Philosophy* **8**, pp. 259–286; reprinted in Samsó 2007, Trace VIII.

Sayılı, A., 1960/1988, *The Observatory in Islam*, 2nd edn., Ankara: Türk Tarih Kurumu Basimevi.

Sezgin, F., 1978, *Geschichte des arabischen Schrifttums, Band VI: Astronomie bis ca. 430 H*, Leiden: Brill.

Sheynin, O. B., 1973, "Mathematical treatment of astronomical observations (A historical essay)", *Archive for History of Exact Sciences* **11**, pp. 97–126.

Sheynin, O. B., 1992, "Al-Bīrūnī and the mathematical treatment of observations", *Arabic Sciences and Philosophy* **2**, pp. 299–306.

Sheynin, O. B., 1993, "The treatment of observations in early astronomy", *Archive for History of Exact Sciences* **46**, pp. 153–192.

Simon, J.-L., *et al.*, 1994., "Numerical expressions for precession formulae and mean elements for the Moon and the planets", *Astronomy and Astrophysics* **282**, pp. 663–683.

The Súrya Siddhánta: An Ancient System of Hindu Astronomy, followed by the Siddhánta Śiromani, 1861/1974. Translated into English by P. B. Deva Sastri and L. Wilkinson, Amsterdam: Philo Press.

The Súrya Siddhánta: A Textbook of Hindu Astronomy, 1860/1997, Gangooly, P. (ed.), Burgess, E. (En. tr.), Delhi: Motilal Banarsidass.

Swerdlow, N. M. and Neugebauer, O., 1984. *Mathematical Astronomy in Copernicus's* De Revolutionibus. New York-Berlin-Heidelberg-Tokyo: Springer-Verlag.

Toomer, G. J., 1969, "The solar theory of az-Zarqāl: A history of errors", *Centaurus* **14**, pp. 306–336.

Toomer, G. J., 1987, "The solar theory of Az-Zarqāl: An epilogue", in: King and Saliba 1987, pp. 513–519.

Toomer, G. J. (ed.), 1998, *Ptolemy's Almagest*, Princeton: Princeton University Press.

Van Der Waerden, B. L., 1987a, "The heliocentric system in Greek, Persian and Hindu astronomy", in: King and Saliba 1987, pp. 525–545.

Van Der Waerden, B. L., 1987b, "The astronomical system of the Persian tables II", *Centaurus* **30**, pp. 197–211.

Yabuuti, K., 1997, "Islamic astronomy in China during the Yuan and Ming Dynasties", Translated into English and partially revised by B. van Dalen, *Historia Scientiarum* **7**, pp. 11–43.

Yang, Q., 2017, "From the West to the East, from the sky to the Earth: A biography of Jamāl al-Dīn", *Asiatische Studien – Études Asiatiques* **71**, pp. 1231–1245.

9

ASTRONOMICAL OBSERVATIONS AT THE MARĀGHA OBSERVATORY IN THE 1260s–1270s[1]

9.1 Introduction

In this chapter, we present and analyze the systematic astronomical observations carried out by Muḥyī al-Dīn al-Maghribī (d. June 1283) at the Marāgha observatory (northwestern Iran, active *ca.* 1260–1320 CE). This information has been documented in his *Talkhīṣ al-majisṭī* (*Compendium of the* Almagest), extant today in a unique copy. No detailed investigation of these observations has been carried out to date. A quantitative analysis of these data can shed light on the precision attained in astronomical observations in the late Islamic period and also on the accuracy of the observational instruments constructed at the Marāgha observatory.

This chapter is organized as follows. First, in Section 9.2, we present a concise overview of the astronomical activities at the Marāgha observatory. Then, in Section 9.3, we outline the main features of Muḥyī al-Dīn's scientific career. In Section 9.4, his observations are classified into three categories and are translated. Section 9.5 is devoted to the determination of the precision of his observational data with regard to the instruments used. In Section 9.6, we explore his method for deriving the equatorial and ecliptical coordinates of a celestial object from its meridian altitude and the time of its culmination and the correlation between these three sets of data; we also compare the precision of his stellar observations with the star table of the *Īlkhānī zīj* produced by his colleagues at the Marāgha observatory. Finally, in Section 9.7, we compare the accuracy of Muḥyī al-Dīn's observational data with that of other astronomical observations made in the medieval Middle East, and offer some concluding remarks.

1 Original publication: S. M. Mozaffari, "Astronomical observations at the Maragha observatory in the 1260s–1270s," *Archive for History of Exact Sciences* **72** (2018), pp. 591–641. © 2018 Springer, and republished by permission.

DOI: 10.4324/9781003481966-13

9.2. The Marāgha Observatory

The Marāgha observatory was built in 1259 CE by Hülegü (d. 1265 CE), the founder of the Īlkhānīd dynasty of Iran.[2] For the 58 years of its activity, the observatory represented the acme of Islamic astronomy, and its influence on the later Islamic observatories founded in Samarqand in the first part of the fifteenth century, in Istanbul in the 1570s, and by Jai Singh II (1688–1743 CE) in India in the early eighteenth century can hardly be overestimated. The intellectual and financial conditions seem to have been ideal for producing an atmosphere of serious, dedicated research.

During the first two decades of the observatory, two *zījes* were written: Naṣīr al-Dīn al-Ṭūsī's (d. 1274 CE) *Īlkhānī zīj*, in Persian, and Muḥyī al-Dīn al-Maghribī's *Adwār al-anwār*, in Arabic. According to the colophon of MS. T of the *Īlkhānī zīj*, this work was completed around the end of Rajab 670 H/end of February 1272.[3] The majority of the underlying parameter values adopted in it are either Ptolemaic or are borrowed from earlier *zījes*, but some others are not known in any earlier texts and appear to be the results of the observational program carried out by the staff of the observatory in the 1260s; a notable example is the non-Ptolemaic star table in the *Īlkhānī zīj*, in which the ecliptical coordinates of 16 bright stars are tabulated along with the updated longitudes from the star tables of Ibn Yūnus and the *Mumtaḥan zīj* (see Section 9.6).[4]

Shams al-Dīn Muḥammad al-Wābkanawī al-Bukhārī (*fl. ca.* 1254–1320 CE), the outstanding figure of the second period of the Marāgha observatory (after 1283 CE), conducted a project to test the accuracy of the theoretical data derived from the *Īlkhānī zīj* and *Adwār al-anwār*, the two products of the first period of the observational activities at Marāgha. These data were the times of the synodic phenomena, such as the planetary conjunctions and eclipses, and the longitudes at which they occur, which he checked against his own observations apparently carried out over a long period of 40 years. The first documented observation in his *Zīj al-muḥaqqaq al-sulṭānī* (*The verified zīj for the sultan*) is a simple measurement of the lunar altitude at Marāgha early on the night of December 3, 1272.[5] Wābkanawī also observed the annular solar eclipse of January 30, 1283,[6] the conjunction between Jupiter and Saturn in 1286 CE, and the triple conjunctions between these two planets in 1305–1306 CE.[7] Furthermore, in the instructions to Ibn al-Fahhād's *'Alā'ī zīj* he made for Gregory

---

2 Sayılı [1960] 1988, pp. 187–223; some necessary corrections to Sayılı's historical arguments have already been given in Mozaffari and Zotti 2013. It appears that the performance of astronomical observations in Marāgha predates the foundation of the observatory there by around a century; in his treatise on the stereographic projection of the celestial sphere (the fundamental basis of the astrolabe), Ibn al-Ṣalāḥ al-Hamadhānī (d. 1153 CE) said that at Marāgha he found a value of 23;35° for the obliquity of the ecliptic (see Lorch 2000, p. 401, Mozaffari and Zotti 2013, p. 51, note 10).

3 *Īlkhānī zīj*, T: Suppl. P: f. 31v.

4 See Chapter 10. For the two star tables in the *Mumtaḥan zīj* and their relation to Ibn al-A'lam (d. 985 CE), see Mozaffari 2016–2017.

5 Wābkanawī, *Zīj*, T: f. 89v, Y: f. 155r, P: 135r.

6 See Chapter 3.

7 Wābkanawī, *Zīj*, T: ff. 2r–v, 134v–135r, Y: ff. 2v–3r, 235v–236r, P: 2v–3r, ff. 205r–v.

Chioniades, Wābkanawī computes the parameters of the solar eclipse of July 5, 1293, and the lunar counterpart of May 30, 1295, from this work for the longitude of Tabriz (northwestern Iran), and Chioniades incorporates these data into its Greek translation.[8] These worked examples were to replace the ones given by Ibn al-Fahhād.[9] In the *Revised Canons* – which was, perhaps, a part of a preliminary version of the *Muḥaqqaq zīj* – Chioniades explains the computation of the parameters of the lunar eclipse of May 30, 1295, and the solar eclipse of October 28, 1296, on the basis of the *Īlkhānī zīj* from the "oral teaching of Shams Bukhārī."[10] In light of his experiments over the period of 40 years (as reflected in his dated observations), Wābkanawī severely criticized the *Īlkhānī zīj* for its obvious errors in the planetary ephemerides, the timing and magnitudes of the eclipses, and the times of the planetary conjunctions computed from this official product of the Marāgha observatory. Instead, Wābkanawī preferred to identify Muḥyī al-Dīn's observational program as the "Īlkhānid observations" (*raṣad-i Īlkhānī*). The most egregious error occurred in the case of the triple conjunctions of Saturn with Jupiter in 1305–1306 CE, of which the *Īlkhānī zīj* could predict only the first; in contrast, all three could be predicted on the basis of Muḥyī al-Dīn's last planetary theory established at Marāgha. This difference appears to have played a decisive role in Wābkanawī's final conclusion.

9.3 Muḥyī al-Dīn al-Maghribī

Little is known about Muḥyī al-Dīn except that, according to Ibn al-Fuwaṭī, the librarian of the Marāgha observatory, his full name was Muḥyī al-Dīn Abu al-Shukr Yaḥyā b. Muḥammad b. Abī al-Shukr b. Ḥamīd al-Tūnisī (of Tunis) al-Maghribī (of the Maghrib). Muḥyī al-Dīn had learned Islamic jurisprudence (*fiqh*), according to the *Mālikī* school, in his native city. He spent some years in the service of al-Sulṭān al-Malik al-Nāṣir Yūsuf b. al-ʿAzīz b. Ghāzī b. al-Malik al-Nāṣir Yūsuf b. Ayyūb (reign: 1237–October 2, 1260 CE) in Damascus, before the king was killed by the Mongols; Muḥyī al-Dīn was then sent to the Marāgha observatory. Other than a short stay in Baghdad in the latter part of the 1270s, he lived in Marāgha until his death in Rabīʿ I 682 H/June 1283 CE.[11] He taught some students at the Marāgha observatory,[12] and his works (some 26 in all) spanned the disciplines of mathematics, astronomy, and astrology.[13]

---

8 Pingree 1985–1986, vol. 1, Chapters 32–36: pp. 131–169.

9 In *ʿAlāʾī zīj* I.35 and I.36: pp. 30–35, Ibn al-Fahhād shows how to compute the parameters of the solar eclipse of April 11, 1176, and the lunar eclipse of April 25, 1176. On their accuracy, see Chapter 5, note 18.

10 Pingree 1985–1986, vol. 1, Chapters XVII–XXII: pp. 307–333.

11 Ibn al-Fuwaṭī, vol. 5, p. 117. A quotation from Mālik b. Anas can be found in Muḥyī al-Dīn's *ʿUmda*, f. 24r (above the table of the anomaly of Saturn).

12 Of them, Ibn al-Fuwaṭī (vol. 1, 146–147) mentions ʿIzz al-Dīn al-Ḥasan b. al-Shaykh Muḥammad b. al-Shaykh al-Ḥasan al-Wāsiṭī al-ʿaṭṭār Shaykh Dār Sūsīyān.

13 See Suter 1900, p. 155, Brockelmann 1937–1942, vol. 1, p. 626, 1943–1949, S$_1$, p. 868, Sarton 1927–1948, vol. 2, Part 2, pp. 1015–1016, Sezgin 1978, p. 292, Rosenfeld and İhsanoğlu 2003, p. 226. Some of his mathematical works were studied; for example, see Hogendijk 1993. S.

In Syria, he wrote his first *zīj*, the *Tāj al-azyāj wa-ghunyat al-muhtāj* (*The crown of the zījes, sufficient for the needy*), for the longitude of Damascus *ca.* 1258 CE. Interestingly, in this *zīj* he presents non-Ptolemaic values for the orbital elements of Jupiter, Mars, and Mercury together with a set of new values for the solar, lunar, and planetary mean motions in longitude and anomaly;[14] in this work, we also encounter the most precise value measured for the rate of precession throughout the ancient and medieval periods (1°/72 years).[15] Shortly after his arrival at Marāgha, he prepared his second *zīj*, the *'Umdat al-hāsib wa-ghunyat al-tālib* (*Mainstay of the astronomer, sufficient for the student*), preserved in a unique manuscript (Cairo, Egyptian National Library, MM 188), *ca.* 1262 CE.[16] The *'Umda* is a collection of various materials taken from different sources, so it is very difficult to ascertain which parts are originally from an independent work by Muhyī al-Dīn. Moreover, more than 40% of the folios are missing from the codex. In fact, its preserved form has very few characteristics in common with Muhyī al-Dīn's observational activities in Syria and Marāgha or with his other two *zījes* written on the basis of their results – so few, in fact, that if his name were not mentioned on its first folio, there would be no reason for considering him as its possible author. This work appears to have come into being during a course of astronomy at the Marāgha observatory and was compiled by one of Muhyī al-Dīn's students there. In the prologue,[17] we are told:

> [O]ur master, the learned, the model of the lords of teaching, the prince of the geometricians, Yahyā b. Muhammad b. Abī al-Shukr al-Maghribī, said: "a group of the friends trying to learn the science of mathematics asked me to lay down a *zīj* for them in order to obtain the ephemerides of the planets, so that its understanding might be easy for a student, its sources might be close at hand for a practitioner, and that it might be a benefit for a beginner and a reminiscence for a professional."

So it was, in essence, a tutorial work for learning mathematical astronomy and the basic structure of the astronomical tables and how to work with them. The similarities between the *'Umda* and *Īlkhānī zīj* are persuasive enough to conclude that, in agreement with van Dalen's hypothesis, the *'Umda* proves to be a side project of working on the *Īlkhānī zīj*. Support for this hypothesis comes from the common values for the stellar coordinates in Ibn Yūnus's star table as preserved in the *'Umda* and the *Īlkhānī zīj*, which are not found in other source of Ibn Yūnus's

---

Tekeli's short entry on al-Maghribī in *DSB* (Gillipsie *et al.*, vol. 9, p. 555) covers only his mathematical works. See also M. Comes's entry in *BEA* (Hockey *et al.* 2007, pp. 548–549).

14 See Dorce 2002–2003, 2003.

15 See Chapter 8, p. 181.

16 On the basis of the date of a Ptolemaic star table found in it for the end of 630 Y/January 9, 1262; al-Maghribī, *'Umda*, f. 137r.

17 al-Maghribī, *'Umda*, f. 1v.

star table.[18] If this is indeed the case, then the *Īlkhānī zīj* evolved from a tutorial work by Muḥyī al-Dīn, and it is very surprising that his significant contribution to this *zīj* was totally ignored by (or was unknown to) al-Ṭūsī.

Muḥyī al-Dīn successfully performed an extensive systematic observational program at the Marāgha observatory from 1262 to 1274 CE, independently of any other astronomical activity conducted there. He explains in detail his observations and measurements of the solar, lunar, and planetary parameters in his *Talkhīṣ al-majisṭī* (*Compendium of the* Almagest). The parameter values resulting from these activities were incorporated into his third *zīj*, *Adwār al-anwār mada 'l-duhūr wa-'l-akwār* (*Everlasting cycles of lights*, also called *Zīj al-kabīr, Great zīj*).[19] Muḥyī al-Dīn's systematic observations at the Marāgha observatory established him as an outstanding figure, so much so that his contemporaries and immediate successors referred to him by unique honorific titles praising his skill in performing astronomical observations. Ibn al-Fuwaṭī, for instance, calls him *al-muhandis al-raṣadī*, the "geometrician of the observations."[20] His observational program is referred to as the "Īlkhānid observations," or the "new Īlkhānid observations" (*raṣad al-jadīd al-Īlkhānī*).[21] His fame was so widespread that his astrological doctrines were widely trusted (nine of his treatises are on astrology). A prime example of it is the interpretation of the appearance of the comet C/1402 D1 on the basis of his astrological doctrines, which triggered a major war in the Middle East at the turn of the fifteenth century.[22] None of Muḥyī al-Dīn's new parameter values were used in the *Īlkhānī zīj*.[23]

The *Talkhīṣ* has been preserved in a unique copy (Leiden, Universiteitsbibliotheek, no. Orientalis 110) in al-Maghribī's handwriting. According to the table of contents given on f. 2r, the treatise consists of ten books (*maqāla*). The books discuss plane and spherical trigonometry (books I and II), spherical astronomy (III), solar theory (IV), lunar theory (V), lunar parallax and theory of eclipses (VI), stellar astronomy (VII), planetary theory in longitude (VIII), retrograde motion

18 See Chapter 10, pp. 277–278.
19 Kamālī, *Ashrafī zīj*, F: ff. 231v, 232r, G: f. 248v; Kāshī, *Khāqānī zīj*, IO: f. 104r. Kāshī refers to Muḥyī al-Dīn as the sage/wise (*ḥakīm*).
20 Ibn al-Fuwaṭī, vol. 5, p. 117.
21 Mozaffari and Zotti2013 introduces all the indications of these terms, as found in the works written either during the lifetime of the Marāgha observatory or afterward.
22 See Mozaffari 2016.
23 Except, perhaps, for his value of 23;30° for the obliquity of the ecliptic, resulting from the measurements of the maximum and minimum annual solar noon altitudes performed on three successive days after the two dates of June 12 and December 7, 1264 (see Section 9.4.1.1, nos. **1** and **2**; Section 9.5.1; and Table 9.1); Muḥyī al-Dīn's altitude values, 76;9,30° and 29;9,30°, strictly result in the value 23.5° for the obliquity of the ecliptic. In the *Īlkhānī zīj* (C: p. 203, T: f. 102v, P: f. 59v, M1: f. 104v, M2: f. 89v), al-Ṭūsī remarks, "[O]n the basis of *our* observations, the obliquity of the ecliptic *exceeds 23;30° by a small amount*, and we estimated it to be 23;30°" (emphasis is added). Also, in his *Risāla fī kayfiyyat al-irṣād* (*The treatise on how to make* [*astronomical*] *observations*) (P: f. 7v, N: f. 41r), al-'Urḍī says that the same value was known from the continuous observations at Marāgha.

216

and latitude of the planets (IX), and the stereographic projection of the celestial sphere on the plane tangential to its north pole (X). The manuscript is incomplete and corrupt; when the author finishes his computations of Mars, the reader would expect him to commence the computations related to the inferior planets. It is probable that the author never managed to write any more of the treatise. The last two books are also missing from this copy, but the contents of the last book would probably have been adopted from (possibly a brief survey of) his treatise on the astrolabe, which deals with the same topic. Muḥyī al-Dīn dedicated the *Talkhīṣ* to Ṣadr al-Dīn Abū al-Ḥasan ʿAlī b. Muḥammad b. Muḥammad b. al-Ḥasan al-Ṭūsī,[24] the son of Naṣīr al-Dīn al-Ṭūsī, who was appointed director of the observatory after the death of his father.[25] As described in the *Talkhīṣ*, Muḥyī al-Dīn's period of observations at the Marāgha observatory extended from March 7, 1262 (lunar eclipse), to August 12, 1274 (the meridian transit of Jupiter).

A manuscript of the *Adwār al-anwār* which was apparently written under Muḥyī al-Dīn's supervision is preserved in Iran (Mashhad, Holy Shrine Library, no. 332), in which our author records the dates of the completion of its explanatory parts (canons) and of its tables, respectively, as Dhu al-qaʿda 674 H (April–May 1276)[26] and Rajab 675 H (December 1276–January 1277).[27] Thus, he seems to have finished his observations between August 12, 1274, and April–May 1276 and was busy with the construction of the tables in the time span between April and December 1276. In the prologue of this copy of the *Adwār*, he also mentions that he began to write it after completing a (now lost) treatise titled *Manāzil al-ajrām al-ʿulwiyya* (*The mansions of the upper bodies*). Thus, we can safely assume that Muḥyī al-Dīn wrote the *Talkhīṣ* after the *Adwār*, that is, after 1276 CE. According to Ibn al-Fuwaṭī,[28] Muḥyī al-Dīn left the observatory and spent a while in the service of al-Ṣāḥib Sharaf al-Dīn b. al-Ṣāḥib Shams al-Dīn in Baghdad. The date of his departure is not given, but it very likely occurred after he finished writing the *Adwār*. Ibn al-Fuwaṭī's statements give the strong impression that his departure was due to certain difficulties in Marāgha after the death of al-Ṭūsī, because he states that after Muḥyī al-Dīn's return to Marāgha, he was honored and supplied with substantial funding. It is quite probable that he wrote the *Talkhīṣ* after his return to Marāgha from Baghdad, when the observatory was under the direction of Ṣadr al-Dīn, and its dedication to Ṣadr al-Dīn was regarded as an act of thanksgiving.

The *Talkhīṣ*, its characteristics, and its place in the history of Islamic observational astronomy were already described in three papers by George Saliba in the 1980s, and the solar and lunar theories in it have been thoroughly studied by the

---

24 al-Maghribī, *Talkhīṣ*, f. 2r.
25 Sayılı 1960 [1988], p. 205.
26 al-Maghribī, *Adwār*, M: f. 55v.
27 al-Maghribī, *Adwār*, M: f. 124v.
28 Ibn al-Fuwaṭī, vol. 5, p. 117.

present author.[29] It resembles the analogous treatises belonging to the category of the "*Almagest* revision/rewritten/exposition . . ." only in the title and subject. These treatises, sometimes accompanied by criticisms of Ptolemy, constituted a genre with its own particular characteristics and played a pivotal role in medieval Islamic theoretical astronomy. However, in spite of its name, the *Talkhīṣ* is neither a rewriting nor an abridgement of the contents of the *Almagest*; a significant feature that sets it apart from other similar treatises is the abundance of its observational recordings, whereas not a single report of an observation can be found in the others. In this respect, the *Talkhīṣ* has no counterpart in the medieval astronomical corpus of the Middle East. Like Ptolemy in the *Almagest*, Muḥyī al-Dīn explains how he systematically established his parameter values, starting from the measurement of the latitude of Marāgha, the length of the solar year, the solar mean daily motion in longitude, up to the planetary parameters. There is no denying that the treatise left its mark on other developments in late Islamic astronomy; indeed, it led Taqī al-Dīn Muḥammad b. Maʿrūf, the director of the short-lived observatory at Istanbul (1526–1585 CE), and the owner of its surviving manuscript, to register the reports of his observations carried out in Istanbul about two centuries later.[30]

9.4 Muḥyī al-Dīn's Observations at the Marāgha Observatory

Muḥyī al-Dīn's observation reports can be divided into three distinct categories:

I. Observations of the meridian transits of the celestial bodies (the Sun, Moon, eight bright clock stars, and superior planets)
II. Observations of three lunar eclipses
III. Observations of five near appulses of the superior planets to Regulus (α Leo)

The observational records are presented later in the order in which they appear in the treatise. Hereafter, we refer to them by the celestial object involved and the number assigned to each of them. In chronological order, Moon-1 is the first and Jupiter-3 is the last. The accounts contain dates, times, and observational data, including durations measured by the water clock, altitudes of the celestial bodies, magnitudes of the lunar eclipses (Moon-1–3), and angular distances in latitude between the superior planets and Regulus (in Saturn-4,5, Jupiter-4, Mars-4). In the observations of type I, which constitute the bulk of his empirical activities, Muḥyī al-Dīn measured

---

29 The contents of the treatise were introduced in Saliba 1983. The computations related to the eccentricity of the Sun and of Jupiter were addressed in the two critical studies by G. Saliba (1985, 1986). For al-Maghribī's solar theory, see Chapter 1; Mozaffari 2018, esp. pp. 229, 235. For his lunar measurements, see Mozaffari 2014. For his determination of the parameters of Mars, see Mozaffari 2018–2019.
30 Taqī al-Dīn's only comment on *Talkhīṣ* can be found on f. 50v. On his observations, see Chapter 5.

the meridian altitude of a celestial object and the time of its meridian transit as counted from a specific moment when a reference celestial body, whose equatorial and ecliptical coordinates are already known, is located in a particular position with respect to the horizon, for example, the meridian transit of the Sun (true noon) or of a bright clock star (in Saturn-**3**, Jupiter-**3**, Mars-**1,3**) or the Sun being located at a low altitude a short while before sunset or after sunrise. Then, the celestial and ecliptical coordinates of that celestial object can be computed from the already-known coordinates of the reference body and the observational data.[31] In the sequel, only the parts of the accounts that contain the dates, times, and observational data are translated, which are sufficient for our purpose in the present study; in a few exceptional cases, some other quantities are given in order to indicate a quantity missing from the text (in Saturn-**3**) or to correct a date (in Jupiter-**4**). The quantitative data in the records of types **I** and **II** are summarized in Tables 9.1 to 9.4.

9.4.1 Categories I and II: The Meridian Transits and Lunar Eclipses

9.4.1.1 The Sun

1. I say: We measured the maximum altitude of the Sun in the circle of the meridian when it was in the end of the [zodiacal] sign Gemini by the high copper quadrant (*al-rubʿ al-nuḥās al-aʿlā*) available at the auspicious, blessed Īlkhānid observatory laid down in the suburb of Maragha for the three consecutive days, the first of which was the fifth [day] of *Shahrīwar-māh* [i.e., the month of *Shahrīwar*, the 6th month] in the year 633 Yazdigird, and found it to be 76;9,30°, which is the altitude of the head of Cancer. Thereafter, it [i.e., the solar noon altitude] decreased.[32]

2. We found its [i.e., the solar] maximum altitude in the circle of the meridian when it was in the end of the [ecliptic] sign Sagittarius for the three successive days, the first of which was the third [day] of *Isfandhārmadh-māh* [i.e., the month of *Isfandhārmadh*, or *Isfand*, the 12th month] [633 Yazdigird] as 29;9,30°, which is the altitude of the head of Capricorn. Afterwards, it [i.e., the Sun's meridian altitude] increased.[33]

3. [The observation of the autumnal equinox:] I say: We measured the maximum altitude of the Sun at midday on Tuesday, 11 *Ādhar-māh* [i.e., the month of *Ādhar*, the 9th month] in the year 633 Yazdigird, and found it to be 52;21°.[34]

---

31 In *Talkhīṣ* III.11: ff. 49r–50v, Muhyī al-Dīn explains in detail how to compute the ecliptical coordinates of a celestial body from its meridian altitude and the time of its meridian transit or from its altitude and azimuth.

32 al-Maghribī, *Talkhīṣ*, f. 31r.

33 al-Maghribī, *Talkhīṣ*, f. 31r.

34 al-Maghribī, *Talkhīṣ*, f. 58r.

4. The observation of the vernal [equinox]: We measured the maximum altitude of the Sun at midday on Saturday, the fifth [day] of *Khurdhādh-māh* [i.e., the month of *Khurdhādh* or *Khurdād*, the 3rd month] in the year 634 Yazdigird, and found it to be 53;8,30°.[35]

5. We say: We measured the maximum altitude of the Sun by the high quadrant erected in [the plane of] the meridian circle laid down at the blessed Īlkhānid observatory which flourished in the suburb of the city of Maragha on Monday, 3 *Ādhar-māh* [i.e., the month of *Ādhar*, the 9th month] in the year 633 Yazdigird, and found it with utmost precision to be 55;29°.[36]

6. Also, we measured its [i.e., the Sun's] maximum altitude in the meridian circle on Sunday, 21 *Diy-māh* [i.e., the month of *Diy*, the 10th month] [633 Yazdigird], and found it to be 37;35°.[37]

7. Also, we measured its [i.e., the solar] maximum altitude at midday on Thursday, 26 *Urdībihisht-māh* [i.e., the month of *Urdībihisht*, the 2nd month] in the year 634 [Yazdigird], and found it to be 49;36,30°.[38]

8. We measured the maximum altitude of the Sun in the meridian circle on Thursday [read: Wednesday], 16 *Khurdhādh-māh* [i.e., the month of *Khurdhādh* or *Khurdād*, the 3rd month] in the year 634 Yazdigird, and found it to be 57;25°.[39]

Comment: The solar noon altitude on Thursday, 17–3–634 Y = 26–3–1265 CE (JDN 2183184), was equal to ~57;45°. The large error of −20′ is incompatible with the errors we encounter in Muḥyī al-Dīn's first seven solar observations, which do not exceed ±4′ (see Section 9.5.1 and Table 9.1). Thus, it is probably the day of the week that is in error, rather than the date.

9.4.1.2 The Moon

1. The trio of lunar eclipses we dealt with, observing them with extreme accuracy:

The first occurred on the night of Wednesday, the twenty-eight of *Urdībihisht-māh* [i.e., the month of *Urdībihisht*, the 2nd month] in the year 631 Yazdigird. It was eclipsed totally with a perceptible duration (*makth*, "staying"). The altitude of *Qalb al-asad* [i.e., Regulus, α Leo] was equal to 51° west at the beginning of its staying in darkness [i.e., start of totality]. The altitude of *al-Samāk al-aʿzal* [i.e., Spica, α Vir] was equal to 17° east at the beginning of its emersion [i.e., end of totality]. It was [observed] in the city of Maragha.

35 al-Maghribī, *Talkhīṣ*, f. 58r.
36 al-Maghribī, *Talkhīṣ*, f. 59r; also translated in Saliba 1985, p. 118.
37 al-Maghribī, *Talkhīṣ*, f. 59r; also, translated in Saliba 1985, p. 118.
38 al-Maghribī, *Talkhīṣ*, f. 59r; also, translated in Saliba 1985, pp. 118–119.
39 al-Maghribī, *Talkhīṣ*, f. 60r.

Table 9.1 Sun

I	II	III		IV	
1	3 days after 5–6–633 Y June 12–14, 1264 JDN 2182897–9	76; 9,30° **76; 8,30**	+1.0′	———	—
2	3 days after 3–12–633 Y December 7–9, 1264 JDN 2183075–7	29; 9,30 **29;10,30**	−1.0	———	—
3	Tuesday 11–9–633 Y September 16, 1264 JDN 2182993	52;21 **52;17**	+4.0	180;46,25° **180;50,25**	−4′
4	Saturday 5–3–634 Y March 14, 1265 JDN 2183172	53; 8,30 **53; 5,0**	+4.0	1;12,45 **1;10,17**	+3
5	Monday 3–9–633 Y September 8, 1264 JDN 2182985	55;29 **55;25**	+4.0	172;53,59 **172;58,16**	−4
6	Sunday 21–10–633 Y October 26, 1264 JDN 2183033	37;35 **37;31**	+4.0	220;42,44 **220;44,25**	−2
7	Thursday 26–2–634 Y March 5, 1265 JDN 2183163	49;36,30 **49;32;30**	+4.0	352;19,54 **352;16,51**	+3
8	[Wednesday] 16–3–634 Y March 25, 1265 JDN 2183183	57;25 **57;22**	+3.0	12; 0,27 **11;58,45**	+2

Note: For **8**, text has "Thursday."

The total time of the mid-eclipse was 630 [years] 1 [month] 27 [days] 8;18 [hours].[40]

2. The second took place on the night of the first of *Tīr-māh* [i.e., the month of *Tīr*, the 4th month] in the year 639 [Yazdigird]. Approximately, a half and a third of its diameter was eclipsed from the south. The altitude of *al-Simāk al-rāmiḥ* [i.e., Arcturus, α Boo] was equal to 42° east at its beginning. The altitude of *Qalb al-asad* [i.e., Regulus, α Leo] was equal to 35° west at its complete emersion. The total time of the mid-eclipse was 638 [years] 3 [months] 0 [day] 10;13 [hours].[41]

3. The third occurred on the night of Wednesday, the eighteenth of *Farwardīn-māh* [i.e., the month of *Farwardīn*, the first month] in the year 643 [Yazdigird].

40 al-Maghribī, *Talkhīṣ*, f. 69v.
41 al-Maghribī, *Talkhīṣ*, f. 69v.

Nearly four-fifths of its diameter was eclipsed from the south. It was removed from the ascending node. The altitude of *al-Simāk al-rāmiḥ* [i.e., Arcturus, α Boo] was equal to 35° east at its beginning and 68° east at its end. The total time of the mid-eclipse was 642 [years] 0 [month] 17 [days] 14;0 [hours]. [42]

4.   I say: We measured the maximum altitude of the Moon in the meridian circle on the morning of Sunday, the twelfth of *Tīr-māh* [i.e., the month of *Tīr*, the 4th month] in the year 633 Yazdigird, and found it to be 28;34°. The altitude of the Sun at that time was equal to 2° [east]. . . . *Dā'ir al-irtifā'*: 2;37,20°.[43]

Comment: Dā'ir al-irtifā' is the time measured from the altitude of a celestial body (in this observation, the Sun), as expressed in terms of the revolution of the celestial sphere (see Section 9.5.2).

5.   We measured the maximum altitude of the Moon in the meridian circle, and found it to be 28;48°. The time was four and a fifth equal hours after midday on Friday, the twenty-third of the month (*shahr*) of *Ābān-māh* [i.e., the month of *Ābān*, the 8th month] in the year 633 [Yazdigird]. The altitude of the Sun at that time was equal to 25;12° from the direction of west. . . . *Dā'ir al-sā'āt*: 63;0°.[44]

Comment: Dā'ir al-sā'āt is the time expressed in terms of the revolution of the celestial sphere: $4^{1}/_{5}$ hours × 15°/hour = 63;0° (see Section 9.5.2).

6.   I say: We measured the maximum altitude of the Moon in the meridian circle, and found it to be 22;48°. It was at the time of sunrise on Wednesday, the fifth of *Khurdhādh-māh* [i.e., the month of *Khurdhādh* or *Khurdād*, the 3rd month] in the year 631 Yazdigird.[45]

9.4.1.3 Fixed Stars

1.   *Al-Simāk al-rāmiḥ* (Arcturus, α Boo)

I say: I measured its [i.e., Arcturus'] greatest altitude by the high quadrant installed in [the plane of] the meridian circle. I found it to be equal to 75;48°. The clepsydra (*al-mankām*) was in service (*khadama*) for three *turns* (*qalbāt*) minus the two-fifth *minutes* from its meridian transit until the altitude of the Sun became equal to 8°. It was in the morning on Friday, the twentieth [day] of *Farwardīn-māh* [i.e., the month of *Farwardīn*, the first month] in the year 636 Yazdigird. . . . The time (*zamān*) of the clepsydra is 15;24,44,26° and the time of *the minute* (*al-daqīqa*): 0;15,2,11°; *Dā'ir al-mankām*: 46;8,12°.[46]

---

42 al-Maghribī, *Talkhīṣ*, f. 69v.
43 al-Maghribī, *Talkhīṣ*, f. 76r.
44 al-Maghribī, *Talkhīṣ*, f. 78v.
45 al-Maghribī, *Talkhīṣ*, f. 85v.
46 al-Maghribī, *Talkhīṣ*, ff. 111r–111v. All the emphases in the quotations are ours.

Comment: Dāʾir al-mankām is the duration measured by the clepsydra, as expressed in terms of the revolution of the celestial sphere, time-degree (see Section 9.5.2).

2. *Qalb al-asad* (Regulus, α Leo)

We released (*aṭlaqnā*) the clepsydra in the evening of Friday, 23 *Kuhrdhādh-māh* [i.e., the month of *Khurdhādh* or *Khurdād*, the 3rd month] in the year 636 [Yazdigird, when] the altitude of the Sun was equal to four degrees. It was in service for 2 [*turns*] 14;48 [*minutes*] until its [i.e., Regulus'] meridian transit. Its maximum altitude: 67;55°. . . . *Dāʾir al-mankām*: 34;32,1°.[47]

3. *Qalb al-ʿaqrab* (Antares, α Sco)

We released (*aṭlaqnā*) the clepsydra in the evening of Thursday, 23 *Shahrīwar-māh* [i.e., the month of *Shahrīwar*, the 6th month] in the year 636 [Yazdigird, when] the altitude of the Sun was equal to five degrees. It was in service for two *turns* minus 6 [*minutes*] 40 *seconds* until its [i.e., Antares'] meridian transit. Its maximum altitude: 28;22°. . . . *Dāʾir al-mankām*: 29;*4*,14° [recomputed:. . .;9,. . .].[48]

4. *Shiʿrā al-yamāniya* (Sirius, α CMa)

Its maximum altitude in the meridian circle was equal to 36;35°. The clepsydra was in service for two complete *turns* from its meridian transit until the altitude of the Sun became equal to 6;26°. It was in the morning of Saturday, 3 *Diy-māh* [i.e., the month of *Diy*, the 10th month] in the year 636 [Yazdigird]. . . . *Dāʾir al-mankām*: 30;49,29°.[49]

5. *Shiʿrā al-shāmiya* (Procyon, α CMi)

Its maximum altitude in the meridian circle was equal to 59;20°. The clepsydra was in service for three complete *turns* from its meridian transit until the altitude of the Sun became equal to 8;42°. It was in the morning of Thursday, 22 *Diy-māh* [i.e., the month of *Diy*, the 10th month] in the year 636 [Yazdigird]. . . . *Dāʾir al-mankām*: 46;14,13°.[50]

6. *Simāk al-aʿzal* (Spica, α Vir)

We released the clepsydra in the evening of Monday, 14 *Murdhādh-māh* [i.e., the month of *Murdhādh* or *Murdād*, the 5th month] in the year 637 [Yazdigird, when] the altitude of the Sun was equal to 6°. It was in service for two *turns*

47 al-Maghribī, *Talkhīṣ*, f. 112r.
48 al-Maghribī, *Talkhīṣ*, ff. 112r–112v.
49 al-Maghribī, *Talkhīṣ*, f. 112v.
50 al-Maghribī, *Talkhīṣ*, ff. 112v–113r.

minus 16 [*minutes*] 30 *seconds*. Its [i.e., Spica's] maximum altitude in the middle of the heaven was equal to 45;23°. . . . the time (*zamān*) of the clepsydra by means of which the observation was made was equal to 15;45,12°, and the time of *the minute* was equal to 0;15,37,23°, because the first clepsydra had broken. . . . *Dā'ir al-mankām*: 27;12,37°.[51]

Comment: In later computations, our author makes use of 27;12,*17*° instead of 27;12,37°.

7.   *Al-Nasr al-ṭā'ir* (Altair, α Aql)

We released the clepsydra in the evening of Monday, 6 *Ādhar-māh* [i.e., the month of *Ādhar*, the 9th month] in the year 637 [Yazdigird, when] the altitude of the Sun was equal to 6°. It was in service until its [i.e., Altair's] meridian transit minus 9 [*minutes*] 45 *seconds*. Its maximum altitude in the middle of the heaven was equal to 59;54. . . . *Dā'ir al-mankām*: 28;58,*20*° [recomputed:. . .;. . .,*4.5*°].[52]

8.   *'Ayn al-thawr* (Aldebaran, α Tau)

Its maximum altitude in the middle of the heaven was equal to 67;30°. The clepsydra was in service for 2 [*turns*] and 15 [*minutes*] from its [i.e., Aldebaran's] meridian transit until the altitude of the Sun became equal to two degrees. It was in the morning of Monday, 13 *Ādhar-māh* [i.e., the month of *Ādhar*, the 9th month] in the year 637 [Yazdigird]. . . . *Dā'ir al-mankam*: 35;24,45°.[53]

9.4.1.4 The Superior Planets

§ SATURN

1.   We released the clepsydra at midday on Thursday, 19 *Diy-māh* [i.e., the month of *Diy*, the 10th month] in the year 632 Yazdigird. It was in service for 11 [*turns*] 50 [*minutes*] until the meridian transit of Saturn on the meridian circle. Its maximum altitude in the middle of the heaven was equal to 65°. The time (*zamān*) of the clepsydra was equal to 15;24,44,26° and that of its *minute*: 0;15,2,11°. *Dā'ir al-mankām*: 182;3,58°.[54]

2.   We released the clepsydra at midday on Thursday, 5 *Isfandhārmadh-māh* [i.e., the month of *Isfandhārmadh* or *Isfand*, the 12th month] in the year 635 [Yazdigird]. It was in service for 11 [turns] 30 [minutes] 40 [seconds] until

51 al-Maghribī, *Talkhīṣ*, f. 113r.
52 al-Maghribī, *Talkhīṣ*, f. 113v.
53 al-Maghribī, *Talkhīṣ*, ff. 113v–114r.
54 al-Maghribī, *Talkhīṣ*, f. 123r.

224

its [i.e., Saturn's] meridian transit. *Dā'ir al-mankām*: 177;13,16°. Its [i.e., Saturn's] maximum altitude was equal to 74;41°.[55]

3. The maximum altitude of Saturn in the middle of the sky was equal to 60;50°. The clepsydra was in service for 1 [*turn*] 38 [*minutes*] 30 [*seconds*] from its [i.e., Saturn's] meridian transit (*tawassuṭ*) to the meridian transit of *al-Samāk al-aʿzal* [i.e., Spica, α Vir]. It was in the night of Monday, [22] *Urdībihisht-māh* [i.e., the month of *Urdībihisht*, the 2nd month] in the year 642 [Yazdigird]. *Dā'ir al-mankām*: [25;3,38°]. The [right] ascension of the transit degree (*al-mamarr*) [i.e., polar longitude] of Spica: 282;12°; we subtracted *al-dā'ir* from it, and the remainder is equal to the [right] ascension of the transit degree [of Saturn]: 257;8,22°.[56]

Comment: The value of *dā'ir al-mankām* is missing from the text, and a value of 25;*4,12°* has been written along the same line in the left margin; but it does *not* seem to be correct, since the *dā'ir* taken as equal to the difference between the ascensions of the transit degrees of Saturn and Spica given in the text, that is, 25;3,38°, is in complete agreement with the duration computed from the value given for the *turns* of the operation of the clepsydra and the conversion formulas for transforming it into the corresponding interval of time (see earlier, the observation reports of Arcturus and Saturn-1, and later, Section 9.5.2 and Table 9.5).

§ JUPITER

1. I say: We released the clepsydra on Sunday, 14 *Diy-māh* [i.e., the month of *Diy*, the 10th month] in the year 633 Yazdigird [when] the altitude of the Sun was equal to 3° west. It was in service for 6 [*turns*] 52 [*minutes*] 30 [*seconds*] until its [i.e., Jupiter's] meridian transit. Its [i.e., Jupiter's] maximum altitude in the circle of meridian was equal to 64;33°. . . . *Dā'ir al-mankām*: 106;0,25°.[57]

2. We released the clepsydra on Tuesday, 24 *Isfandhārmadh-māh* [i.e., the month of *Isfandhārmadh* or *Isfand*, the 12th month] in the year 635 [Yazdigird, when] the altitude of the Sun was equal to 3;20° west. It was in service for 7 [*turns*] 19 [*minutes*] until its [i.e., Jupiter's] meridian transit. Its [i.e., Jupiter's] maximum altitude was equal to 75;40°. . . . *Dā'ir al-mankām*: 112;38,52°.[58]

3. We released half of the clepsydra (*niṣf al-mankām*) [?] in the night of Sunday, the eight [day] of *Ābān-māh* [i.e., the month of *Ābān*, the 8th month] in the year 643 [Yazdigird] for 5 [*turns*] 15 [*minutes*] 30 [*seconds*] from the meridian transit of *al-Nasr al-ṭā'ir* [i.e., Altair, α Aql] to the meridian transit of the planet. Its *dā'ir*: 42;13,38°. . . . Its [i.e., Jupiter's] maximum altitude was equal to 38;56°.[59]

55 al-Maghribī, *Talkhīṣ*, f. 123r.
56 al-Maghribī, *Talkhīṣ*, f. 123v.
57 al-Maghribī, *Talkhīṣ*, f. 128r.
58 al-Maghribī, *Talkhīṣ*, f. 128v
59 al-Maghribī, *Talkhīṣ*, ff. 128v–129r.

§ MARS

1. We released *the minute (al-daqīqa)* from its [i.e., Mars's] meridian transit to the meridian transit of *'Ayn al-thawr* [i.e., Aldebaran, α Tau]. It was in service for 33 *minutes*. Its [i.e., Mars's] maximum altitude in the circle of the meridian was equal to 72;2°. It was on the night of Wednesday, the first [day] of *Bahman-māh* [i.e., the month of *Bahman*, the 11th month] in the year 633 Yazdigird. *Dā'ir al-daqīqa*: 8;16°.[60]

Comment: Dā'ir al-daqīqa is the duration measured by *the minute*, as expressed in terms of the revolution of the celestial sphere (see Section 9.5.2).

2. We released the clepsydra at the end of the daylight on Sunday, 15 *Isfandhārmadh-māh* [i.e., the month of *Isfandhārmadh* or *Isfand*, the 12th month] in the year 635 [Yazdigird, when] the altitude of the Sun was 3° west. It was in service for 7 [*turns*] 18 [*minutes*] until its [i.e., Mars's] meridian transit. Its [i.e., Mars's] maximum altitude was equal to 79;42°. . . . *Dā'ir al-mankām*: 112;36,22°.[61]

3. We released the clepsydra in the night of Thursday, the 20th [day] of *Urdībihisht-māh* [i.e., the month of *Urdībihisht*, the 2nd month] in the year 640 [Yazdigird] from the meridian transit of *Qalb al-asad* [i.e., Regulus, α Leo]. It was in service for 1 [*turn*] 32 [*minutes*] 30 [*seconds*] until its [i.e., Mars's] meridian transit. Its maximum altitude in the circle of the meridian was equal to 62;33°. *Dā'ir al-mankām*: 23;33,25°.[62]

In Tables 9.1 to 9.4, (1) the true modern values are in bold,[63] and (2) Cols. **I** and **II** indicate, respectively, the numbers assigned to the observations and their dates in the Yazdigird era, the Julian calendar, and Julian Days Number (JDN). The Yazdigird era, June 16, 632, is used with the Egyptian/Persian year, consisting of 12 months of 30 days plus 5 epagomenal days which, in the early Islamic period, were put after the eighth month but, in the late Islamic period, were transferred to the end of the year. Also, Muḥyī al-Dīn expresses the time intervals in terms of the revolution of the celestial sphere (except for Moon-2), the so-called "*dā'ir*," which are first divided by 15 and are then listed in Tables 9.1 to 9.4.

---

60 al-Maghribī, *Talkhīṣ*, f. 132v.
61 al-Maghribī, *Talkhīṣ*, ff. 132v.
62 al-Maghribī, *Talkhīṣ*, f. 133r.
63 In the analysis of these observations, allowance should be made for atmospheric refraction and parallax, especially when a celestial body is setting, rising, or at a low altitude. Marāgha is 1,550 m above sea level, where the average atmospheric pressure is ~850 mbar. The changes in temperature at the different times are also taken into account, based on weather reports for the 2000s. Throughout this chapter, the true modern values are derived from the software Alcyone Ephemeris 4.3, which is well suited for historical investigations.

Table 9.2 Moon

A. Lunar Eclipses

I	II	III	IV	V	VI	VII	VIII	IX
1	Night of Wednesday 28–2–631 Y March 7, 1262 JDN 2182069	Sun, true noon	**12:10**	**20:23**	$8;18^h$ $+5^m$	Total **1.77**	Regulus: 51° W **51**	Spica: 17° E **18**
2	Night of Tuesday 1–4–639 Y April 7, 1270 JDN 2185022	Sun, true noon	**12: 0**	**22: 8**	$10;13^h$ $+5$	≈1/2 + 1/3 from south **0.823**	Arcturus: 42° E **43**	Regulus: 35° W **36**
3	Night of Wednesday 18–1–643 Y January 24, 1274 JDN 2186410	Sun, true noon	**12:15** (-1^d)	**2:12**	$14;0^h$ $+3$	≈4/5 from north **0.77**	Arcturus: 35° E **34**	Arcturus: 68° E **68**

B. Lunar Altitudes

I	II	III	IV	V	VI	VII	VIII
4	Morning of Sunday 12–4–633 Y April 20, 1264 JDN 2182844	Sun, sunrise	**5:10**	2; 0° [E] **1;45** **2;36**	28;34° **28;53** **28;54**	 **5:21** (1) **5:25** (2)	$0;10,30^h$ -4^m
5	Afternoon of Friday 23–8–633 Y August 29, 1264 JDN 2182975	Sun, true noon	**11:58**	25;12 W **25;20** **24;30**	28;48 **28;30** **28;30**	 **16:10** (1) **16:14** (2)	 4;12 −4
6	Sunrise of Wednesday 5–3–631 Y March 15, 1262 JDN 2182077	Sun, sunrise	**6: 3**	0 **0** **1;29**	22:48 **22;57** **22;59**	 **6: 3** (1) **6:11** (2)	 0 —

Note:

4. (1) 10.5 minutes after sunrise, when azimuth of the Moon ≈359°; (2) the time when the Moon transited the meridian. **5**. (1) 4;12 hours after true noon, when azimuth of the Moon ≈359°; (2) the time when the Moon transited the meridian. **6**. (1) At sunrise, when azimuth of the Moon ≈358°; (2) the time when the Moon transited the meridian.

Table 9.2 (Continued)

C. The horizontal coordinates of the Moon and the stars at the specific contact times of the triple lunar eclipses.

1	Start of totality (19:31 MLT)		End of totality (21:15 MLT)	
	Moon	Regulus	Moon	Spica
Altitude	17°	51°	36°	18°
Azimuth	281	294	300	295°
2	Beginning (20:38 MLT)		End (23:39 MLT)	
	Moon	Arcturus	Moon	Regulus
Altitude	24°	43°	42°	36°
Azimuth	304	272	352	81
3	Beginning (0:43 MLT)		End (3:40 MLT)	
	Moon	Arcturus	Moon	Arcturus
Altitude	69°	34°	41°	68°
Azimuth	22	265	80	304

Table 9.3 Stars

I	II	III	IV	V		VI	VII		VIII		IX		X		XI	
Arcturus (α Boo)	Morning, Friday 20–1–636 Y January 28, 1267 JDN 2183857	Sun, 8;0° E	7:50	75;48° **75;44.0**	+ 4.0'	**4:46**	3; 5ʰ	+1ᵐ	+23; 8.5° **+23; 7.5**	+ 1.0'	205;50 **205;34**	+16'	+31;22° **+31;14**	+ 8'	194;19° **194; 1**	+18'
Regulus (α Leo)	Evening, Friday 23–3–636 Y April 1, 1267 JDN 2183920	Sun, 4;0 W	**18: 5**	67;55 **68; 0.0**	− 5.0	**20:22**	2;18	+1	+15;15.5 **+15;23.5**	− 8.0	142;21 **142;14**	+ 7	+ 0;22 **+ 0;26**	− 4	139;49 **139;40**	+ 9
Antares (α Sco)	Evening, Thursday 23–6–636 Y June 30, 1267 JDN 2184010	Sun, 5;0 W	**18:50**	28;22 **28; 8.0**	+14.0	**20:43**	1;56	+3	−27;17.5 **−24;30.0**	+12.5	236;44 **236;17**	+27	− 4;11 **− 4;28**	+17	239;58 **239;33**	+25
Sirius (α CMa)	Morning, Saturday 3–10–636 Y October 8, 1267 JDN 2184110	Sun, 6;26 E	**6:45**	36;35 **36;40.5**	− 5.5	**4;39**	2; 3	−3	−16; 4.5 **−15;57.0**	− 7.5	94; 4 **93;13**	+51	−39,30 **−39;26**	− 4	94;59 **94; 0**	+59
Procyon (α CMi)	Morning, Thursday 22–10–636 Y October 27, 1267 JDN 2184129	Sun, 8;42 E	**7:19**	59;20 **59;27.0**	− 7.0	**4:12**	3; 5	−2	+ 6;40.5 **+ 6;50.0**	− 9.5	105;55 **105;12**	+43	−15;55 **−15;52**	− 3	106;28 **105;42**	+46
Spica (α Vir)	Evening, Monday 14–5–637 Y May 21, 1268 JDN 2184336	Sun, 6;0 W	**18:34**	45;22 **45;22.0**	0	**20:20**	1;49	+3	− 7;17.5 **− 7;15.0**	− 2.5	192; 9 **191;46**	+23	− 1;53 **− 2; 0**	+ 7	194; 2 **193;39**	+23
Altair (α Aql)	Evening, Monday 6–9–637 Y September 10, 1268 JDN 2184448	Sun, 6;0 W	**17;30**	59;54 **59;48.5**	+ 5.5	**19;26**	1;56	0	+ 7;14.5 **+ 7;11.5**	+ 3.0	288;26 **288;46**	−20	+29;24 **+29;20**	+ 4	291;10 **291;28**	−18
Aldebaran (α Tau)	Morning, Monday 13–9–637 Y September 17, 1268 JDN 2184455	Sun, 2;0 E	**6; 2**	67;30 **67;21.5**	+ 8.5	**3;41**	2;22	+1	+14;50.5 **+14;45.0**	+ 5.5	59; 0 **58;36**	+24	− 5;29 **− 5;31**	+ 2	60; 0 **59;35**	+25

Table 9.4 Planets

A. Saturn

I II	III	IV	V	VI	VII	VIII	IX	X	XI
1 Thursday 19–10–632 Y October 25, 1263 JDN 2182666	Sun, true noon	**11:44**	65; 0° −9.5′ **65; 9.5**	**23:51**	12;8^h +1^m	+12;20.5 −12.5′ **+12;33.0**	39; 7 +18′ **38;49**	−2;51° −15′ **−2;36**	40;35° +9′ **40;26**
2 Thursday 5–12–635 Y December 9, 1266 JDN 2183807	Sun, true noon	**11:57**	74;41 −6.5 **74;47.5**	**23:46**	11;49 0	+22; 1.5 − 9.5 **+22;11.0**	82;16 + 5 **82;11**	−1;17 − 8 **−1; 9**	82;54 +8 **82;46**
3 Night of Monday 22–2–642 Y February 27, 1273 JDN 2186079	Spica, meridian transit	**1:51**	60;50 +6.0 **60;44.0**	**0:13**	1;40 +2	+ 8;10.5 + 3.5 **+ 8; 7.0**	167; 8 − 1 **167; 9**	+2;27 + 4 **+2;23**	165; 0 −1 **165; 1**
4 Morning, Friday 29–9–639 Y October 3, 1270 JDN 2185201	Appulse to Regulus See Section 9.4.2								
5 Beginning of the night of Sunday 22–6–640 Y June 27, 1271 JDN 2185468	Appulse to Regulus See Section 9.4.2								

(*Continued*)

Table 9.4 (Continued)

B. Jupiter

I	II	III	IV	V	VI	VII	VIII	IX	X	XI
1	Sunday 14–10–633 Y October 19, 1264 JDN 2183026	Sun, 3;0° W	**16:49**	64;33° −1.0' **64:34.0**	**23:50**	7; 4^h $+3^\mathrm{m}$	+11;53.5 −3.5' **+11;57.0**	33;52 +31' **33;21**	−1;38° −13' **−1;25**	35;37° +27' **35;10**
2	Tuesday 24–12–635 Y December 28, 1266 JDN 2183826	Sun, 3;20 W	**16:34**	75;40 −4.0 **75;44.0**	**0: 3** ($+1^\mathrm{d}$)	7;31 +2	+23; 0.5 −7.0 **+23; 7.5**	105;21 +17 **105; 4**	+0;16 − 3 **+0;19**	104; 8 +18 **103;50**
3	Night of Sunday 8–8–643 Y August 12, 1274 JDN 2186610	Altair, meridian transit	**21:26** (-1^d)	38;56 +3.0 **38;53.0**	**0:10**	2;49 +5	−13;43.5 +1.0 **−13;44.5**	330;45 +47 **329;58**	−1;37 −15 **−1;22**	327;59 +43 **327;16**
4	Midnight of Sunday 2–[11]–636 Y November 6, 1267 JDN 2184139	Appulse to Regulus See Section 9.4.2								

(*Continued*)

231

Table 9.4 (Continued)

C. Mars

I II	III	IV	V		VI	VII		VIII		IX		VIII		IX	
1 Night of Wednesday 1–11–633 Y November 5, 1264 JDN 2183043	Aldebaran, meridian transit	**0:28**	72; 2° **72; 6.5**	−4.5′	**23:59** (−1^d)	$0;33^h$	$+4^m$	+19;22.5° **+19;30.0**	−7.5′	50;40° **51;15**	−35′	+0;45° **+0;43**	+ 2′	53;19° **53;51**	−32′
2 Sunday 15–12–635 Y December 19, 1266 JDN 2183817	Sun, 3; 0° W	**16:29**	79;42 **79;46.0**	−4.0	**0: 0** (+1^d)	7;30	−1	+27; 2.5 **+27; 9.5**	−7.0	95; 6 **95;20**	−14	+3;37 **+3;43**	− 6	94;41 **94;46**	− 5
3 Night of Thursday 20–2–640 Y February 26, 1271 JDN 2185347	Regulus, meridian transit	**22:39** (−1^d)	62;33 **62;35.5**	−2.5	**0:18**	1;34	−5	+ 9;53.5 **+ 9;59.0**	−5.5	165;57 **166;56**	−59	+3;34 **+4; 1**	−27	163;15 **164; 6**	−51
4 14;30 hours after true noon of Tuesday 10–10–639 Y October 15, 1270 JDN 2185213	Appulse to Regulus See Section 9.4.2														

Table 9.1 summarizes the solar observations: Col. **III** presents Muḥyī al-Dīn's and the true values for the solar meridian altitudes in the left-hand side of the column, and their differences in the right-hand side (NB: we also keep this arrangement in the subsequent tables); Col. **IV** gives his values for the solar longitudes (which were computed from the observed noon altitudes), the modern values at the times, and their differences.

Table 9.2(A) shows the trio of the lunar eclipses. The first eclipse is total, with a perceptible duration; the two others are partial. Col. **III** gives the origins of time, from which the instants of the maximum phases of the eclipses were measured, and Col. **IV** the corresponding mean local times (MLT); Col. **V** presents the times of the maximum phases of the eclipses computed on the basis of modern theories.[64] Col. **VI** contains (on the left-hand side) our author's values for times of the maximum phases of the eclipses measured from the origins mentioned in Col. **III** and their differences from the true time intervals, that is, the differences between Cols. **IV** and **V** (on the right-hand side). Col. **VII** presents the magnitudes of the eclipses and the directions of the obscuration of the lunar disk. The magnitudes of the lunar eclipses may have been naked-eye estimates; however, two optical devices for the direct measurements of the magnitudes of the eclipses were invented and constructed at the Marāgha observatory (see Section 9.5.3). As we can see, their measurements were accurate. Cols. **VIII** and **IX** give, respectively, the altitudes and horizontal directions of some reference stars at the start and end of totality (in the first eclipse) or at the beginning and end of the eclipse (in the two others). We are not told precisely what the directions refer to. It seems at first sight that they refer to the meridian; the horizontal coordinates of a celestial body can be given by means of its altitude along with its direction with respect to the meridian in the same manner that Muḥyī al-Dīn expresses the solar altitude in Moon-**5**. Nevertheless, in the cases of the stellar altitudes in Moon-**1–3**, the directions appear to refer to the lunar disk. The modern horizontal coordinates of the Moon and the stars in question at the specific contact times of the triple lunar eclipses (rounded to the nearest degree) are given in Table 9.2(C). It can clearly be seen that the direction of west for Regulus in the first eclipse must refer to the lunar disk, simply because this star was in the east of the meridian of Marāgha at that time. However, in five other stellar altitudes, it makes no difference at all whether the directions he gives refer to the meridian or the lunar disk, since the stars were located in the same direction with respect to both.

Based on what our author says in Moon-**1**, this trio of lunar eclipses were the ones that he observed carefully, and therefore, he was able to rely on his observations and be confident about the accuracy of the data obtained from them. During the period of his observations, nine other lunar eclipses were observable at their maximum phases from Marāgha, which Muḥyī al-Dīn may have witnessed as well.

---

64 The modern times and magnitudes of the eclipses are based on 5MCLE, nos. 07878, 07897, and 07907.

Table 9.2(B) shows the observations of the lunar meridian transits: Col. **III** gives the origins of time, from which the times of the lunar meridian transits were computed, and Col. **IV**, the corresponding MLTs. Col. **V** contains the altitudes of the Sun at the times of the observations, and Col. **VI** the meridian altitudes of the Moon. Col. **VII** indicates the times of the meridian transits of the Moon. Col. **VIII** contains our author's values for the times of the lunar meridian transits measured from the origins mentioned in Col. **III** and their differences from the true time intervals, that is, the differences between Cols. **IV** and **VII**. For reasons that will be mentioned later (see Section 9.5.1), we present the two sets of the modern values in Cols. **V–VII**: the one (in the second line) for the times given by our author (= Col. **IV** + Col. **VIII**) and the other (in the third line) for the true modern times of the lunar meridian transits (see notes to the table).

Table 9.3 presents Muḥyī al-Dīn's stellar observations: Col. **III** gives the origins of time, from which the times of the meridian transits of the stars were computed, and Col. **IV** the corresponding MLTs. Col. **V** presents Muḥyī al-Dīn's and true modern values for the meridian altitudes of the stars and their differences. Col. **VI** contains the true times of the meridian transits of the stars. Col. **VII** gives our author's values for the times of the meridian transits of the stars, measured from the origins mentioned in Col. **III**, and their differences from the true time intervals, that is, the differences between Cols. **IV** and **VI**. Cols. **VIII** and **IX** give, respectively, his values for the declinations and right ascensions of the stars in comparison with the corresponding modern values at that time. Note that he does not calculate the declinations since they were not necessary in his computations; we have derived them simply from $\delta = h_{max} + \varphi - 90°$ with the values of h_{max} in Col. **V** and $\varphi = 37;20,30°$, and have rounded the resulting figures to the nearest 0.5'. Cols. **X** and **XI** indicate, respectively, his computed values for the stellar latitudes and longitudes together with the corresponding modern values at that time and the differences between them.

Table 9.4 lists Muḥyī al-Dīn's planetary observations; the columns are the same as in Table 9.3.

9.4.2 Category III: Near Appulses of the Superior Planets to Regulus

Muḥyī al-Dīn reports four near appulses of the superior planets to Regulus. The observations of the appulses of the planets to the bright zodiac stars (especially Regulus) appear to have been common in early Islamic astronomy,[65] as

---

65 Making observations of this kind seems to have been a characteristic of planetary research ever since antiquity (for Babylonian observations of the appulses of the planets to the stars, see Jones 2004). Even in the telescopic era, the astronomers found that "of all the celestial observations that have hitherto been made, none are so capable of perfect exactness, as the near appulses of the Moon and planets to the fixed stars" (Halley 1720–1721, p. 209).

documented in Ibn Yūnus's *Ḥākimī zīj*.[66] They were deployed in order to test the ecliptical coordinates computed from the *Mumtaḥan zīj*.

Muḥyī al-Dīn's observations are as follows:

§ Saturn

4. We took an observation of Saturn and Regulus on the morning of Friday, 29 *Ādhar-māh* [i.e., the month of *Ādhar*, the 9th month] in the year 639 Yazdigird. We found that the circle of latitude approximately aligned the two. Then, we expected the conjunction to take place at midday on this day. [At the time of the observation,] Saturn was in the north of Regulus by approximately half a degree. The longitude of Regulus at the time of the observation was Leo 19;52° [= 139;52°] which was equal to the longitude of Saturn [as well].[67]

The date is October 3, 1270. Saturn rose at 1:22 MLT. The observation must have been made between this time and the start of the civil twilight at 5:38 MLT (Saturn transited the meridian of Marāgha at 8:14 MLT, after sunrise at 6:04 MLT). Muḥyī al-Dīn's statement about the time of the observation gives the impression that the observation was made some time before civil twilight. At this time, the apparent longitude of Saturn was about **139;43°**, and that of Regulus **139;41°**, and thus, Saturn was located about **2′** in longitude *in front of* Regulus. Since our author estimated that the conjunction occurred at noon on the date in question, and that the daily motion of the planet was 0;5°,[68] he appears to have observed

66 For example, Alī b. Amājūr's single observations of Mercury and Venus with Antares on December 24, 918, and Mars with Procyon on January 1, 919 (Ibn Yūnus, *Zīj*, L: p. 99; Caussin 1804, pp. 108–111, Delambre 1819, p. 83), and periodic observations of Jupiter with Vega and Mars with Sirius from July 13 to September 9, 918 (Ibn Yūnus, *Zīj*, L: pp. 98–99; Caussin 1804, pp. 104–107, Delambre 1819, p. 83). The majority of other planetary observational records are the closeness of the planets to Regulus. Examples: A conjunction of Jupiter with Regulus was observed by Ḥabash on September 6, 864 (Ibn Yūnus, *Zīj*, L: p. 108; Caussin 1804, pp. 155–156, Delambre 1819, p. 87); Māhānī found Saturn 2/3° in longitude behind Regulus on August 28, 858 (Ibn Yūnus, *Zīj*, L: p. 96; Caussin 1804, pp. 94–97, Delambre 1819, p. 80); an occultation of Regulus by Venus on September 10, 885 (Ibn Yūnus, *Zīj*, L: p. 109, F1: f. 10r (the only observation of the Banū Amājūr that is mentioned in MS. F1 of the *Ḥākimī zīj*); Caussin 1804, pp. 157–158, Delambre 1819, p. 87), and a conjunction of Mars with Regulus on September 20, 909 (Ibn Yūnus, *Zīj*, L: p. 109; Caussin 1804, pp. 161–162, Delambre 1819, p. 89) were observed by Alī b. Amājūr; and eight conjunctions of Regulus with Venus and five with Mars were observed by Ibn Yūnus during 987–1002 CE (Ibn Yūnus, *Zīj*, L: pp. 113–120; Caussin 1804, pp. 179–211, Delambre 1819, pp. 90–92). The other two conjunctions of the planets with Regulus are reported by a thirteenth-century Yemenite astronomer, al-Kawāshī: Jupiter–Regulus on October 5, 1279, and Mars–Regulus on June 16, 1282 (see King and Gingerich 1982, pp. 124, 126–127).

67 al-Maghribī, *Talkhīṣ*, f. 127r.

68 Muḥyī al-Dīn could determine the daily motion in longitude of a planet by deriving its longitudes from its horizontal coordinates measured at the same time on two successive nights, according to his method. From his values for the motional parameters and orbital elements of Saturn, the longitude of Saturn at noon on October 2 and 3, 1270, in Marāgha were, respectively, 139;44,11° and

Saturn ~1.5′ in longitude *behind* Regulus at that time; accordingly, he must have committed an error of ~+3.5′ in the measurement of the difference in longitude between Saturn and Regulus. He gives their longitudes as 139;52°, which was, in fact, derived from adding the increment of ~3′ in the longitude of the star in three years (on the basis of his value of 1° in 66 Persian years for the rate of the precession) to the value 139;49° measured for it in 1267 CE (Table 9.3). The modern value at that time was **139;42°**. The difference between the apparent latitudes of the planet (**+1;9°**) and Regulus (**+0;27°**) was around **42′**, and so Muḥyī al-Dīn's error is ~−12′.

5.   The circle of latitude aligned Saturn and Regulus early on the night of Sunday, 22 *Shahrīwar* [i.e., the 6th month] in the year 640 [Yazdigird]. It was supposed that the conjunction occurred at midday on Saturday. The longitude of Regulus and of Saturn was Leo 19;53° [= 139;53°].[69]

The date is equivalent to June 28, 1271, but since night precedes day, "early on the night of Sunday" means a short while after sunset on Saturday, June 27, 1271. Sunset came at 19:21 MLT. The observation must have been made between 19:54 MLT (the end of the civil twilight) and 21:34 MLT when Saturn set. At the first time, the apparent longitude of the planet was equal to **140;12°**, and that of Regulus **139;44°**; thus, Saturn was ~**28′** in longitude ahead of Regulus, and with the daily motion of 0;6°,[70] our author appears to have found the longitude of the planet more than that of Regulus by only ~2′. Thus, it seems that he committed an error of ~+26′ in the estimation of the longitudinal difference between Saturn and Regulus. Saturn (with a latitude of **+1;29°**) was located around 1° in latitude above Regulus (at an apparent latitude of ~**+0;29°**).

The two preceding conjunctions occurred in the period of the direct motion of the planet; meanwhile, another conjunction of Saturn with Regulus took place on January 20, 1271, when the planet was in the retrograde motion; it was also observable in Marāgha.

§ *Jupiter*

4.   I say that the circle of latitude aligned the centers of Jupiter and Regulus at midnight on Sunday, the second of *Ādhar-māh* [i.e., the month of *Ādhar*, the

---

139;49,28° (modern: **139;39,55°** and **139;45,11°**); both sets of values result in a daily longitudinal motion of 0;5°.

69 al-Maghribī, *Talkhīṣ*, f. 127r.

70 From Muḥyī al-Dīn's parameter values, the longitudes of the planet at the noon of the two days June 27 and 28, 1271 were, respectively, 139;53,42° and 140;00,28° (modern: **140;7,57°** and **139;14,28°**).

9th month] [*sic*] in the year 636 Yazdigird. Jupiter was in the north of Regulus approximately as much as one and one half of its [i.e., Jupiter's] body (*jirm*). The longitude of Regulus was 4 [signs] 19;49° [= 139;49°] which was equal to the longitude of Jupiter [as well]. At the mentioned midnight, the mean longitude of the Sun: 231;50,24°, the mean longitude of Jupiter: 124;58,22°, its [mean] anomaly: 106;52,2°. . . .[71]

The date 2nd of *Ādhar-māh* in the year 636 Y is equal to Wednesday, September 7, 1267; the year is correct, because of the value given for the longitude of Regulus (cf. Table 9.3), but the day is incorrect, since at midnight on September 7, 1267, Jupiter was in excess of 8° in longitude behind Regulus (Muḥyī al-Dīn's parameter values and epoch positions result in a value of 131;24° for the longitude of Jupiter at that time; modern: **131;13°**). The date should be read as "the 2nd of *Bahman* (the 11th month) in the year 636 Y" = November 6, 1267, which was a Sunday, because on the basis of al-Maghribī's mean motions and epoch mean positions, it can easily be verified that all values he gives for the mean longitudes of the Sun and Jupiter and the mean anomaly of Jupiter in the preceding passage are, indeed, for midnight on this date. Jupiter rose on November 5 at 23:10 MLT and transited the local meridian on November 6, at 6:01 MLT; sunrise came at 6:39 MLT. When passing across the meridian, the apparent longitude of Jupiter was equal to **139;44°**, and that of Regulus **139;40°**; at midnight, when they were located at an altitude of ~9°, their longitudes were equal, respectively, to **139;38°** and **139;35°**, but Muḥyī al-Dīn found Jupiter in conjunction with Regulus and, consequently, must have committed an error of ~+3′ in the estimation of the distance in longitude between Jupiter and Regulus. The difference between the apparent latitudes of the planet (**0;53°**) and Regulus (**0;28°**) was ~**25′**; our author gives this distance in terms of the "body" of Jupiter. Ibn Yūnus also used the body of Jupiter to express the apparent angular separation of Mars from Jupiter on December 15, 989, about 1 hour after sunset (17:04 MLT in Cairo); the modern value is ~**10′**.[72] The use of the angular diameters of the luminous celestial bodies such as the Moon in order to express the distance between the heavenly objects goes back to the ancient observations recorded in the *Almagest*; for example, an observation of the near appulse of Mercury to the stars β and δ Sco dated from 265 BCE in *Almagest* IX.10.[73] Nevertheless, in both Ibn Yūnus's and Muḥyī al-Dīn's observations, *body* does not refer to the actual angular diameter of Jupiter, simply because its apparent diameter is less than 1 arc minute, and thus, it is completely undetectable using the optical instruments available in medieval times. However, a partly physiological optical effect enhances the apparent diameter of the planets

71 al-Maghribī, *Talkhīṣ*, f. 131v.
72 Ibn Yūnus, *Zīj*, L: p. 114; Caussin 1804, pp. 181–183.
73 Toomer [1984] 1998, p. 464.

and stars, meaning, that the brighter the star, the larger it is.[74] Thus, since Jupiter is the brightest of the superior planets and much brighter than Regulus (at the time of this observation, the planet was about 24 times as luminous as Regulus), it appears as a large shining point, which medieval astronomers could use as reference to estimate the narrow distances between Jupiter and another object. However, it was less frequently used than standard units such as the cubit (= 2° or 2;20°), the span (= 1/2 cubit), and the finger (= 1/24 cubit).

§ Mars

4. I say that the circle of latitude had aligned the centers of Mars and Regulus when 14;30 hours had elapsed since midday on Tuesday, the tenth of *Diy-māh* [i.e., the month of *Diy*, the 10th month] in the year 639 Yazdigird, when nearly one-third of [the ecliptic sign] Virgo rose. [Mars] was in the north of Regulus as much as 40 arc minutes. The longitude of Regulus and of Mars was 4 [ecliptic signs] 19;5°2 [= 139;52°].[75]

The time of this observation is given as 14;30^h after true noon on October 14, 1270. Mars rose on October 15 at 0:34 MLT. At 2:30 MLT, the apparent longitude of Mars was about **139;42,30°**, and that of Regulus **139;40,0°** (both at an altitude of **22°**), whereas Muḥyī al-Dīn found them to be in conjunction with each other; consequently, he appears to have committed an error of ~+2.5′ in the estimation of the distance in longitude between Jupiter and Regulus. The apparent conjunction took place between 0:30 MLT and 1:0 MLT. The difference between the apparent latitudes of Mars (**+1;37°**) and Regulus (**+0;27°**) was ~**70′**, and thus, his error is ~−30′.

It is worth noting that Muḥyī al-Dīn's style of describing the near appulse of the planets to Regulus as "the circle of latitude aligned" differs from the common standard terminology adopted in Islamic astronomy, for example, the roots *q-r-n*, "to be in conjunction"; *k-s-f*, "to occult"; or *l-ṣ-q*, "to associate/to be close."

---

74 See Clark and Stephenson 1977, p. 133; in the *Planetary Hypotheses*, Ptolemy presents Hipparchus's estimates of the apparent diameters of the planets, in which the diameter of Jupiter at its mean distance from the Earth is given as 1/12th of that of the Sun (Goldstein 1967, p. 11). Some theoretical estimates of the other sort were proposed for the practical purposes in Islamic astronomy; for instance, a table for the angular diameters of the planets can be found in al-Fahhād's *ʿAlāʾī zīj* (composed *ca.* 1172 CE), p. 182, which was also copied in Wābkanawī's *Zīj*, T: f. 170v, the *Ashrafī zīj*, f. 92v, and the anonymous *Sulṭānī zīj*, f. 172v. The table gives the diameters for the true epicyclic anomaly of the planets at the mean distance of the center of the epicycle from the Earth. The tabular values for the diameter of Jupiter run from a minimum of 2′56″ to a maximum of 4′21″.

75 al-Maghribī, *Talkhīṣ*, f. 135r.

238

9.5 Muḥyī al-Dīn's Applied Instruments and the Precision of His Observational Data

The meridian altitudes in the observations of type **I** were measured by the mural quadrant of the observatory, called the "high copper quadrant," and the durations by a clepsydra called the *mankām* (in the observations of both types **I** and **II**). In the first two sections that follow, we discuss the precision attained in Muḥyī al-Dīn's observations in relation to these two instruments. Although he does not mention any other instrument in his reports, he probably used an optical device in order to estimate the magnitudes of the lunar eclipses (type **II**) and an armillary sphere in the observations of type **III**, which are briefly considered in the other two sections.

9.5.1 Quadrant (the Observations of Type I)

As described in al-ʿUrḍī's *Fī kayfiyyat al-irṣād*,[76] the mural quadrant al-ʿUrḍī constructed at the Maratha observatory was made of copper, with dimensions of 3 fingers (1 finger = 1/24 cubit) in width and 1 finger in thickness, and was installed in the center of a wooden quadrant-shaped frame with a radius of 5 cubits (1 cubit ≈ 66.5 cm)[77] and a width of 1/4 cubit. The copper quadrant was graduated to 90°, and each degree was subdivided into 60′. Consequently, its radius would have been ~324.19 cm, and thus, every degree on the quadrant corresponded to a length of ~5.66 cm, and each arc minute ~0.94 (~1) mm.[78] The quadrant was probably the central instrument of the Marāgha observatory, built in the middle of its principal building, where the remains of a long arc in stone appears to have survived until the modern period;[79] it was partially reconstructed in the past century (the precise time is unknown to us). Figure 9.1 displays its current situation and shape. Muḥyī al-Dīn appears to have been so interested in the instrument that he composed a poem during the observations of

76 al-ʿUrḍī, *Fī kayfiyyat al-irṣād*, P: ff. 2v–4r, N: ff. 38r–39v; Seemann 1929, pp. 28–32. For a reconstruction of it, see Sezgin and Neubauer 2010, vol. 2, p. 38.

77 See Mozaffari and Zotti 2013, p. 82.

78 al-ʿUrḍī was aware of the fact that the smallest subdivision marked off on an instrument, easily readable by an unaided eye, should correspond to a length of ~1 mm. In his description of the solstice armilla ("Two Circles" in *Almagest* I.12: Toomer [1984] 1998, pp. 61–62), al-ʿUrḍī makes an interesting remark about the subdivisions, which confirms our estimation; its inner diameter is equal to 5 cubits (about half of the diameter of the mural quadrant) and 4 fingers in both width and thickness: "If the diameter of the smallest circle drawn on the two [lateral] surfaces of the ring is equal to 5 cubits, the circumference of the greatest of these circles will not be less than 16 2/3 cubits. Thus, 1/16 of it becomes greater than 3 spans [NB: 1 cubit = 3 spans, as also noted by al-ʿUrḍī in the description of the mural quadrant; al-ʿUrḍī, *Fī kayfiyyat al-irṣād*, P: f. 3r, N: f. 38v], which contains 22.5° of the circumference of a great circle; each degree becomes larger than one finger, and thus it will be possible to divide it into 60 parts or 30 distinct parts (al-ʿUrḍī, *Fī kayfiyyat al-irṣād*, P: f. 11r, N: f. 43v)." By an outer diameter of 5 cubits + 4 fingers (~210.58 cm), the circumference of the ring amounts to ~16.23 (≈ 16 1/4) cubits (~1079.40 cm), and thus, each 1° corresponds to a length of around 1.1 fingers (~3.0 cm) on it; consequently, in al-ʿUrḍī's estimate, a distance of 30 mm can be subdivided to 30 "distinct" parts.

79 For the figures of it, see Sezgin and Neubauer 2010, vol. 2, p. 31.

Figure 9.1 The remains of the base of the central quadrant of the Marāgha observatory in its present shape. The historical core is enveloped by a twentieth-century brick reconstruction.

Source: Photo captured by the author.

1265–1266 CE in its praise, and a certain astronomer, Majd al-Dīn Abū Muḥammad al-Ḥasan b. Ibrāhīm b. Yūsūf al-Baʿalbakī (modern: Baalbek in Lebanon), engraved the poem on the quadrant.[80]

The use of this quadrant is explicitly mentioned three times in the observations of Sun-**1,5** and Arcturus. Muḥyī al-Dīn expresses the solar altitudes to a precision of 1/2′, which was probably an estimated value between two successive arc minutes.[81]

In the case of the observations of the solar noon altitudes (Table 9.1), the modern values are rounded to half a minute in order to accord with the degree of precision intended in Muḥyī al-Dīn's values. For the modern values in Sun-**1,2**, the solar meridian altitudes on June 14, 1264 (the day of the summer solstice), and December 9, 1264, are taken into account. Note that December 9, 1264, was four days prior to the day of the winter solstice (December 13, 1264). It is surprising that the

---

80 Ibn al-Fuwaṭī, vol. 4, pp. 413–414; the poem reads:

أنا ربع دائرة الفلك / طوبى لمن مثلي ملك / بى تدرك الأوقات حقًّا / ويقيناً دون شك

A tentative translation of it is as follows. "I am a quadrant of the circle of the orb./Good for everyone such as me as an angel!/By me the times are known truly/and securely, without any doubt."

81 Noted earlier in Saliba 1985, p. 115.

decrease of the solar meridian altitude in the days after December 9 escaped Muḥyī al-Dīn's attention.[82] An error of ±1′ occurred in Sun-**1,2**, but in the other solar observations, we are confronted with a constant error of +4′. The mean absolute error (i.e., the average of the absolute values of the errors) in the solar meridian altitudes is $MAE = 3.1'$; the mean error $\mu = +2.9'$ with the standard deviation $\sigma = 1.9'$.

Muḥyī l-Dīn does not specify the instrument used to measure the altitudes in his planetary observations as well as in Moon-**4,5,6**. It can be safely assumed that the meridian altitudes of the planets were measured by the central quadrant. Except for Antares, the errors in his stellar and planetary meridian altitudes (Tables 9.3 and 9.4) are within ±10′. In these cases, we encounter solely random/irregular errors. For the stars: $MAE = 6.2'$, $\mu = +1.8'$ with $\sigma = 7.5'$. And for the planets: $MAE = 4.6'$, $\mu = -2.6'$ with $\sigma = 4.7'$.

The sighting of a brighter heavenly object through the tiny holes in the pinnulas of the alidade of the quadrant would have been more precise than observing

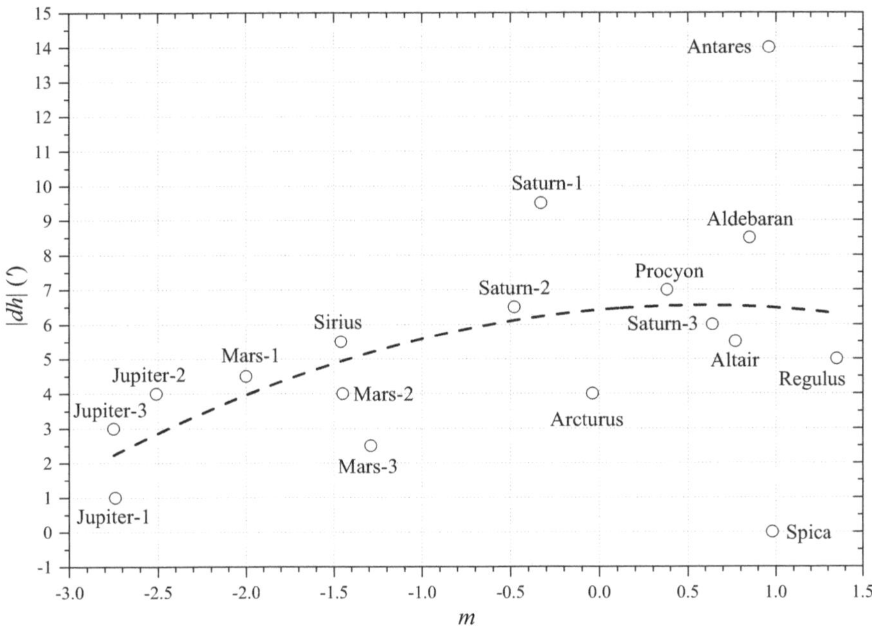

Figure 9.2 The correlation between the absolute values of the errors in altitude $|dh|$ and magnitude m in Muḥyī al-Dīn's observations. The dashed curve displays a polynomial fit of degree 2.

82 The solar noon altitudes at Marāgha on December 7–14, 1264, were as follows (rounded to 30″):

| 7: | 29;15,30° | 9: | 29;10,30° | 11: | 29;7,0° | 13: | 29;5,30° |
| 8: | 29;12,30 | 10: | 29; 8,30 | 12: | 29;6,0 | 14: | 29;6, 0 |

a fainter celestial body. In Figure 9.2, the absolute values of the errors in altitude $|dh|$ are plotted against the visual magnitudes m together with a polynomial fit of degree 2 in order to show the general trend of the relation between the two variables in Muḥyī al-Dīn's observations. As can be seen, the two variables are, in general, weakly correlated. There is a significant relation between $|dh|$ and m in his planetary observations, so that the errors in altitude in the case of Jupiter ($MAE = 2.7'$), the brightest of the three superior planets ($m < -2.5$ in the three observations), are significantly less than those in the case of Saturn ($MAE = 7.3'$; $-0.5 < m < 1$) and Mars: $MAE = 3.7'$ ($-2 < m < -1$); but both maximum and minimum values of $|dh|$ occur in the cases of the two stars of magnitude 1.

In Moon-**4,5,6**, Muḥyī al-Dīn had to measure the altitudes of the Sun and Moon at the same time. Although four large instruments constructed at the Marāgha observatory were used specifically to measure the horizontal coordinates, only one of them, known as the "Two Quadrants," could be used for the simultaneous observations of the altitudes and azimuths of the two celestial objects that appear at any angular distance from each other in the sky, as is necessary in Moon-**4,5**.[83] We may therefore suggest that this instrument was used in Moon-**4,5,6**. Based on the time intervals that our author gives for these observations (Table 9.2(B), Col. **VIII**), it is evident that the Moon had still one or 2° of azimuth to travel to reach the meridian of Marāgha (see note to Table 9.2(B)). This is why we give the two sets of the modern values for the altitudes at these observations: the first for the times mentioned by Muḥyī al-Dīn, and the second for the true times of the meridian transit of the Moon. This also gives the impression that, if the Two Quadrants had indeed been used in these observations, this systematic error could have been due to a misalignment of 1° or 2° toward the east in the meridian line marked on the azimuth ring of the instrument. The errors in the altitude of the Sun and of the Moon in the observations Moon-**4,5** are, respectively, within $\sim\pm1/4°$ and $\pm1/3°$, and the error in the altitude of the Moon in the observation Moon-**6** is $\sim-10'$. These errors are appreciably larger than those in the other observations which were made by the mural quadrant, suggesting, perhaps, that a different instrument was used in these lunar observations.

The stellar altitudes related to the lunar eclipses (Table 9.2(A), Cols. **VIII** and **IX**), as well as the solar altitudes in the vicinity of the horizon (Tables 9.3 and 9.4, Col. **III**), are almost always given in integer degrees, except in the cases of the solar altitudes in the observations of Sirius, Procyon, and Jupiter-**2**. We are not

---

83 al-ʿUrḍī, *Fī Kayfiyyat al-irṣād*, P: ff. 15r–17v, N: ff. 45v–47v; Seemann 1929, pp. 72–81. This instrument consisted of the two quadrants pivoted on an iron axis, which could move freely on a circular wall. The altitude was determined by the quadrants, and the azimuth was read from a graduated copper ring installed on the top of the wall. The other two instruments could be used to simultaneously measure the horizontal coordinates of the two celestial objects with the diametrically opposed azimuths. The last instrument may only have been used to measure the altitude and azimuth of one object at a given time; of course, al-ʿUrḍī does not mention that he constructed it at the Marāgha observatory (al-ʿUrḍī, *Fī Kayfiyyat al-irṣād*, P: ff. 19v–25r, N: ff. 48v–52v; Seemann 1929, pp. 87–88, 92–104). For a reconstruction of these instruments, see Sezgin and Neubauer 2010, vol. 2, pp. 44, 46–51.

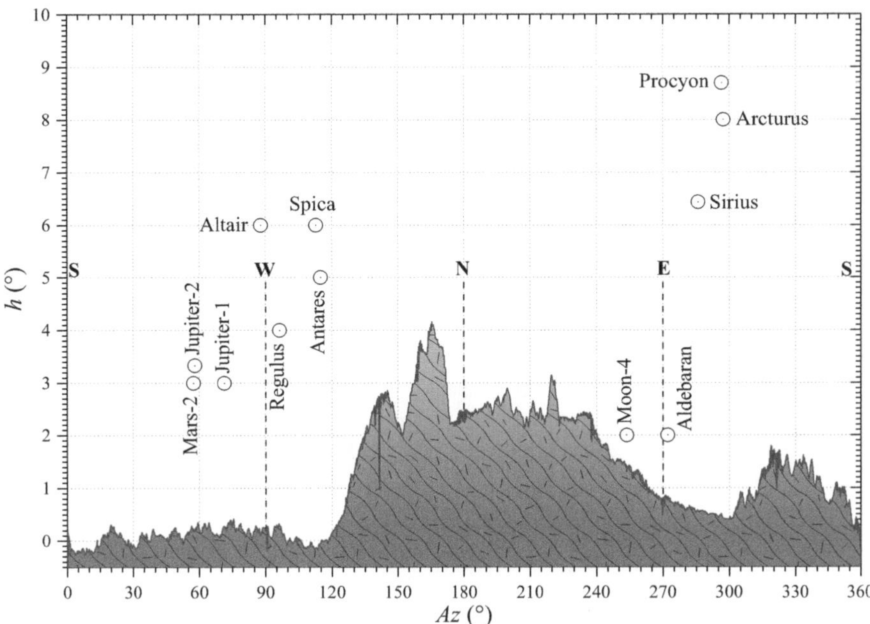

Figure 9.3 The limitation of the altitude observations around the central building of the Marāgha observatory. The open circles display the positions of the Sun at the origin of Muḥyī al-Dīn's time measurements in the lunar, stellar, and planetary observations (Tables 2(B), 3, 4(B), and 4(C)).

told which instrument was used to measure these solar and stellar altitudes (an astrolabe, the Two Quadrants, or another device?). The profile exhibited in Figure 9.3 presents an approximate estimation of the limitation in the altitude observations near the horizon around the hill of the Marāgha observatory. As can be seen, the limitation of the altitudes in the southern half of the horizon (azimuths from 270° to 90°) does not exceed 2°. The open circles display the positions of the Sun in the vicinity of the horizon at the start of Muḥyī al-Dīn's time measurements in the lunar, stellar, and planetary observations.

The atmospheric refraction in Muḥyī al-Dīn's measurement of the meridian altitudes scarcely exceeds 2′ (with Moon-**6** the only exception). In Figure 9.4, the values for the atmospheric refraction R at the apparent meridian altitudes in Muḥyī al-Dīn's observations are shown along the general graph of R plotted against the apparent altitude h.[84] We shall return to this in our discussion of the correlation between the errors in altitude and declination (see Section 9.6).

84 Drawn on the basis of the formula given in Meeus 1998, p. 106.

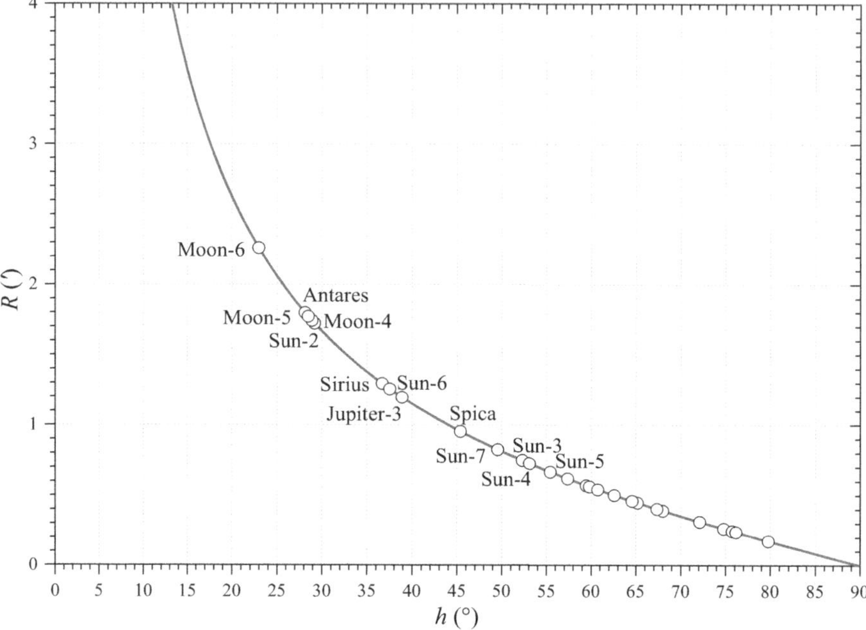

Figure 9.4 The atmospheric refraction R with respect to the apparent altitude h in Muḥyī al-Dīn's observations.

9.5.2 Clepsydra (Observations of Types I and II)

Muḥyī al-Dīn refers to the application of the clepsydra in his systematic observations, to allow the measurement of all the times of the culmination of the planets and stars. He called the clepsydra *mankām* (or *minkām*), a corruption of the term *bankām* (or *binkām*), which comes from (in fact, is the Arabicized form of) the Persian term *pangān*. This Persian term appears to have been converted into Arabic in two different forms in the early Islamic period: *fankān* (or *finkān*), which meant "clock," in general, and *bankām*, which meant "water clock," in particular, as can be understood from the work on mechanical devices by al-Jazarī, the twelfth-century Islamic engineer and craftsman.[85]

Pangān was originally a simple Persian clepsydra in the shape of a floating bowl (*ṭās*), with a hole in its apex and two graduated scales (usually drawn with the aid of an astrolabe) for both the equal/equinoctial and unequal/seasonal hours on its peripheral surface. The bowl was placed into a vessel of water. From the level of water drained into it, one could read off the amount of time that had elapsed since a particular moment. An early description of the instrument in the Islamic period appears in al-Ṣufī's (903–986) *Book on the Astrolabe*.[86] This model

---

85 al-Jazarī 1973, p. 17.
86 See al-Ṣufī 1995, Chapters 354–357: pp. 299–302.

of clepsydra goes back to Babylonian and Indian texts of the first millennium BC,[87] and archaeological excavations have unearthed its earliest models in India, apparently from the same period.[88] In his *India*, Bīrūnī (973–1048 CE) mentions witnessing the popular use of these clepsydras in Purshūr (today, Peshawar).[89]

Clepsydras were often used for both everyday timekeeping and astronomical time-measuring in antiquity and the Middle Ages, so that the references to them in both primary and secondary sources are more abundant than can be summarized here. Owing to certain specific historical considerations which will be discussed in the sequel, it is worth noting that, in Chinese astronomy, clepsydras were used for various functions, such as the timing of eclipses.[90] The accuracy attained is ~15 minutes; of course, there is evidence that the Chinese astronomers could measure the times of sunrise/sunset by clepsydras with an accuracy of ~5 minutes.[91] The use of clepsydras with compound mechanical components had been well established in Chinese astronomy since, at least, the eleventh century.[92]

The time-measuring equipment and, especially, various types of clepsydras were so commonly used in medieval Islam that timekeeping without the need to know the positions of the celestial objects was called " *'ilm* (science of) *al-bankāmāt* (plural form of *bankām*)." It was part of a longtime tradition of mechanical devices in the Eastern Mediterranean and the Middle East which came from Greek and Indian sources and was developed in the works of Islamic engineers like the Banū Mūsā (ninth century), Hibat-Allāh b. al-Ḥusayn (d. 1139–1140 CE), Yūnus al-Asṭurlābī, and al-Jazarī.[93] In *On experimental astronomy* I.4–8, al-Khāzinī mentions the time-measuring devices as the fifth type of the observational instrument:

> The instruments with the aid of which the motions of the celestial sphere in time are measured: Many of them work with water; a large amount of water, the weight of which is known, is poured into them in one time.

87 Pingree 1973, pp. 3–4. It is highly probable that Babylonians used clepsydras for astronomical purposes, for example, to determine the times of eclipses (Stephenson 1997, p. 59). On Babylonian clepsydras, see, especially, Neugebauer 1947, Michel-Nozières 2000. Indian sources, for example, *Sūrya Siddhánta* XIII.23 ([1860] 1997, p. 264, [1861] 1974, p. 91), refer to this type of clepsydra. For more details, see Sarma 1994a, 1994b, pp. 512–514, 2001, p. 54; and especially Sarma 2004, Sharma 2000, pp. 241–243, Ōhashi 2008, Pandey 2011; and other articles cited therein.

88 See Rao 2005, pp. 205–206, Ōhashi 2008, p. 270.

89 See al-Bīrūnī 1910, vol. 1, pp. 337–338.

90 See Cullen 1996, p. 42, Stephenson 1997, Chapter 9.

91 Stephenson 1997, p. 278.

92 Needham 1981, p. 136; more on Chinese clepsydras can be found in the classic study Needham *et al.* 1986, Chapters 6 and 7. The best example of the compound mechanism of Chinese clepsydras is Su Song's water-powered mechanical clock implemented with an escapement regulator from about 1088 CE (cf. Yan 2007, pp. 163–198, 2009).

93 Hill 2008, p. 130. For example, Pseudo-Archimedes's *On the construction of water-clocks* is preserved in an Arabic translation (Archimedes 1976, Sezgin and Neubauer 2010, vol. 3, pp. 94–95), and the names of al-Jazarī's three clocks ("peacock", "man," and "monkey") are also mentioned in *Sūrya Siddhánta* XIII.21 ([1860] 1997, pp. 263–264, [1861] 1974, p. 90). For illustrations of some Islamic clepsydras, see Sezgin and Neubauer 2010, vol. 3, Chapter 4.

> Some of them work with sand, into which a large amount of sand, the weight of which is known, is poured.[94]

Nevertheless, there is explicit evidence that not all the astronomers of the classical medieval Islamic period (prior to 1050 CE) relied on water or sand clocks. For instance, in his *Taḥdīd*, Bīrūnī mentions (in the context of using lunar eclipses in order to determine the differences in longitude between cities) that "some measure time with care and attention by continuous motions which are empirically equal in equal times and, as a rule, this has been done by the use of water"; but he rejects this practice, because "[water] is subject to variation in many respects. For instance, [1] the purity and density depend on its sources. . . . Also, [2] it is subject to accidental variations, by variation in the quality of the air." "For these and similar reasons," Bīrūnī follows, "man has preferred the motions of sand to it."[95] He then asserts that the measurement of time by the altitudes of the stars has a definite advantage over the use of water or sand clocks: the time can be computed from the altitudes of certain stars so that "the corroboration of evidence obtained from some of them with that obtained from others leads to better accuracy."[96] Also, the Islamic astronomers' reluctance to use water clocks may have been rooted in the Ptolemaic context in which they worked, since the use of the clepsydra is rejected as a means of measuring the apparent diameter of the Sun in both *Almagest* V.14 and *Tetrabiblos* III.2, for the reason that they are "capable of error . . . by stoppages and irregularities in the flow of the water from different causes and by mere chance."[97] Moreover, for example, from al-Jazarī's description of his six clepsydras, it is evidently understood that they were *representational* timekeeping instruments for announcing the seasonal or equinoctial hours and sometimes for showing the ecliptic longitude of the Sun and the Moon, but a practitioner of astronomy needs reliable time-measuring devices precise enough to be used as the *empirical* instruments in astronomical queries.[98] Only one of al-Jazarī's

---

94 al-Khāzinī, *Kayfiyyat al-i'tibār*, in *Zīj*, V: ff. 6r–7r. The other four instruments are the alidade, the triquetrum, the dioptra, and the triangle. See also the following note.

95 The use of sand instead of water is seen in the automatic celestial globe invented by al-Khāzinī, about one century later, in which a sand reservoir provided the power required for the rotation of the globe (see Lorch 1980). Sand clocks are also mentioned in *Sūrya Siddhánta* XIII.21 and 22 ([1860] 1997, pp. 263–264, [1861] 1974, pp. 90–91).

96 al-Bīrūnī 1967, pp. 155–156 (with a change in the translation: in the first sentence, "care and attention" has replaced "precision," which may be somewhat misleading).

97 Toomer [1984] 1998, p. 252; Ptolemy, *Tetrabiblos* [1940] 1956, p. 231. The method of measuring the Sun's apparent diameter using the clepsydra was to count how many times the vessel of clepsydra is filled during the time from one sunrise to the next. The Egyptian value 750, for example, corresponds to the Sun's angular diameter being equal to $1/_{750}$-th part of its entire orbit, that is, 28′ 48″ (see Neugebauer 1975, vol. 2, pp. 865–868).

98 This classification of the Islamic astronomical instruments is adopted from Charette 2006, p. 123. The other two examples of mechanical astronomical devices for representational purposes are al-Khāzinī's automatic celestial globe and his balance clock; on these instruments, see Lorch 1980, 1981.

clepsydras, called the "beaker water clock," could measure the time ideally with a precision of 4 minutes.[99] So these appear to be the reasons that time-measuring devices were not used in Islamic astronomy – especially for timing eclipses, which were usually measured directly from the altitude of the Sun (in the case of solar eclipses) or of the reference stars (in the case of lunar eclipses).[100] However, from the time of al-Maghribī onward, a combination of the use of the two methods (altitude–clepsydra) appears to have been established as a standard for measuring the times of eclipses in the second period of the Marāgha observatory; for example, in his *Muḥaqqaq zīj* IV.15.8–9, Wābkanawī emphasizes that this is done to ensure the correctness of the times measured for the phases of the eclipses, since "the times derived from both [methods] should be equal."[101]

Based on the information given by al-Maghribī, as we will see later, one might speculate about the kinds, models, and calibrations of the clepsydras used at the Marāgha observatory, but nothing more can be concluded regarding their structure and application. Unfortunately, he did not give a physical description of the instrument. Of course, it appears to have been so accurate that it could measure the time intervals with errors never more than 5 minutes (Tables 9.3 and 9.4, Col. **VII**); so it does not appear to have been a simple inflow/sinking or outflow clepsydra. Of course, at that time, there were important cultural exchanges between Iran and China, the two realms of the Mongol Empire; Khubilai Khān, the first emperor of the Yuan dynasty of China (d. 1294 CE), founded two astronomical institutes in his capital, Beijing, contemporary to the Marāgha observatory: the Islamic Astronomical Bureau (1271 CE) and an astronomical bureau for revising the calendar (1278 CE). The first bureau was led by a person referred to as Zhamaluding in the Chinese annals from the year 1267 CE onward, who seems to have been an Iranian astronomer named Jamāl al-Dīn Muḥammad b. Ṭāhir b. Muḥammad al-Zaydī of Bukhārā (in modern Uzbekistan). He appears to have authored a *zīj* in Persian, preserved in a Chinese translation, known as *Huihuili*, prepared in Nanjing in 1383 CE; the parameter values underlying this work have been embedded in an Arabic *zīj* written by a certain Sanjufīnī in Tibet in 1366 CE. The astronomical bureau for revising the calendar was directed by Zhang Wenqian (1217–1283 CE), and Gue Shoujing (1231–1316 CE) and Wang Xun (*ca.* 1235–1281 CE) were its co-directors. In 1276 CE, when Khubilai Khan captured the Southern Song capital, Linan, Gue and Wang had a new calendar created. The new official astronomical system of the Mongol dynasty, *Shoushili*,

---

99 al-Jazarī 1973, pp. 17–82.

100 In mid-latitudes and for mid-altitudes, these measurements might be precise to some 5–6 minutes (Stephenson 1997, p. 466).

101 Wābkanawī, *Zīj*, T: ff. 92r–v, Y: ff. 159r–160r, P: ff. 139r–140r. The passage in question can also be found in the *Ghāzānid treatise on observational instruments*, written *ca.* 1294–1305 CE (about it, see Sections 9.5.3 and 9.5.6). In it, the time-measuring device is called the "time glass" (*shīsha-i sāʿat*) (see Chapter 11, p. 329; Mozaffari and Zotti 2013, pp. 128–130). Interestingly, in the seventeenth century, European astronomers still preferred to time eclipses by measuring altitudes rather than by relying on mechanical clocks (Stephenson and Said 1991, p. 207, note 26).

was completed in 1280 CE.[102] Khubilai Khan's astronomical bureaus in Beijing were equipped with some clepsydras (both *representational*, such as "The Lamp Water Clock for the Palace of Supernal Brightness," and *empirical*) and with the water-powered armillary sphere and celestial globe.[103] Also, Chinese astronomers (at least Fu Mengchi or Fu Muzhai) were in the service of the Īlkhānid dynasty of Iran and worked at the Marāgha observatory.[104] Marāgha was also a place where the native Islamic Persian architecture met the Chinese Mongol one: among a few broken tiles discovered in the ruins of the observatory, one bears an ornament in the shape of a dragon (see Figures 9.5(a) and 9.5(b)), obviously indicating that the craftsmen and artists of the observatory copied and applied the Chinese and Mongolian forms and motifs in their works. (An ornament of this kind, which is not found in the Marāgha tiles before the Mongol period,[105] would have served as a medium of cultural and social experience in a general sense and as a mode of visual address through which Mongols apparently proclaimed and projected their definable social, political, ideological, and cultural values and aspects.)[106] These findings give the impression that, perhaps, there was also a connection between the clepsydras of the Marāgha observatory and the elaborate Chinese technology for making time-measuring devices.[107]

Returning to Muḥyī al-Dīn's *Talkhīṣ*, in what follows we gather together all the information we can glean from the statements in his observation reports.

1. A time interval measured by the clepsydra is called *sāʿāt al-mankām*, "hours of the clepsydra," or, if it is expressed in terms of the revolution of the celestial sphere (time-degree), *dāʾir al-mankām*. This is in order to distinguish it from time computed from the altitude, which is called *sāʿāt al-irtifāʿ*, "hours of the altitude", or *dāʾir al-irtifāʿ*.
2. The use of the clepsydra is mentioned in the stellar and planetary observations. In all likelihood, it was also used to measure the times of the conjunctions of the superior planets with Regulus (especially in the case of Mars-**4**). In the triple lunar eclipses (Moon-**1–3**), its use was not indicated. The times might have been measured from the star altitudes (Table 9.2(A), Cols. **VIII** and **IX**); however, it is not unreasonable to assume that the clepsydra was also used in these observations. In Moon-**4**, we are told that the time interval was measured from the solar altitude. In Moon-**6**, no measurement of time

102 On the relationship of Islamic and Chinese astronomy in the Mongol period, see Johnson 1940, Hartner 1950, Yabuuti 1987, 1997, van Dalen 2002a, 2002b, Peng Yoke 2008, Sivin 2009.
103 See Sivin 2009, Chapter 5, esp. pp. 174–176 and Appendixes A and B.
104 It seems that he introduced the Chinese–Uighur calendar incorporated in *Īlkhānī zīj*, which is also found in Persian *zījes* from the Mongol period onward (see van Dalen *et al.* 1997).
105 For example, see Makovicky 1992. For more on the dragon imagery as a symbol of cross-cultural exchange during the Īlkhānid period, see Kuehn 2011, pp. 209f.
106 The basic idea has been taken from Morrall 2009.
107 Curiously, Lorch (1980, esp. pp. 291–294, 1981) embarks on a lengthy discussion of the Chinese counterparts of al-Khāzinī's astronomical representational devices (see earlier, note 98).

(a)

(b)

Figure 9.5 The broken tiles discovered in the ruins of the Marāgha observatory (courtesy of the Īlkhānid Museum in Marāgha: (a) On one of the samples, a three-dimensional letter *Jīm* (ج) can apparently be seen; (b) the tile with an ornament in the shape of dragon.

Source: Photo captured by the author.

was required. In Moon-**5**, despite our author's habit of expressing the times measured either by the clepsydra or from the altitudes in degrees, the time interval is first given in the equinoctial hours (4 $\frac{1}{5}$ hours), and then the corresponding arc of the revolution of the celestial sphere is mentioned as 63°; our author adopts a different term here and calls it the *dā'ir al-sā'āt* (neither *dā'ir al-mankām* nor *dā'ir al-irtifā'*). It is not clear whether this duration was measured by the clepsydra or from the altitude.

3. It appears that the two mechanisms were embedded in the clepsydras used in Marāgha, in order to measure hours and minutes independently of each other. The first is given the general name of the clepsydra, that is, *al-mankām*, which was used for measuring *hours*, while the other is called *the minute, al-daqīqa*, which was used to count *minutes*. There are two features that suggest this. First, in Mars-**3**, the time interval is less than 1 hour, and al-Maghribī's statements make it clear that only *the minute* was used to measure it; the resulting time-degree is called *dā'ir al-daqīqa* (cf. *dā'ir al-mankām*). Second, when the first model of the clepsydra was broken (see following, no. 5), *the minute* of it could no longer be used either; this indicates that they were not two separate instruments but connected to each other in some way.

4. It is clear that the *hours* and *minutes* measured, respectively, by *the mankām* and *the minute* do not correspond exactly to the hours and minutes of time; rather, there was a deviation between them. We are told in the observational records of Arcturus and Saturn-**1** that

> 1 *hour* (or *turn*) of *the mankām* = 15;24,44,26° ≈ 1;2 hours
> 1 *minute* of *the minute* = 0;15,2,11° ≈ 0;1 hours (1)

Examples: The duration from the meridian altitude of Arcturus to the moment when the Sun was located at an altitude of 8° after sunrise (see earlier, 4.1.3) is equal to three *turns* of the clepsydra minus 2/5 *minute*; consequently, 3 × 15;24,44,26° − $^2/_5$ × 0;15,2,11° = 46;8,12°, as Muḥyī al-Dīn mentions, or ~3;5 hours (Table 9.3); and the time interval from the meridian transit of Mars to that of Aldebaran in Mars-**1** (see earlier, 4.1.4) is 33 *minutes*; then, 33 × 0;15,2,11 = 8;16° or ~0;33 hours (Table 9.4(C)).

5. At least two models of this kind of clepsydra were available at the Marāgha observatory. The first six stellar observations were performed with the aid of the first clepsydra (1). When explaining the observation of Spica, Muḥyī al-Dīn notes that before this observation (May 1268), the first clepsydra had been broken, and he therefore used the other model in the observations of Spica, Altair, and Aldebaran (note that there was a 7-month gap between the observations of Procyon and Spica). Our author mentions that in the second model,

> 1 *hour* (or *turn*) of *the mankām* = 15;45,12° ≈ 1;3 hours
> 1 *minute* of *the minute* = 0;15,37,23° ≈ 0;1 hours. (2)

6. We list Muḥyī al-Dīn's values for the times of the clepsydra and the corresponding time-degrees in the stellar and planetary observations in Table 9.5.

Note that our reading of his values, expressed in the *Abjad* numerals, for the time-degrees (Col. **5**) is confirmed, because he uses the same values in the computation of the ecliptical coordinates of the stars and planets.

A. For the stars, there is a deviation in the value of the minutes in the case of Antares (9 → *4*), and another in the value of the seconds in the case of Altair (4 → *20*). Also, as noted in the observation report of Spica, Muḥyī al-Dīn correctly gives the time-degree as 27;12,*37*° in the text but uses 27;12,*17*° in his calculations. These deviations do not seem to be due to the probable confusions between the *Abjad* numerals with similar forms: 9/ط → 4/د; 4/د → 20/ک; 37/لز → 17/یز).

B. In the case of the planets:

a. One would have expected the second model of the clepsydra to be used in the third observations of the superior planets, which were performed after May 1268. Nevertheless, only the conversion formulas of the first model of the clepsydra, (1), are mentioned at the beginning of the planetary observations (Saturn-**1**). Moreover, the time-degrees given in Saturn-**3** and Mars-**3** are in close agreement with the conversion formulas (1). It is not known whether Muḥyī al-Dīn failed to allow for the change in clepsydras when presenting his data, or whether the clepsydras used at Marāgha had been repaired, and so the first model was available again after February 1271 (if not earlier), when the observation of Mars-**3** was made.

b. In the cases of Mars-**2** and Jupiter-**1,3**, Muḥyī al-Dīn's values for the time-degrees do not agree with either those computed from (1) or those computed from (2). In the cases of Mars-**2** and Jupiter-**1**, as expected, they are closer to the first clepsydra than to the second. For Jupiter-**3**, our author says that "half of the clepsydra" was deployed for 5 *turns* + 15;30 *minutes*, which corresponds to either 80;56,46° (first model) or 82;48,9° (second model); both are nearly twice the value of 42;13,38° he gives in the text. This passage is very difficult to interpret: Was the upper part of the water reservoir of the clepsydra badly damaged due to an accident after February 1273, when the observation of Saturn-**3** was made (or after January 1274, when the third lunar eclipse was observed, assuming that the clepsydra was also used in the observations of the lunar eclipses), so that each *turn* of water pouring into or out of it now lasted roughly half the time it would normally take to perform its function? We do not know.

To sum up, assessing the accuracy of Muḥyī al-Dīn's time-measuring using the clepsydras available at Marāgha, it can be said that they were able to fix the time intervals with the precision of ±5 minutes. In the case of the fixed stars, $MAE = 1.75^m$, $\mu = 0.5^m$ with $\sigma = 2.1^m$, and in the case of the planets, $MAE = 2.6^m$, $\mu = 1.2^m$ with $\sigma = 3.0^m$. The two models obtained very nearly the same degree of accuracy.

Table 9.5 The durations measured by the clepsydra

1	2	3	4		5	
	Hours (turns) of al-mankām	*Minutes of al-daqīqa*	*Model of the used clepsydra*		*Time-degrees*	
					Recomputed	*Al-Maghribī*
Arcturus	3	− 2/5	(1)		46; 8,12°	46; 8,12°
Regulus	2	+14;48	(1)		34;32, 1	34;32, 1
Antares	2	− 6;40	(1)		29; 9,14	29; **4**,14
Sirius	2	0	(1)		30;49,29	30;49,29
Procyon	3	0	(1)		46;14,13	46;14,13
Spica	2	−16;30	(2)		27;12,37	27;12,37 (1)
						27;12,**17** (2)
Altair	2	− 9;45	(2)		28;58, 4.5̲	28;58,**20**
Aldebaran	2	+15	(2)		35;24,45	35;24,45
Saturn-**1**	11	+50	(1)		182; 3,58	182; 3,58
Saturn-**2**	11	+30;40	(1)		177;13,16	177;13,16
Saturn-**3**	1	+38;30	(1)	(1):	25; 3,38	25; 4,12 (1)
				(2):	25;46,44	25; 3,38 (2)
Jupiter-**1**	6	+52;30	*Unknown*	(1):	105;37,51	106; 0,25
				(2):	108;11,25	
Jupiter-**2**	7	+19	(1)		112;38,52.5	112;38,52
Jupiter-**3**	5	+15;30	*Unknown*	(1):	80;56,46	42;13,38
				(2):	82;48, 9	
Mars-**1**	0	+33	(1)		8;16	8;16
Mars-**2**	7	+18	*Unknown*	(1):	112;23,50	112;36,22
				(2):	114;57,37	
Mars-**3**	1	+32;30	(1)		23;33,25	23;33,25

Note:

Spica: (1) The value given in the text on f. 113r; (2) the value used in the calculations.
Saturn-**3**: (1) The value given in the margin on f. 123v; (2) the value computed from the relevant quantities explicitly given in the text (see the observational record).

9.5.3 Dioptra/Pinhole Image Device (Observations of Type II)

In *Almagest* V.14, Ptolemy describes a dioptra used by Hipparchus which was 4 cubits in length ($\approx$ 185.28 cm).[108] This dioptra has a fixed lower pinnula on which there is a hole for sighting, and a movable outer one, which is placed in front of the Sun. The solar and lunar angular diameters are calculated from the width of the movable pinnula and the distance between the two pinnulas. In his *Fī kayfiyyat 'l-irṣād*, al-ʿUrḍī presents an addition for the antique dioptra in order to determine the eclipsed area/diameter of the Sun or Moon.[109] Indeed, Muḥyī al-Dīn had at his disposal a specific instrument for measuring the magnitude of eclipses, which he may have applied to his observations of the lunar eclipses. Interestingly, a pinhole image device able to measure the magnitude of the solar eclipses is introduced in Wābkanawī's *Muḥaqqaq zīj* IV.15.8,[110] as well as in the *Risāla al-Ghāzāniyya fī 'l-ālāt al-raṣadiyya* (*The Ghāzānid treatise on observational instruments*).[111] The latter work presents the physical descriptions and applications of 12 unprecedented observational instruments constructed in the second period of the Marāgha observatory, invented by Ghāzān Khān, the seventh ruler of the Īlkhānid dynasty of Iran (reg. October 21, 1295–May 17, 1304).

9.5.4 Armillary Sphere (Observations of Type III)

In Muḥyī al-Dīn's four observations of the near appulses of the superior planets to Regulus, there are two systematic errors:

1. A systematic positive error in the longitudinal distance between each superior planet and Regulus, which amounts to a mean value of +3' in Saturn-**4**, Jupiter-**4**, and Mars-**4**, but which rises to a relatively large value of ~+26' in Saturn-**5**. Note that Muḥyī al-Dīn had more time to perform the first three observations than the latter.

2. A systematic negative error in his values for the separation in latitude between Regulus and every superior planet in Saturn-**4** (~−12'), Jupiter-**4** (note that Muḥyī al-Dīn takes a relatively sizable angular distance of ~**25'** as equal to 3/2 of Jupiter's body, while Ibn Yūnus takes a distance of ~**10'** as equal to Jupiter's body), and Mars-**4** (~−30'). This quantity has not been given in Saturn-**5**.

---

108 Toomer [1984] 1998, p. 56.

109 al-ʿUrḍī, *Fī Kayfiyyat al-irṣād*, P: ff. 12r–14v, N: ff. 44r–45v; Seemann 1929, pp. 61–71, Sezgin and Neubauer 2010, vol. 2, p. 43.

110 Wābkanawī, *Zīj*, T: ff. 92r–v, Y: ff. 159r–v, P: ff. 139r–v. For the measurement of the apparent angular diameter of the Moon and the magnitude of lunar eclipses, Wābkanawī recommends the use of an instrument called *dhāt al-misṭaratayn*, "having two rulers," which is also used to measure the angular distance of the two heavenly bodies near their conjunction/occultation (Wābkanawī, *Zīj* IV.15.7,9: T: ff. 91r–92v, Y: ff. 158r–160r, P: ff. 138r–140r).

111 See Section 11.4.12; Mozaffari and Zotti 2013, pp. 127–135.

Nothing is said about the instrument(s) applied in these observations. However, assuming that his derived values for the ecliptical coordinates of Regulus are correct, it is quite probable that Muḥyī al-Dīn used an armillary sphere to continuously measure the distances in longitude and latitude of a superior planet from Regulus as a reference point to detect when they are in conjunction with each other, as well as to estimate the approximate separation in latitude between them at that time. Support for this idea comes from the fact that the only passage in the *Talkhīṣ* in which the use of the armillary sphere is mentioned is a quotation from the *Almagest* related to the conjunctions of the planets with the fixed stars (see Section 9.6). In addition, "the circle of latitude" in the four observational accounts in question may indicate a reference to the latitude ring of an armillary sphere. Al-'Urḍī had constructed a model of the armillary sphere (with five rings) at the Marāgha observatory. The inner diameter of its ecliptic ring was equal to 3 cubits, and its width and thickness both measured 4 fingers. Thus, the outer diameter of the latitude ring of the instrument was equal to 3 cubits (note that the convex surface of the latitude ring closely fits the concave surface of the ecliptic one). By an analogy drawn from al-'Urḍī's statement on the subdivisions,[112] each degree marked on the ecliptic and latitude rings of the armillary sphere corresponded to a length of slightly less than 3/4 of a finger (respectively, ~18.4 mm and 17.4 mm), and thus, they could be divided into, at the most, 20 distinct parts, each standing for 3'. Of course, al-'Urḍī only speaks of the degree divisions for both rings, while the possibility of small subdivisions of a degree is mentioned in the case of the meridian ring.[113] Thus, if Muḥyī al-Dīn indeed used this instrument, his values for the angular distance in latitude between a superior planet and Regulus may be rough estimations made with the aid of the two adjacent degree marks on the latitude ring, separated from each other by a distance < 2 cm; his errors in the determination of the alignment of the two celestial bodies correspond to a distance of 1 mm (in Saturn-**4**, Jupiter-**4**, and Mars-**4**) and of 8 mm (in Saturn-**5**) on the ecliptic ring, and the fact that he made two systematic errors may indicate the existence of structural defects in the construction of the components of the instrument and/or joining them together.

9.6 The Accuracy of Muḥyī al-Dīn's Observational-Computational Method for the Derivation of the Celestial and Ecliptical Coordinates

According to al-Maghribī's method, the data obtained from the observations type **I**, that is, (1) the meridian altitude and (2) the duration, together with (3) the already-known right or oblique ascension of a reference body, are converted to the celestial and ecliptical coordinates, which then served as the input data for other computations. The sources of the errors in the declination and right ascension and

---

112 See earlier, note 78.
113 al-'Urḍī, *Fī Kayfiyyat al-irṣād*, P: ff. 4r–10v, N: 39v–43r; Seemann 1929, pp. 33–53, Sezgin and Neubauer 2010, vol. 2, pp. 39–40.

the latitude and longitude of a celestial body are these three quantities, together with the errors in (4) the basic parameters, mainly the obliquity of the ecliptic to the celestial equator (ε = 23;30,0°; modern: ~23;32,10°; note that the mean obliquity ε_0 decreased slightly from **23;32,4.9°** to **23;32,4.7°**, and nutation in the obliquity from **+5.4″** to **+4.3″** in the time interval from Sun-1 to Sun-2) and the geographical latitude of the observatory (φ = 37;20,30° N; modern: ~37;23,46° N, longitude L = **46;12° E**), and (5) the computational inadequacy (the errors in the process of calculations, auxiliary trigonometric or astronomical tables, etc.). The effects of these factors on the celestial and ecliptical coordinates are briefly considered in what follows.

1. The errors in declination $d\delta$ and ecliptical latitude $d\beta$ are, to a great extent, dependent upon those in meridian altitude dh and geographical latitude, $d\varphi$ ≈−3′, together with the atmospheric refraction R (Figure 9.4). The geocentric declination δ of a celestial body is derived from its apparent (topocentric) meridian altitude h by $\delta = h + \varphi - R - 90°$; thus, $d\delta = dh + d\varphi + R$. Consequently, if $dh < 0$, the negative error $d\varphi$ ≈−3′ enlarges the corresponding negative error $d\delta$, but R reduces it (e.g., in the case of Sirius: dh ≈−5.5′ and R ≈ 1′, and so $d\delta$ = −7.5′). In contrast, if $dh > 0$, $d\varphi$ decreases $d\delta$, but R increases it (e.g., in the case of Antares: dh ≈ +14′ and R ≈ 2′, and thus, $d\delta$ ≈ +13′, or Jupiter-3: dh ≈ +3′ and R ≈ 1′, and so $d\delta$ ≈ +1′). The correlation between the statistical results of dh, $d\delta$, and $d\beta$ is shown in the following tabulation.

Stars	MAE	μ	σ	Planets	MAE	μ	σ
dh	6.2′	+1.8′	7.5′	dh	4.6′	−2.6′	4.7′
$d\delta$	6.2	−0.7	7.6	$d\delta$	6.3	−5.3	5.0
$d\beta$	6.1	+3.4	7.3	$d\beta$	10.3	−9.0	9.7

2. The degree of precision attained in the right ascension RA and ecliptical longitude λ is greatly dependent upon the precision in the right ascension/longitude of the reference body and timekeeping, which is strongly linked to the accuracy of the water clock. Note that an error of 1 minute in counting time corresponds to an error of 15′ in the right ascension and to an error of approximately the same amount in longitude. The solar longitudes in Sun-**3–8** are directly derived from its noon altitudes. Once the solar theory is established, it serves as the principal reference object for the measurement of the celestial and ecliptical coordinates of the other celestial bodies. As discussed in detail elsewhere, Muḥyī al-Dīn constructed an elegant solar theory from his experimental activities at Marāgha, with the errors between a lower limit >−5′ (occurring around mid-summer) and an upper limit <+8′ (occurring around mid-winter).[114] Accordingly, the errors in RA and λ are probably due to errors

114 See Mozaffari 2018, esp. pp. 229, 235.

in the durations measured rather than to an imprecise solar theory. Of course, in the four planetary observations, the four different stars are used as the reference body, and as we shall see later, the errors in their right ascensions noticeably diminish the errors in the right ascensions and longitudes of the planets. The statistical results of the errors in duration dt, dRA, and $d\lambda$ are presented in what follows.

Stars	MAE	μ	σ	Planets	MAE	μ	σ
dt	1.75^m	0.5^m	2.1^m	dt	2.6^m	1.2^m	3.0^m
dRA	26.4'	21.4'	21.8'	dRA	25.2'	1.0'	33.0'
$d\lambda$	27.9	23.4	23.1	$d\lambda$	21.6	1.8	28.9

As can be seen, in the case of the stars, a mean absolute error of 1.75 minutes = 26.3' in duration is strongly correlated to the preceding values for the mean absolute errors in RA and λ. In the case of the planets, we would expect at the outset that a mean absolute error of 2.6 minutes would result in a large mean absolute error of ~40' in RA and λ, but in fact, we find errors that are significantly smaller. The reason is that Muḥyī al-Dīn used stars as the reference bodies, instead of the Sun, in the four planetary observations, and so the errors in their right ascensions compensate those in duration, as explained later. In Jupiter-**3** and Mars-**3**, the stars passed across the meridian of Marāgha before the planets, and hence $RA = RA^* + t$, in which RA indicates the right ascension of the planet, RA^* that of the star, and t stands for the duration expressed in time-degree; in both cases, dRA^* and dt are of the opposite signs (see Tables 9.3, 9.4(B) and 9.4(C)); as a result, $dRA < dt$. In contrast, in Saturn-**3** and Mars-**1**, the planets transited the meridian before the stars, and so $RA = RA^* - t$; in both cases, dRA^* and dt are of the same sign, and thus, again, $dRA < dt$.

In the four observations of type **III**, the longitudes of the planets were simply taken as equal to that of Regulus (i.e., its longitude as observed in 1267 CE corrected for the precession).

Muḥyī al-Dīn never explains why he preferred to use a lengthy computational process of spherical astronomy rather than an armillary sphere, with which he could read off the ecliptical coordinates directly. The use of the armillary sphere is mentioned only in one passage in *Talkhīṣ* VII.4, which is, in fact, a quotation from *Almagest* IX.2.[115] His preference for observing the Sun, Moon, planets, and stars in their culminations might also have something to do with avoiding errors due to refraction, which Ptolemy had already referred to in *Optics* V.28.[116]

---

115 al-Maghribī, *Talkhīṣ*, f. 117r, corresponding to Toomer [1984] 1998, p. 423: lines 10–13.

116 Smith 1996, p. 241: "when the star rises to position H [= a point on the meridian], it reaches a point where the visual ray is refracted without any perceptible difference between apparent and true location."

It is worth considering that in *Talkhīṣ* VI.6, Muḥyī al-Dīn explicitly refers to the *Book on the optics* (*Kitāb al-manāẓir*) when using the inverse relation between the angular diameter of a heavenly object and its distance from the Earth in order to derive the apparent diameters of the Sun and Moon.[117] Nevertheless, it is difficult to determine whether this was the main reason for Muḥyī al-Dīn's appeal to spherical astronomy to derive the ecliptical coordinates from the observational data.[118] The other (and, indeed, more likely) reason is that he saw some difficulty in working with the armillary sphere. It seems that in medieval Middle Eastern astronomy, there was a tendency to replace it with altitude-azimuthal instruments; Abu al-ʿAbbās al-Lawkarī (d. 1071–1072 CE), for instance, devised an altitude-azimuthal instrument and worked out a method of spherical astronomy, which he described as a manual for its use, especially for the sake of "making unnecessary [the use of] the armillary sphere" (*mughniyya ʿan dhāt al-ḥalaq*), "because," we are told, "it is hard to take the armillary sphere and it is too expensive, whether we take a Complete or a Sufficient model (*nawʿ*) of it" (by the *Complete model*, he means an armillary sphere with nine rings, and by the *Sufficient model*, an armillary sphere with six armillas).[119] Also, when explaining the progress of astronomical

117 al-Maghribī, *Talkhīṣ*, ff. 94r–v. He does not identify the book on optics he has in mind. The inverse relation between the distance and apparent diameter is discussed, for example, in Ibn al-Haytham's *Optics* II.3 (vol. 1, pp. 273–295; Smith 2001, vol. 1, pp. 164–191, vol. 2, pp. 475–494; Sabra 1989, vol. 1, pp. 173–190).

118 A problem arises from the fact that in *Talkhīṣ* VII.4 (f. 117r), Muḥyī al-Dīn quotes a passage from *Almagest* IX referring to the difference in the angular distance of two heavenly objects between the horizon and near the zenith (Toomer [1984] 1998, p. 421). In the other parallel passage in *Almagest* I.3 (Toomer [1984] 1998, p. 39), Ptolemy refers to a similar phenomenon, that is, the enlargement of the apparent diameters of the Sun and the Moon in the vicinity of the horizon; there he treats the problem as the effect of the atmospheric refraction, but later in *Planetary Hypotheses* I (Goldstein 1967, pp. 9, 34–35) and *Optics* III.59 (Smith 1996, p. 151), he explains it as merely an optical illusion. This is relevant to the problem mentioned earlier of the angular separation of two celestial bodies near the horizon (also, see Goldstein 1997, p. 5); if the relation between the enlargement of the luminaries and the increase in the angular distance between two objects near the horizon was correctly understood, the latter would no longer be referred to as an observational fact. Muḥyī al-Dīn does not seem to have seen any clear relation between them. Interestingly, in his *Taḥrīr al-majisṭī*, al-Ṭusī does not comment upon either of the passages from the *Almagest* in question (P1: pp. 5, 284, P2: f. 82v, P3: ff. 18v, 107v).

119 Lawkarī was a peripatetic philosopher who wrote an encyclopedia titled *Bayān al-ḥaqq bi-ḍimān al-ṣidq*, including an epitome of Ptolemy's *Almagest*, which was well-known to the later Islamic astronomers (e.g., Quṭb al-Dīn al-Shīrāzī referred to it in the prologue on astronomy of his *Durrat al-tāj li qurrat al-Dibāj*; see al-Shīrāzī 1944, vol. 2, p. 1); however, the surviving manuscripts and published editions of this work do not contain its astronomical part. A chapter from it, dealing with al-Lawkarī's invented instrument, has been partially preserved in MS. Utrecht, Universiteitsbibliotheek, 1442, pp. 23–26, titled "On the description of the instrument making unnecessary [the use of] the armillary sphere and of the method (*ṭarīqa*) by which it becomes possible to attain knowledge of the position of any star investigated without the use of the armillary sphere." In it, the *Book on [astronomical] observations* (*Kitāb al-irṣād*) is attributed to al-Lawkarī, and he also mentions that he had already written a treatise on the instrument in question; neither work is extant today. The detailed description of the instrument is missing from the Utrecht MS, but this text is

systems over the course of history in the introduction to his treatise *On experimental astronomy* (*Kayfiyyat al-i'tibār*), 'Abd al-Raḥmān al-Khāzinī (*fl. ca.* 1120 CE) considers "dispensing with the armillary sphere" (*al-istighnā' 'an dhāt al-ḥalaq*) as a third (seemingly, particular) aspect of the developments made by his Islamic predecessors (*al-muta'akhkhirūn*), along with the two other (apparently, general) ones: "The investigation into [astronomical] observations" (*al-taḥqīq fī al-raṣad*) and "the improvement of the precision of [observational] instruments" (*al-tadqīq fī al-ālāt*).[120] Consequently, he does not mention the armillary sphere among his five observational instruments described in *On experimental astronomy* I.4–8.[121] In addition, in the first period of the Marāgha observatory, al-'Urḍī mentions that his Two Quadrants replaces the armillary sphere.[122] Furthermore, in the *Ghāzānid treatise on observational instruments*, which deals with the 12 new altitude-azimuthal instruments constructed in the second period of the astronomical activities at Marāgha, the classic observational instruments are rejected for various reasons and, instead, a new approach to the design and construction of the observational instruments is proposed: the use of long and straight rulers instead of rings. We are told that, in general, if the instruments consisting of the rings are small, it will be impossible to divide them into the minutes and seconds of arc, and so the data obtained will inevitably be approximate; and if they are large, it will be impossible to make them completely circular, as they ought to be, and then their defects will outweigh their benefits. Specifically in the case of the armillary sphere, we are told that (1) it is extremely difficult to use the instrument, and (2) its determination of the longitude of a given heavenly object is dependent on that of a reference celestial body; such a measurement is "approximate" (we are not told why). A third reason is also given, which is slightly ambiguous and can be interpreted in one of two ways: either the principal reference body must be located on the ecliptic (i.e., it can only be the Sun, and hence, all measurements are dependent upon the solar theory) or, as a special case, the latitude ring of the instrument may not be perfectly perpendicular

---

very likely the source of al-Marrākushī (d. 1262 CE) in his *Jāmi' al-mabādī wa-'l-ghāyāt* II.7.7 (I: pp. 116–119, N: ff. 155r–156r, P: ff. 207r–v), in which he describes al-Lawkarī's instrument in the category of observational instruments: It consists of a single azimuthal ring, an altitudinal quadrant of another ring of the same diameter, both made of copper, and an alidade with two pinnulas with tiny holes. The azimuthal ring is installed on a round hollow booth (*dakka*) on a steady flattened horizontal ground. The quadrant is installed in a cross (*ṣalīb*) which is erected in the center of the circular wall, in such a manner that the cross smoothly rotates in the hollow inside the azimuthal ring and the quadrant steadily and smoothly rotates on its circumference toward any direction. It is worth noting that as he himself clarifies, al-Marrākushī's source for the description and application of the armillary sphere in *Jāmi' al-mabādī wa-'l-ghāyāt* II.7.5,6 (I: pp. 113–116, N: ff. 154r–155r, P: ff. 206r–207r) was also al-Lawkarī's *al-Bayān*.

120 al-Khāzinī, *Kayfiyyat al-i'tibār*, in *Zīj*, V: ff. 4v–5r. This treatise comes as an introduction to his *zīj*, in which Khāzinī deals with the principal features of observational astronomy and explains the technical experiments for testing and re-measuring the astronomical quantities and parameter values from a methodologically consistent point of view.

121 al-Khāzinī, *Kayfiyyat al-i'tibār*, in *Zīj*, V: ff. 6r–7r.

122 al-'Urḍī, *Fī Kayfiyyat al-irṣād*, P: f. 15r, N: f. 45v.

to its ecliptic ring. The negative consequence is that the longitude and latitude of the celestial object cannot be known "with certainty."[123]

Of course, Muḥyī al-Dīn's specific method placed some restrictions on his systematic observations. For instance, when dealing with the stellar measurements, he complains that

> it is not possible for us to observe either *al-Nasr al-wāqi'* [i.e., Vega, α Lyr] or *al-'Ayyūq* [i.e., Capella, α Aur], both of which transit the circle of the [local] meridian in its northern direction, because there is no northern quadrant [established] on the meridian line in this auspicious, blessed observatory.[124]

For a comparative view of the observational activities made at the Marāgha observatory, we compare Muḥyī al-Dīn's errors in the ecliptical coordinates of the 8 stars he observed at the Marāgha observatory (Table 9.3) with those in the ecliptical coordinates of 16 bright stars his colleagues observed there in the 1260s, as recorded in a non-Ptolemaic star table in the *Īlkhānī zīj*.[125] The statistical results of the errors in the latitudes and longitudes of the 8 stars in this table, which are in common with Table 9.3, are as follows.

	MAE	μ	σ
$d\beta$	10.8'	+4.0'	11.8'
$d\lambda$	13.1	+5.4	16.5

Taking the ecliptical coordinates of all 16 stars into account, we have:

	MAE	μ	σ
$d\beta$	12.5'	−0.4'	16.8'
$d\lambda$	13.5	−3.4	17.5

Comparing these results with Muḥyī al-Dīn's errors given earlier in this section, it is clear that his values for the stellar latitudes are more precise than those measured in the parallel observational program carried out at the Marāgha observatory, while the opposite is true in the case of the stellar longitudes. With regard to the accuracy of the armillary sphere constructed by al-'Urḍī, as discussed in the

---

123 See Chapter 11, p. 303; Mozaffari and Zotti 2013, pp. 70–71, 72–73.

124 al-Maghribī, *Talkhīṣ*, f. 114v. The declinations of Vega and Capella were, respectively, about **+44;51.5°** and **+38;17.5°** at that time, and both thus transited the meridian of Marāgha in its northern half. The non-Ptolemaic star table of *Īlkhānī zīj* includes the ecliptical coordinates of both Vega and Capella.

125 See Chapter 10.

previous section, it is highly improbable that it was used in the measurement of the ecliptical longitudes of the 16 stars in question.

9.7 Comparison and Conclusion

1. The precision of Muḥyī al-Dīn's first two observations of the solar noon altitudes in the days near the summer and winter solstices of 1264 CE can be compared with the best solar noon altitude observations performed by the early Islamic astronomers in the days of the solstices, with the mean error of ±1′, which is near the limit of resolution of the unaided eye;[126] for example, Khālid b. ʿAbd al-Malik al-Marwarūdhī (*fl. ca.* 832 CE), who had used a quadrant with radius ~10 m; al-Ṣūfī (903–986 CE), who had used a solstice ring with diameter ~5 m;[127] al-Khujandī (945–1000 CE), who had constructed a sextant with radius ~20 m;[128] and so on. In the other solar meridian altitude observations, Muḥyī al-Dīn committed a consistent error of +4′, the source of which is unknown to us. From the early Islamic period, the most accurate solar observations recorded at times other than solstices were made by Bīrūnī with a mean absolute error of ~2′.[129]

2. In the systematic measurements of the meridian altitudes of the stars and planets, Muḥyī al-Dīn's errors are irregular and, except for Antares, never exceed ±10′; their mean absolute value is ~5.3′ (for the superior planets: ~4.6′ and for the stars: ~6.2′). We do not know of any comparable observational data of this kind in Islamic astronomy. However, Ibn al-Shāṭir's (Damascus, 1306–1375/1376) star table in his *Jadīd zīj*, which lists the celestial/equatorial coordinates of 80 stars for 765 H (which began on October 9, 1363; JDN 2219175), must have been based on similar observations to Muḥyī al-Dīn's.[130] The mean absolute errors in the values given for the declinations

---

126 Said and Stephenson 1995, pp. 122–123, 125, 129–130.

127 Kennedy 1961, p. 105. The instrument used was likely a Two Circles (*Almagest* I.12: Toomer [1984] 1998, pp. 61–63), in which the inner circle was replaced by an alidade.

128 al-Khujandī's treatise *Fī tashīh al-mayl wa ʿard al-balad* (*On the correction of the obliquity of the ecliptic and the latitude of place*), containing the description of his gigantic instrument, was edited in Khujandī 1908 and translated into German in Schirmer 1926, pp. 63–79. A surviving tract on this instrument, titled *Ḥikāyat al-ālat al-musammāt al-suds al-Fakhrī* (*Information on the instrument called the* Fakhrī *Sextant*), is attributed to Bīrūnī (edited in al-Bīrūnī 1908; translated into German in Wiedmann 1910 and into English in Sezgin and Neubauer 2010, vol. 2, p. 25; see also Schirmer 1926, pp. 43–46). On al-Khujandī's sextant, see also al-Bīrūnī 1967, pp. 70–77; Kennedy 1973, pp. 44–48; Bīrūnī's *al-Qānūn al-masʿūdī*, a separate chapter appended to VI.6: 1954–1956, vol. 2, p. 643; al-Marrākushī's *Jāmiʿ al-mabādī waʾl-ghāyāt fī ʿilm al-mīqāt* II.7.2: I: ff. 55v–56r (Sezgin's facsimile edition, vol. 2, pp. 110–111), N: ff. 152v–153r, P2: 205r.

129 Said and Stephenson 1995, p. 126.

130 Ibn al-Shāṭir, *Jadīd zīj*, O: ff. 147v–148r, L1: ff. 107v–109r, L2: ff. 137r–v. Of course, in Section 19 of this work (O: f. 61v, L1: ff. 23r–v, L2: ff. 28v–29r, PR: ff. 30v–31r), we are told that he observed many of the fixed stars and laid down their *ecliptical* coordinates in a table for the year 760 H/December 2, 1358. The present author is currently preparing a detailed study of this table.

and right ascensions of the eight stars of this table that are in common with Table 9.3 are, respectively, 20.3' (μ = +13.5', with σ = 22.6') and 10.5' (μ = +6.4', with σ = 11.9'). As discussed in Section 9.6, the corresponding errors found in Muḥyī al-Dīn's observations are, respectively, 6.2' and 26.4'. Therefore, Muḥyī al-Dīn obtained more precise values for the stellar meridian altitudes/declinations than Ibn al-Shāṭir, while the latter achieved more accurate measurements for the culmination times/right ascensions.

3. Muḥyī al-Dīn's use of the water clock seems to be a characteristic of his own observations. To the best of our knowledge, no other uses of it for astronomical purposes were recorded in the Islamic period. As already noted, it appears that the medieval Islamic astronomers did not use water clocks in their observations – partly, it seems, because of Ptolemy's rejection of their use in the astronomical/astrological queries, and partly due to their awareness of the possible deficiencies of the instrument. Nevertheless, thanks to the accuracy of the clepsydras available at the Marāgha observatory, Muḥyī al-Dīn attained a remarkable degree of precision in measuring time (maximum absolute error = 5 m; mean absolute error = 2.5 m). The clepsydras at Marāgha were probably Chinese models. As understood from Muḥyī al-Dīn's *Talkhīṣ*, they could measure hours and minutes independently of each other, and at least two models were available at the observatory.

4. A quantitative analysis of the surviving reports of the near appulses of the planets to the stars from the medieval Islamic period is necessary for a reliable comparative study.[131] There is a relatively tolerable systematic error of ~+3' in three of Muḥyī al-Dīn's observations of the near appulses of the superior planets to Regulus (the fourth has an error of ~+0.5°) in estimating the distance in longitude between each planet and the star.

5. A comparison of the accuracy of Muḥyī al-Dīn's observations with those of his colleagues who produced the *Īlkhānī zīj*, with reference to the values for the ecliptical coordinates of the fixed stars, shows that Muḥyī al-Dīn's values for the latitudes are more precise, while the *Īlkhānī zīj* has better values for the longitudes.

6. No observation reports from the Beijing and Samarqand observatories are available, other than the two large star catalogues compiled there.[132] We compare the precision of Muḥyī al-Dīn's observations with that achieved by Taqī al-Dīn Muḥammad b. Maʿrūf in his observations made at the Istanbul observatory in the 1570s. The latter describes some mechanical clocks for astronomical use, seemingly influenced by European sources and models;[133]

---

131 A detailed study of the early Islamic planetary observations recorded in Ibn Yūnus's *Ḥākimī zīj* (see earlier, note 66) is under preparation by the author.

132 See Knobel 1917, Shevchenco 1990, Krisciunas 1993, Verbunt and van Gent 2012, van Dalen 2000.

133 Taqī al-Dīn, *Sidra*, K: f. 90r; see also Sezgin and Neubauer 2010, vol. 3, pp. 118–122, Ben-Zaken 2011.

it is not known whether he applied them to his astronomical observations. The accuracy he attained in his measurement of the solar noon altitudes and longitudes and the equinox times is comparable to that achieved by Muḥyī al-Dīn, but his estimated times of the maximum phases of the trio of the lunar eclipses that he and his colleagues observed in 1576–1577 CE suffer from a relatively large negative systematic error of between −1/2 and −1 hour.[134]

References

5MCLE: Espanek, F. and Meeus, *NASA's Five Millennium Catalog of Lunar Eclipses*, retrieved from https://eclipse.gsfc.nasa.gov/LEcat5/LEcatalog.html.

Anonymous, *Sulṭānī zīj*, MS. Iran, Parliament Library, no. 184.

Archimedes, 1976, *On the Construction of Water-Clocks: An Annotated Translation from Arabic Manuscripts of the Pseudo-Archimedes Treatise: Kitāb Arshimīdas fī ʿamal al-binkamāt*, Hill, D. R. (ed.), London: Turner and Devereux.

Ben-Zaken, A., 2011, "The revolving planets and the revolving clocks: Circulating mechanical objects in the Mediterranean", *History of Science* **49**, pp. 125–148.

al-Bīrūnī, Abū al-Rayḥān, 1908, *Ḥikāyat al-ālat al-musammāt al-suds al-Fakhrī* [*Information on the instrument called the Fakhrī Sextant*], Shaykhu, L. (ed.), *al-Mashriq* **11**, pp. 68–69.

al-Bīrūnī, Abū al-Rayḥān, 1910, *Alberuni's India*, Sachau, E. C. (En. tr.), 2 Vols., London: Kegan Paul, Trench, Trübner & Co.

al-Bīrūnī, Abū al-Rayḥān, 1954–1956, *al-Qānūn al-masʿūdī* (*Masʿūdīc canons*), 3 Vols., Hyderabad: Osmania Bureau.

al-Bīrūnī, Abū al-Rayḥān, 1967, *Taḥdīd nahayāt al-amākin li-taṣḥīḥ masāfāt al-masākin* (*Determination of the Coordinates of Positions for the Correction of Distances between Cities*), Ali, J. (En. tr.), Beirut: American University of Beirut.

Brockelmann, K., 1937–1942, *Geschichte der arabischen Literatur*, 2 Vols., 2nd edn., Leiden: Brill, 1943–1949, Supplementbände 1–3, Leiden: Brill.

Burnett, C., *et. al.* (eds.), 2004, *Studies in the History of the Exact Sciences in Honour of David Pingree*, Leiden-Boston: Brill.

Caussin de Perceval, J.-J.-A., 1804, "Le livre de la grande table hakémite, Observée par le Sheikh,. . ., ebn Iounis", *Notices et Extraits des Manuscrits de la Bibliothèque nationale* **7**, pp. 16–240.

Charette, F., 2006, "The locales of Islamic astronomical instrumentation", *Journal of History of Science* **44**, pp. 123–138.

Clark, D. and Stephenson, F. R., 1977, *The Historical Supernovae*, Oxford: Pergamon Press.

Cullen, C., 1996, *Astronomy and Mathematics in Ancient China: The Zhou bi suan jing*, Cambridge: Cambridge University Press.

van Dalen, B., 2000, "A non-ptolemaic Islamic Star table in Chinese", in: Folkerts, M. and Lorch, R. (eds.), *Sic Itur Ad Astra; Studien zur Geschichte der Mathematik und Naturwissenschaften*, Wiesbaden: Harrassowitz.

van Dalen, B., 2002a, "Islamic and Chinese astronomy under the Mongols: A little-known case of transmission", in: Dold-Samplonius, Y., *et al.* (eds.), *From China*

---

134 See Table 5.2 on p. 121.

to Paris: 2000 Years Transmission of Mathematical Ideas, Stuttgart: Franz Steiner, pp. 327–356.

van Dalen, B., 2002b, "Islamic astronomical tables in China: The sources for the Huihui li", in: Ansari, R. (ed.), *History of Oriental Astronomy* (Proceedings of the Joint Discussion 17 at the 23rd General Assembly of the International Astronomical Union, Organised by the Commission 41 (History of Astronomy), Held in Kyoto, August 25–26, 1997), Dordrecht: Kluwer, pp. 19–30.

van Dalen, B., Kennedy, E. S., and Saiyid, M. K., 1997, "The Chinese-Uighur calendar in Ṭūsī's Zīj-i Īlkhānī", *Zeitschrift fur Geschichte der Arabisch-Islamischen Wissenschaften* **11**, pp. 111–152.

Delambre, M., 1819, *Histoire de l'Astronomie du Moyen Age*, Paris: Courcier.

Dorce, C., 2002–2003, "The *Tāj al-azyāj* of Muḥyī al-Dīn al-Maghribī (d. 1283): Methods of computation", *Suhayl* **3**, pp. 193–212.

Dorce, C., 2003, "El Tāŷ al-azyāŷ de Muḥyī al-Dīn al-Maghribī", in: *Anuari de Filologia*, Vol. 25, Secció B, Número 5, Barcelona: University of Barcelona.

[*DSB*:] Gillispie, C. C., *et al.* (eds.), 1970–1980, *Dictionary of Scientific Biography*, 16 Vols., New York: Charles Scribner's Sons.

Goldstein, B. R., 1967, "The Arabic version of Ptolemy's planetary hypotheses", *Transactions of the American Philosophical Society* **57**, pp. 3–55.

Goldstein, B. R., 1997, "Saving the phenomena: The background to Ptolemy's planetary theory", *Journal for the History of Astronomy* **28**, pp. 1–12.

Halley, E., 1720–21, "On the method of determining the Places of the Planets by observing their near Appulses to the fixed stars", *Philosophical Transactions of the Royal Society* **31**, pp. 209–211.

Hartner, W., 1950, "The astronomical instruments of Cha-ma-lu-ting, their identification, and their relations to the instruments of the observatory of Marāgha," *Isis* 41, pp. 184–194.

Hill, D. R., 2008, "Al-Jazarī", in: Selin 2008, pp. 130–131.

[*BEA*:] Hockey, T., *et al.* (eds.), 2014, *The Biographical Encyclopedia of Astronomers*, 2nd edn., New York [etc.]: Springer.

Hogendijk, J. P., 1993, "An Arabic text on the comparison of the five regular polyhedra: 'Book XV' of the 'Revision of the Elements' by Muḥyī al-Dīn al-Maghribī", *Zeitschrift fur Geschichte der Arabisch-Islamischen Wissenschaften* **8**, pp. 133–233.

Ibn al-Fahhād: Farīd al-Dīn Abu al-Ḥasan ʿAlī b. ʿAbd al-Karīm al-Fahhād al-Shirwānī or al-Bākū ʾī, *Zīj al-ʿAlā ʾī*, MS. India, Salar Jung, no. H17.

Ibn al-Fuwatī, Kamāl al-Dīn ʿAbd al-Razzāq b. Muḥammad, 1995, *Majmaʿ al-ādāb fī muʿjam al-alqāb*, Kāẓim, M. (ed.), 6 Vols., Tehran: Ministry of Culture.

Ibn al-Haytham, al-Ḥasan, 1983, *Kitān al-manāẓir, Books I-II-III: On Direct Vision*, Sabra, A. I. (ed.), Kuwait: The National Council for Culture, Arts, and Letters.

Ibn al-Shāṭir, ʿAlāʾ al-Dīn Abu ʾl-Ḥasan ʿAlī b. Ibrāhīm b. Muḥammad al-Muṭaʿʿim al-Anṣārī, *al-Zīj al-Jadīd*, MSS. K: Istanbul, Kandilli Observatory, no. 238, O: Oxford, Bodleian Library, Seld. A inf 30, L1: Leiden, Universiteitsbibliotheek, Or. 65, L2: Leiden, Universiteitsbibliotheek, Or. 530, PR: Princeton, Princeton University Library, no. Yahuda 145.

Ibn Yūnus, ʿAlī b. ʿAbd al-Raḥmān b. Aḥmad, *Zīj al-kabīr al-Ḥākimī*, MSS. L: Leiden, Universiteitsbibliotheek, no. Or. 143, O: Oxford, Bodleian Library, no. Hunt 331, F1: Paris, Bibliothèque Nationale, no. Arabe 2496 (formerly, arabe 1112; copied in 973 H/1565–1566 CE), F2: Paris, Bibliothèque Nationale, no. Arabe 2495 (formerly, arabe 965; the 19th-century copy of MSS. L and the additional fragments in F1).

al-Jazarī, Ibn al-Razzāz, 1973, *The Book of Knowledge of Ingenious Mechanical Devices* (*Kitāb fī ma'rifat al-ḥiyal al-handasiyya*), Hill, D. R. (En. Tr.), Dordrecht-Boston: D. Reidel Publishing Company.

Johnson, M. C., 1940, "Greek, Moslem and Chinese instrument design in the surviving Mongol equatorials of 1279 A.D.", *Isis* **32**, pp. 27–43.

Jones, A., 2004, "A study of Babylonian observations of planets near normal stars", *Archive for History of Exact Sciences* **58**, pp. 475–536.

al-Kamālī, Muḥammad b. Abī 'Abd-Allāh Sanjar (Sayf-i munajjim), *Ashrafī zīj*, MSS. F: Paris: Bibliothèque Nationale, no. 1488, G: Iran–Qum: Gulpāyigānī, no. 64731.

al-Kāshī, Jamshīd Ghiyāth al-Dīn, *Khāqānī zīj*, MSS. IO: London: India Office, no. 430; P: Iran: Parliament Library, no. 6198.

Kennedy, E. S., 1961, "Al-Kāshī's treatise on astronomical observational instruments", *Journal for Near Eastern Studies* **20**.2, pp. 98–108. Rep. Kennedy, 1983, pp. 394–404.

Kennedy, E. S., 1973, *A Commentary Upon Bīrūnī's* Kitāb Taḥdīd al-Amākin, Beirut: American University of Beirut.

al-Khāzinī, 'Abd al-Raḥmān, *al-Zīj al-mu'tabar al-sanjarī*, MSS. V: Vatican, Biblioteca Apostolica Vaticana, no. Arabo 761, L: London, British Library, no. Or. 6669; *Wajīz* [Compendium of] *al-Zīj al-mu'tabar al-sanjarī*, MSS. I: Istanbul, Süleymaniye Library, Hamidiye collection, no. 859; S: Tehran: Sipahsālār, no. 682.

al-Khujandī, Abū Maḥmūd Ḥāmid b. al-Khiḍr, 1908, *Fī tashīḥ al-mayl wa 'arḍ al-balad* [*On the Correction of the Obliquity of the Ecliptic and the Latitude of Place*], Shaykhu, L. (ed.), *al-Mashriq* **11**, pp. 60–67.

King, D. and Gingerich, O., 1982, "Some astronomical observations from thirteenth-century Egypt", *Journal for the History of Astronomy* **13**, pp. 121–128.

King, D. A. and Saliba, G. (eds.), 1987, *From Deferent to Equant: A Volume of Studies on the History of Science of the Ancient and Medieval Near East in Honor of E. S. Kennedy*, Vol. 500, New York: Annals of the New York Academy of Sciences.

Knobel, E. B., 1917, *Ulugh Beg's Catalogue of Stars*, Washington: Carnegie.

Krisciunas, K., 1993, "A more complete analysis of the errors in Ulugh Beg's star catalogue", *Journal for the History of Astronomy* **24**, pp. 269–280.

Kuehn, S., 2011, *The Dragon in Medieval East Christian and Islamic Art*, Leiden-Boston: Brill.

al-Lawkarī, Abu al-'Abbās Faḍl b. Muḥammad, 1995, *Bayān al-ḥaqq bi-ḍimān al-ṣidq*, Dībājī, S. I. (ed.), Tehran: The International Institute of Islamic Thought and Civilization.

Lorch, R., 1980, "Al-Khāzinī's 'Sphere that rotates by itself'", *Journal for the History of Arabic Science* **4**, pp. 287–329. Rep. Lorch 1995, Trace XI.

Lorch, R., 1981, "Alkhāzinī's balance-clock and the Chinese steelyard clepsydra", *Archives Internationales d'Histoire des Sciences* **31**, pp. 183–189. Rep. Lorch 1995, Trace XV.

Lorch, R., 1995, *Arabic Mathematical Sciences: Instruments, Texts, Transmission*, Aldershot: Variorum.

Lorch, R., 2000, "Ibn al-Ṣalāḥ's treatise on projection: A preliminary survey", In: Folkerts, M. and Lorch, R. (eds.), *Sic Itur ad Astra: Studien zur Geschichte der Mathematik und Naturwissenschaften*, Wiesbaden: Harrassowitz Verlag, pp. 401–408.

al-Maghribī, Muḥyī al-Dīn, *Adwār al-anwār mada 'l-duhūr wa-'l-akwār* (*Everlasting cycles of lights*), MSS. M: Iran, Mashhad, Holy Shrine Library, no. 332; CB: Ireland, Dublin, Chester Beatty, no. 3665.

al-Maghribī, Muḥyī al-Dīn, *Talkhīṣ al-majisṭī* (*The compendium of the* Almagest), MS. Leiden: Universiteitsbibliotheek, Or. 110.

al-Maghribī, Muḥyī al-Dīn, ʿUmdat al-ḥāsib wa-ghunyat al-ṭālib (*Mainstay of the astronomer, sufficient for the student*), MS. M: Cairo: Egyptian National Library, no. MM 188.

Makovicky, E., 1992, "800-year-old pentagonal tiling from Marāgha, Iran, and the new varieties of aperiodic tiling it inspired", in: Hargittai, I. (ed.), *Fivefold Symmetry*, Singapore: World Scientific, pp. 67–86.

al-Marrākushī, Abū ʿAlī al-Ḥasan b. ʿAlī b. ʿUmar, *Jāmiʿ al-mabādī waʾl-ghāyāt fī ʿilm al-miqāt* (*Comprehensive Collection of Principles and Objectives in the Science of Timekeeping*), MSS. I: Istanbul, Topkapı Sarayı Museum Library, Ahmet III Collection, no. 3343 (facsimile edition by Sezgin, F., 2 Vols., Frankfurt: Institute for the History of Arabic-Islamic Science at the Johann Goethe University, 1984), N: Istanbul, Süleymaniye Library, Nuruosmaniye Collection, no. 2902, P: Tehran, Parliament Library, no. 37234–10.

Meeus, J., 1998, *Astronomical Algorithms*, Richmond: William-Bell.

Michel-Nozières, C., 2000, "Second millennium Babylonian water clocks: A physical study", *Centaurus* **42**, pp. 180–209.

Morrall, A., 2009, "Ornament as evidence", in: Harvey, K. (ed.), *History and Material Culture*, 1st edn., London-New York: Routledge, pp. 47–66.

Mozaffari, S. M., 2014, "Muḥyī al-Dīn al-Maghribī's lunar measurements at the Maragha observatory", *Archive for History of Exact Sciences* **68**, pp. 67–120.

Mozaffari, S. M., 2016, "Astronomy and politics: Three case studies on the service of astrology to society", in: Rappenglück, M. A., Rappenglück, B., Campion, N., and Silva, F. (eds.), *Astronomy and Power: How Worlds Are Structured; Proceedings of the SEAC 2010 Conference*, Oxford: British Archaeological Reports, pp. 241–246.

Mozaffari, S. M., 2016–2017, "A revision of the star tables in the *Mumtaḥan zīj*", *Suhayl* **15**, pp. 67–100.

Mozaffari, S. M., 2018, "An analysis of medieval solar theories", *Archive for History of Exact Sciences* **72**, pp. 191–243.

Mozaffari, S. M., 2018–2019, "Muḥyī al-Dīn al-Maghribī's measurements of Mars at the Maragha observatory", *Suhayl* **16**, pp. 149–249.

Mozaffari, S. M. and Zotti, G., 2013, "The observational instruments at the Maragha observatory after AD 1300", *Suhayl* **12**, pp. 45–179.

Needham, J., 1981, *Science in Traditional China: A Comparative Perspective*, Cambridge, MA: Harvard University Press.

Needham, J., Ling, W. and de Solla Price, D. J., 1986, *Heavenly Clockwork. The Great Astronomical Clocks of Medieval China*, 2nd edn., with supplement by John H. Combridge, Cambridge: Cambridge University Press.

Neugebauer, O., 1947, "Studies in ancient astronomy: VIII. The water clock in Babylonian astronomy", *Isis* **37**, pp. 37–43. Rep. Neugebauer 1983, pp. 239–245.

Neugebauer, O., 1975, *A History of Ancient Mathematical Astronomy*, Berlin-Heidelberg-New York: Springer.

Neugebauer, O., 1983, *Astronomy and History: Selected Essays*, New York: Springer.

Ōhashi, Y., 2008, "Astronomical instruments in India", in: Selin 2008, pp. 269–273.

Pandey, G. S., 2011, "Divisions of time and measuring instruments of Varāhmihira", in: Yadav, B. S. and Mohan, M. (eds.), *Ancient Indian leaps into Mathematics*, Birkhäuser, Boston: Springer, pp. 75–110.

Peng Yoke, H., 2008, "Guo Shoujing", in: Selin 2008, p. 1041.

Pingree, D., 1973, "The mesopotamian origin of early Indian mathematical astronomy", *Journal for the History of Astronomy* **4**, pp. 1–12.

Pingree, D. (ed.), 1985–1986, *Astronomical Works of Gregory Chioniades, Part 1: Zīj al-'Alā'ī*, 2 Vols., Amsterdam: Gieben.

Ptolemy, 1940/1956, *Tetrabiblos*, Robbins, F. E. (ed.), Cambridge, MA: Harvard University Press, London: Heinemann (Loeb classical library, Greek authors, 350).

Rao, N. K., 2005, "Aspects of prehistoric astronomy in India", *Bulletin of the Astronomical Society of India* **30**, pp. 499–511.

Rosenfeld, B. A. and İhsanoğlu, E., 2003, *Mathematicians, Astronomers, and Other Scholars of Islamic Civilization and Their Works (7th-19th c.)*, Istanbul: IRCICA.

Sabra, A. I., 1989, *The Optics of Ibn al-Haytham*, 2 Vols., London: Warburg Institute.

Said, S. S. and Stephenson, F. R., 1995, "Precision of medieval Islamic measurements of solar altitudes and equinox times", *Journal for the History of Astronomy* **26**, pp. 117–132.

Saliba, G., 1983, "An observational notebook of a thirteenth-century astronomer", *Isis* **74**, pp. 388–401. Rep. Saliba 1994, pp. 163–176.

Saliba, G., 1985, "Solar observations at Maragha observatory", *Journal for the History of Astronomy* **16**, pp. 113–122. Rep. Saliba 1994, pp. 177–186.

Saliba, G., 1986, "The determination of new planetary parameters at the Maragha observatory", *Centaurus* **29**, pp. 249–271. Rep. Saliba 1994, pp. 208–230.

Saliba, G., 1994, *A History of Arabic Astronomy: Planetary Theories During the Golden Age of Islam*, New York: New York University Press.

Sarma, S. R., 1994a, "The bowl that sinks and tells time", *India Magazine, of her People and Culture* **14**, pp. 31–36.

Sarma, S. R., 1994b, "Indian astronomical and time-measuring instruments: A catalogue in preparation", *Indian Journal of History of Science* **29**, pp. 507–528.

Sarma, S. R., 2001, "Measuring time with long syllables: Bhāskara I's Commentary on Āryabhaṭīya, Kālakriyāpāda 2", *Indian Journal of History of Science* **36**, pp. 51–54.

Sarma, S. R., 2004, "Setting up the water clock for telling the time of marriage", in: Burnett *et. al.* 2004, pp. 302–330.

Sarton, G., 1927–1948, *Introduction to the History of Science*, 3 Vols. in 5 Parts, Baltimore: Williams & Wilkins.

Sayılı, A., 1960/1988, *The Observatory in Islam*, 2nd edn., Ankara: Türk Tarih Kurumu Basimevi.

Schirmer, O., 1926/1927, "Studien zur Astronomie der Araber", *Sitzungsberichte der Physikalisch-Medizinischen Sozietät zu Erlangen* **58–59**, pp. 33–88.

Seemann, H. J., 1928/1929, "Die Instrumente der Sternwarte zu Marāgha nach den Mitteilungen von al-'Urḍī", in: *Sitzungsberichte der Physikalisch-medizinischen Sozietät zu Erlangen*, Schulz, O. (ed.), Vol. 60, Erlangen: Kommissionsverlag von Max Mencke, pp. 15–126.

Selin, H. (ed.), 2008, *Encyclopaedia of the History of Science, Technology, and Medicine in Non-Western Cultures*, 2nd edn., Netherlands: Springer.

Sezgin, F., 1978, *Geschichte Des Arabischen Scrifttums*, Vol. 6, Leiden: Brill.

Sezgin, F. and Neubauer, E., 2010, *Science and Technology in Islam*, 5 Vols., Frankfurt: Institut für Geschichte der Arabisch–Islamischen Wissenschaften.

Sharma, V. N., 2000, "Astronomical instruments at Kota", *Indian Journal of History of Science* **35**, pp. 233–244.

Shevchenco, M., 1990, "An analysis of errors in the star catalogues of Ptolemy and Ulugh Beg", *Journal for the History of Astronomy* **21**, pp. 187–201.

al-Shīrāzī, Quṭb al-Dīn, 1944, *Durrat al-tāj li qurrat al-Dibāj*, 5 Vols., Tehran: the Iran Ministry of Culture.

Sivin, N., 2009, *Granting the Seasons: The Chinese Astronomical Reform of 1280, with a Study of Its Many Dimensions and a Translation of Its Records*, New York: Springer.

Smith, A. M., 1996, "Ptolemy's theory of visual perception: an English translation of the Optics with introduction and commentary", *Transactions of the American Philosophical Society* **86**.2, Philadelphia.

Smith, A. M., 2001, "Alhacen's theory of visual perception: A critical edition, with English translation and commentary, of the first three books of Alhacen's De aspectibus, the Medieval Latin version of Ibn al-Haytham's Kitāb al-Manāẓir, 2 Vols.", *Transactions of the American Philosophical Society* **91**.4 and 5.

Stephenson, F. R., 1997, *Historical Eclipses and Earth's Rotation*, Cambridge: Cambridge University Press.

Stephenson, F. R. and Said, S. S., 1991, "Precision of medieval Islamic eclipse measurements", *Journal for the History of Astronomy* **22**, pp. 195–207.

al-Ṣūfī, ʿAbd al-Raḥmān, 1995, *Al-ʿamal bi-l-Asṭurlāb*, Morocco: ISESCO.

Súrya Siddhánta, 1860/1997, *The Súrya Siddhánta: A Textbook of Hindu Astronomy*, Gangooly, P. (ed.) and Burgess, E. (tr.), Delhi: Motilal Banarsidass. 1861/1974, *The Súrya Siddhánta: An Ancient System of Hindu Astronomy, followed by the Siddhánta Śiromani*, Deva Sastri, P. B. and Wilkinson, L. (En. trs.), Amsterdam: Philo Press.

Suter, H., 1900, *Die Mathematiker und Astronomen der Araber und ihre Werke*, Leipzig: Teubner.

Taqī al-Dīn Muḥammad b. Maʿrūf, *Sidrat muntaha 'l-afkar fī malakūt al-falak al-dawwār* (*The lotus tree in the seventh heaven of reflection*) or *Shāhanshāhiyya Zīj*, MSS. K: Istanbul, Kandilli Observatory, no. 208/1 (up to f. 48v; autograph); N: Istanbul, Süleymaniye Library, Nuruosmaniye Collection, no. 2930; V: Istanbul, Süleymaniye Library, Veliyüddin Collection, no. 2308/2 (from f. 10v).

Toomer, G. J. (ed.), 1984/1998, *Ptolemy's Almagest*, Princeton: Princeton University Press.

al-Ṭūsī, Naṣīr al-Dīn, *Īlkhānī zīj*, MSS. C: University of California, Caro Minasian Collection, no. 1462; T: University of Tehran, Ḥikmat Collection, no. 165 + Suppl. P: Iran, Parliament 6517 (Remark: The latter is not actually a separate MS, but contains 31 folios missing from MS. T. The chapters and tables in MS. T are badly out of order, presumably owing to the folios having been bound in disorder), P: Iran, Parliament Library, no. 181, M1: Iran, Mashhad, Holy Shrine Library, no. 5332a; M2: Iran, Qum, Marʿashī Library, no. 13230.

al-Ṭūsī, Naṣīr al-Dīn, *Taḥrīr al-majisṭī* (*Exposition of the* Almagest), MSS. Iran, Parliament Library, P1: no. 3853, P2: no. 6357, P3: no. 6395.

al-ʿUrḍī, Muʾayyid al-Dīn Muʾayyid b. Barmak b. Mubārak, *Risāla fī kayfiyyat al-irṣād* (*The treatise on how to make* [*astronomical*] *observations*), MSS. P: Tehran, Parliament Library, no. 4345/1, ff. 1v–26r (date: Wednesday, 26 Dhu 'l-ḥijja 1243 H/9 July 1828), N: Tehran, National Library, no. 28211/5, ff. 37v–53r.

Verbunt, F. and van Gent, R. H., 2012, "The star catalogues of Ptolemaios and Ulugh Beg: Machine-readable versions and comparison with the modern HIPPARCOS Catalogue", *Astronomy & Astrophysics* **544**, p. A31.

Wābkanawī, Shams al-Dīn Muḥammad, *Zīj al-muḥaqqaq al-sulṭānī ʿalā uṣūl al-raṣad al-Īlkhānī* (*The verified zīj for the sultan on the basis of the parameters of the Īlkhānid observations*), MSS. T: Turkey, Aya Sophia Library, no. 2694; Y: Iran, Yazd, Library of ʿUlūmī, no. 546, its microfilm is available in Tehran university central library, no. 2546; P: Iran, Library of Parliament, no. 6435.

Wiedmann, E., 1910, "Uber den Sextant des al-Chogendi", *Archiv für die Geschichte der Naturwissenschaften und der Technik* **2**, pp. 148–151.

Yabuuti, K., 1987, "The influence of Islamic astronomy in China", in: King and Saliba 1987, pp. 547–559.

Yabuuti, K., 1997, "Islamic astronomy in China during the Yuan and Ming Dynasties", translated and partially revised by Benno van Dalen, *Historia Scientiarum* **7**, pp. 11–43.

Yan, H.-S., 2007, *Reconstruction Designs of Lost Ancient Chinese Machinery* (*History of Mechanism and Machine Science*, Vol. 3), Netherlands: Springer.

Yan, H.-S., 2009, "An approach for the reconstruction synthesis of lost ancient Chinese mechanisms", in: Yan, H.-S. and Ceccarelli, M. (eds.), *International Symposium on History of Machines and Mechanisms, Proceedings of HMM 2008*, Netherlands: Springer.

Part IV

STELLAR ASTRONOMY

10

A MEDIEVAL BRIGHT STAR TABLE

The Non-Ptolemaic Star Table in the *Īlkhānī Zīj*[1]

10.1 Introduction

The Marāgha observatory was founded by Hülegü (d. 1265), the first ruler of the Īlkhānid dynasty of Iran, in northwestern Iran about the beginning of the 1260s. During the first period of the astronomical activities carried out there, two separate and independent observational programs were established: one by the first director, Naṣīr al-Dīn al-Ṭūsī (1201–1274), and the main staff of the observatory, whose names are mentioned in the prologue of the *Īlkhānī zīj*,[2] and the other by Muḥyī al-Dīn al-Maghribī (d. 1283), which altogether lasted until about the middle of the 1270s. Muḥyī al-Dīn explains the details of his observations, the data obtained from them, and the procedure for the derivation of the Ptolemaic fundamental solar, lunar, stellar, and planetary parameters (eccentricities, epicycle radii, mean motions, precession rate, and the like) from them in a unique treatise titled *Talkhīṣ al-majisṭī* (*Compendium of the* Almagest).[3] There are no analogous data from the observations carried out by the main staff of the Marāgha observatory, although undisputable evidence of the existence of a purposeful program of systematic observations can be found in the *Īlkhānī zīj*, the official product of the Marāgha observatory. This chapter concerns the clearest evidence of such observations, that is, the non-Ptolemaic star table of the *Īlkhānī zīj*.

---

1 Original publication: S. M. Mozaffari, "A medieval bright star table: The non-Ptolemaic star table in the *Īlkhānī Zīj*," *Journal for the History of Astronomy* **47** (2016), pp. 294–316. © 2016 Sage, and republished by permission.

2 According to al-Ṭūsī, they were Mu'ayyad al-Dīn al-ʿUrḍī from Damascus, Fakhr al-Dīn al-Marāghī from Mawṣil (Mosul in Iraq), Fakhr al-Dīn al-Akhlāṭī from Tiflīs (Tbilisi, today, capital of Georgia), Najm al-Dīn Dabīrān from Qazwīn (central Iran). See Naṣīr al-Dīn al-Ṭūsī, *Īlkhānī zīj*, C: p. 5, T: f. 2r, P: —, M1: f. 4r, M2: f. 2v.

3 See Chapter 9.

DOI: 10.4324/9781003481966-15

10.2 The Star Tables in the *Īlkhānī zīj*

There are the two star tables in the *Īlkhānī zīj*.[4]

First, a Ptolemaic star table for the epoch of this *zīj*, that is, the beginning of the year 601 Yazdigird (hereafter, Y)/January 18, 1232, that includes the ecliptical coordinates, magnitudes, and temperaments[5] of 60 stars. In it, the longitudes were updated from the star catalogue of *Almagest* VII and VIII by adding 16;45° to Ptolemy's longitude values.[6] This increment is one of the puzzling discrepancies that can be found in the *Īlkhānī zīj*, because the value adopted for the rate of precession in this *zīj* is 1° in 70 Persian years (of 365 days unvarying), which is the value determined by Ibn al-Aʿlam (d. 985 CE); with this value, the increment should amount to 15;39° in the interval of 1,095 years between the epochs of the *Īlkhānī zīj* and the *Almagest* star catalogue. But the tabular longitudes were, instead, updated from the *Almagest* by the use of the precession rate of 1°/66^y together with the false hypothesis that an earlier star catalogue prepared by Menelaus, some 40 years before Ptolemy, had served as the source for the *Almagest* star catalogue. According to this false opinion, Ptolemy added a value of 0;25° to the longitudes in the alleged star catalogue of Menelaus. This hypothesis goes back to early medieval Middle Eastern astronomy (see later text). Alternatively, it can be said that the compiler of this table added 4;3° to the longitudes in al-Ṣūfī's *Ṣuwar al-kawākib al-thābita* (*On the constellations of the fixed stars*) for October 1, 964, which are, in fact, equal to the longitude values in the *Almagest* +12;42°.[7]

Second, a comparative non-Ptolemaic star table that gives the ecliptical coordinates according to Ptolemy, Ibn al-Aʿlam, and Ibn Yūnus (d. 1009 CE) for 18 stars as well as for those 16 stars (the same 18 stars, except for θ Eri and β Per) observed by the Marāgha astronomers (Table 10.1). As with the Ptolemaic star table in the *Īlkhānī zīj*, the epoch of longitudes in this table is the beginning of 601 Y (January 18, 1232), that is, about 30 years before the founding of the Marāgha observatory. This is evident from the fact that the longitudes written down in the column devoted to Ptolemy in it are equivalent to those in the first (Ptolemaic) star table.

---

4 *Īlkhānī zīj*, P: f. 56v, T: f. 100r, M1: f. 100v, M2: f. 86v, C: p. 195, IT: f. 70v, L: f. 89v, F: f. 59v, B: f. 84r, O: f. 114r, Fl: 102r.

5 Temperament (*mizāj*) is an astrological idea, according to which one or some planets are associated with a star and it is assumed that that star has the auspicious or malevolent property of that or those planets.

6 It was edited in Kunitzsch 1964, pp. 382–409.

7 It is noteworthy that below the star tables in MS. T of the *Īlkhānī zīj*, we are told that from the epoch of this *zīj* until 688 Y, the stellar longitudes increased by 1;18,18°, which is in accordance with the yearly precessional motion of 54″. This figure seems to be an approximation of the rate 1°/66^y and equals $1\frac{1}{2}$° in each century, as mentioned, for example, in the Banū Mūsā's *On the solar year*; see Neugebauer 1962, p. 267.

Table 10.1 Non-Ptolemaic star table in the *Īlkhānī Zīj* (epoch: January 18, 1232)

	Star names		Īlkhānid		Ibn Yūnus			Mumtaḥan Ibn/Ibn al-A'lam (IA)				Ptolemy	
			Longitude	Latitude	Longitude		Latitude	Longitude			Latitude	Longitude	Latitude
					Original	Īlkhānī zīj +2;51°		Original	IA +2;36°	Īlkhānī zīj +3;9°		Īlkhānī zīj Alm. +16;45°	
1	ākhir al-nahr	θ Eri	—— 	—— 	13;29° 9;37	16;20° 	−53;28° −53;49	10;10° 6;46	13;33° 	16;42° 	−53;30° −53;49	16;55° 	−53;30° −53;51
2	kaff al-khaḍīb	β Cas	24;15° 24;28	+51;45° +51;13	21;43 21;42	24;34 	+51;50 +51;14	18;24 18;54	21; 0 	24; 9 	+51;45 +51;13	24;35 	+51;40 +51;14
3	ra's al-ghūl	β Per	—— 	—— 	42;33 42;42	45;24 	+22;39 +22;19	39;16 39;53	41;52 	45;01 	+22;45 +22;18	46;25 	+23; 0 +22;13
*4	al-dabarān	α Tau	59;22 59; 4	− 5;13 − 5;31	56;16 56;17	59; 7 	− 5;15 [M.] − 5;10 [Pt.] − 5;32	52;55 53;27	55;31 	58;40 	− 5;15 − 5;33	59;25 	− 5;10 − 5;36
5	rijl al-jawzā'	β Ori	65;30 66; 7	−31;25 −31;13	62;50 63;19	65;[41] 	−31; 4 [M.] −31;15	59;30 60;30	62; 6 	65;15 	−31; 4 −31;17	66;35 	−31;30 −31;22
6	al-'ayyūq	α Aur	71;10 71; 9	+22;40 +22;51	68;41 68;22	71;32 	+22;33 +22;51	65; 5 65;32	67;41 	70;50 	+22;50 +22;51	71;45 	+22;30 +22;50
7	yad al-jawzā'	α Ori	78; 0 78; 2	−16;50 −16; 8	74;55 75;15	77;46 	−16;50 −16;10	71;36 72;26	74;12 	77;21 	−16;45 −16;11	78;45 	−17;00 −16;17
*8	al-shi'rā al-yamāniya	α CMa	93;50 93;31	−39;20 −39;26	91;11 90;46	94; 2 	−39;30 −39;23	87;50 87;58	90;31 	93;40 	−39;20 −39;21	94;25 	−39;10 −39;12
*9	al-shi'rā al-sha'āmiya	α CMi	105;45 105;12	−16; 5 −15;52	102;21 102;27	105;12 	−16; 2 −16;10 [Pt.] −15;49	99; 0 99;40	101;36 	104;45 	−16; 0 −15;47	105;55 	−16;10 −15;39
*10	qalb al-asad	α Leo	139;14 139;10	+ 0;17 + 0;26	136;19 136;24	139;10 	+ 0;10 [Pt.] + 0;26	133; 0 133;35	135;36 	138;45 	+ 0;15 + 0;25	139;15 	+ 0;10 + 0;23
*11	al-simāk al-a'zal	α Vir	193;25 193; 8 ?	− 1;52 − 2; 0	190; 7 190;22	192;58 	− 2;10 − 1;59	186;48 187;33	189;24 	192;33 	− 2; 6 − 1;58	193;25 	− 2; 0 − 1;56
*12	al-simāk al-rāmiḥ	α Boo	193; 0 193;31	+31;25 +31;16	190;58 190;28 190;44	193;49 	+31;33 +31;13 +31;25	187;10 187;54	189;46 	192;55 	+31;12 +31;32	193;45 	+31;30 +32; 0

(*Continued*)

Table 10.1 (Continued)

Star names			Īlkhānid		Ibn Yūnus			Mumtaḥan Ibn/Ibn al-Aʿlam (IA)				Ptolemy	
			Longitude	Latitude	Longitude		Latitude	Longitude			Latitude	Longitude	Latitude
					Original	Īlkhānī zīj +2;51°		Original	IA +2;36°	Īlkhānī zīj +3;9°		Īlkhānī zīj Alm. +16;45°	
*13	qalb	α Sco	239; 2	− 4;10	236;16	239; 7	− 4;25	232;55	235;31	238;40	−4;24	239;25	− 4; 0
	al-ʿaqrab		**239; 3**	**− 4;28**	**236;16**		− 4;27	**233;27**			**−4;25**		− **4;20**
							− **4;26**						
14	raʾs	α Oph	251;15	+35;15	248;43	251;34	+35;59	245;18	247;54	251; 2	+36; 0	251;35	+36;00
	al-ḥawwāʾ			[. . .;**55**?]									
			251;41	**+35;59**	**248;54**		**+36; 2**	**246; 5**			**+36; 3**		**+36;12**
15	al-nasr	α Lyr	274;40	+61;50	272;17	275; 8	+61;45 [M.]	269; 0	271;36	274;45	+61;45	274; 5	+62; 0
	al-wāqiʿ		**274;34**	**+61;46**	**271;47**		**+61;47**	**268;58**			**+61;47**		**+61;50**
*16	al-nasr	α Aql	290;40	+29;15	287;43	290;34	+29;10 [Pt.]	284;18	287;49	290;58	+29;12	290;35	+29;10
	al-ṭāʾir		**290;56**	**+29;20**	**288; 8**		**+29;20**	**285;17**			**+29;21**		**+29;23**
17	al-ridf	α Cyg	324;30	+59;50	320;40	323;31	+59;38	318;24	321; 0	324; 9	+59;36	325;55	+60; 0
			324;46	**+59;57**	**322; 2**		+59;33	**319;15**			**+59;58**		**+60; 0**
							+59;58						
18	mankib	β Peg	348;30	+31;10	345;53	348;44	+31;12	342;20	345; 5	348;14	+31;10	348;55	+31; 0
	al-faras		**348;40**	**+31; 8**	**345;53**		**+31; 8**	**343; 4**			**+31; 7**		**+31; 7**

Notes

- [Pt.]: Ptolemaic; *Alm.: Almagest*; [M.]: Mumtaḥan.
- *Īlkhānī zīj*, P: f. 56v, T: f. 100r, M1: f. 100v, M2: f. 86v, C: p. 195, IT: f. 70v, L: f. 89v, F: f. 59v, B: f. 84r, O: f. 114r, Fl: 102r.
- The true modern values at the times are in bold cases.
- Original epochs: Ptolemy, the beginning of 885 Nabonassar/July 20, 137; Mumtaḥan, the beginning of 214 н/March 11, 829; Ibn al-Aʿlam, the beginning of 380 y/March 14, 1011; Ibn Yūnus, the end of 400 y/March 7, 1032.
- In the columns related to Mumtaḥan/Ibn al-Aʿlam, note that Ibn al-Aʿlam appears to have made the corrections only in the longitudes of the four stars θ Eri, α CMa, α Aql, and β Peg, as highlighted in the table; for the rest, his longitudes are equal to Mumtaḥan +2;36°. The variants in the case of the Mumtaḥan/ Ibn al-Aʿlam's table have already been given in Mozaffari 2016–2017.
- * The stars that were observed in the independent observational program carried out by Muḥyī al-Dīn al-Maghribī at the Marāgha observatory (cf. Table 9.3).

- **Īlkhānid**

Places: 1–3. In MSS. T, C, O, and Fl, the coordinates of β Cassiopeiae were incorrectly written down in the row related to θ Eridani. The coordinates of θ Eridani and β Persei are not presented in any manuscripts.

Latitudes

8. A strange value of −4;20 is given in most of the manuscripts (even in the authoritative MS. F, prepared under the supervision of al-Ṭūsī's son), except for O: −39;10 and B and IT: −39;20.
11. C, P, M2, IT, B: −1;12; F, L, O, Fl, T: −1;52. In M1, the value of −1;12 was corrected in another hand as −1;52.

Longitudes

13. T, O, Fl: 239;–, C: 239;42, B: 239;12, P, IT, F, L, M1, M2: 239;2.

- **Ibn Yūnus**

Latitudes

5. Dropped from all the consulted MSS, but only the integer degrees can be found in M1; Bernard (1684, p. 571) gives −31;35 (presumably a confusion with the Īlkhānid corresponding value).
6. All the consulted manuscripts of the *Īlkhānī zīj* (except for F) follow the same incorrect value of +22;3 as given in the *ʿUmda*; F has +22;53.
14. *Mukhtār zīj*, L: +35;19.
15. All the consulted manuscripts of the *Īlkhānī zīj* have +61;15 (quite probably misreading the value in the *ʿUmda*).

Longitudes

5. Dropped from all consulted MSS, but only the integer degrees can be found in M1; Bernard (1684, p. 571) gives a value of −65;*30*(seemingly a confusion with the Īlkhānid corresponding value).

6. *Īlkhānī zīj*, C: [*1st* ecliptic sign] *9*;32 (a scribal error: ـ → ـ), P, L: *1st* [ecliptic sign] *22*;*50* (confused with the Mumtaḥan value for the latitude of this star in the adjacent column related to Ibn al-Aʿlam).

17. *Mukhtār zīj*, L: 3*08*;40 (a scribal error: ـ → ـ).

This table was first brought to light more than three centuries ago by Edward Bernard (1638–1696 CE) in vol. 14 of the *Philosophical Transactions of the Royal Society*.[8] Bernard apparently made use of the extant manuscripts of the *Īlkhānī zīj* in Oxford.[9] Although the epochs he takes for the star tables of Ibn al-Aʿlam, Ibn Yūnus, and al-Ṭūsī (980, 996, and 1233 CE, respectively) are all erroneous, his reading of the *abjad* numerals are notably precise. This work was followed by E. B. Knobel.[10] Also, an excerpt of the non-Ptolemaic star table of the *Īlkhānī zīj* was published by the late E. S. Kennedy.[11]

The latitude values in the column ascribed to Ibn al-Aʿlam in the non-Ptolemaic star table of the *Īlkhānī zīj* are equal to the ones in the star tables included in the *Mumtaḥan zīj*; this work is a fruit of the observational programs established by early Islamic astronomers working in Baghdad and Damascus in the first half of the ninth century. The two star tables can be found in the extant manuscripts of the *Mumtaḥan zīj*: the first and, in all likelihood, original table contains the ecliptical coordinates of 24 stars; the longitudes are for (presumably the beginning of) the year 214 al-Hijra/March 11, 829/18 Bahman [11th Persian month] 197 ʏ.[12] In the second table, the ecliptical coordinates of 18 out of the 24 stars of the first table are listed for (presumably the beginning of) the year 380 ʏ/March 14, 1011,[13] and the longitudes in it (except for the four stars highlighted in Table 10.1) were updated from the first table by adding the increment of 2;36° due to the precession. This value is certainly the result of adopting the value 1°/70 Persian years for the rate of precession ((380−198)/70 = 2;36°).[14] The longitudes in the column ascribed to

---

8 Bernard 1684, p. 571. This reflects a peculiarity of the epoch when the astronomical data from the preceding centuries were not counted as purely historical materials, but as a part of the scientific literature of the day. On Bernard, see Mercier 1994, pp. 177–191.

9 On the history of the two manuscripts of the *Īlkhānī zīj* preserved in Oxford, the Huntington collection, nos. 143 (= our MS. O) and 144, see Mercier 1994, p. 163. Note that MS. no. 144 is not the original of the *Īlkhānī zīj* but a reworking by Shihāb al-Dīn al-Ḥalabī (d. 1455 CE).

10 Knobel 1875–1877, pp. 8–9, 11, 21–22.

11 Kennedy 1956, p. 170.

12 *Mumtaḥan zīj*, E: p. 188.

13 *Mumtaḥan zīj*, E: pp. 189–190, L: ff. 31v, 153r.

14 Note that the annual precessional motion in the *Mumtaḥan zīj* (E: p. 187) is equal to 0;0,54,44,20° per *Arabic* year, which is approximately equal to 1° in 66 *Arabic* (*not* Egyptian/Persian) years. On the probable Indian origin of this value, see Pingree 1964, p. 138, 1972, p. 29, 1976, p. 113.

Ibn al-Aʿlam in the non-Ptolemaic star list of *Īlkhānī zīj* were, in reality, derived from the second Mumtaḥan star table; the difference in longitude between the two amounts to 3;9°, which is, also, consistent with adopting the rate of precession of 1°/70^y ((601–380)/70 ≈ 3;9°). That al-Ṭūsī and his assistants engaged in compiling the *Īlkhānī zīj* attributed the second star table of the *Mumtaḥan zīj* to Ibn al-Aʿlam seems problematic, but elsewhere, we have settled a highly plausible hypothesis that the second Mumtaḥan star table can quite probably be a refinement made by Ibn al-Aʿlam of the first table on the basis of a few stellar observations made by himself dated to about 976 CE.[15]

Ibn Yunus's star table is preserved in the Cairo manuscript of Muḥyī al-Dīn's first *zīj* written at Marāgha, the *ʿUmdat al-ḥāsib*,[16] and in Abu ʾl-ʿUqūl's *Mukhtār zīj*, written in Yemen *ca.* 1300 CE.[17] These tables ascribed to Ibn Yūnus include the ecliptical coordinates of, respectively, 59 and 62 stars for the end of 400 Y/ March 7, 1032. In total, 18 of these stars are included in the non-Ptolemaic star table of the *Īlkhānī zīj*; the longitudes are converted to the epoch of the *Īlkhānī zīj* by adding the increment of 2;51° to those in the original. This value is in conformity with adopting the precession rate of 1°/70^y, since (601–401)/70 ≈ 2;51°. In seven cases (nos. 4, 6, 9, 12, 13, 14, and 17), the latitude values in all the consulted manuscripts of the *Īlkhānī zīj* and the Cairo manuscript of al-Maghribī's *ʿUmda* are not in agreement with the *Mukhtār zīj*. In one case (no. 15), the *Īlkhānī zīj* does not agree with the two others. It is likely that all these differences are due to simple scribal errors in alphanumerics. In five cases, it is uncertain which value should properly be selected, and hence, we record both values: for nos. 4, 9, 12, 13, and 17, the upper row indicates the value mentioned in the *Īlkhānī zīj* and Cairo manuscript of the *ʿUmda*, and the lower row shows that registered in the *Mukhtār zīj*. For no. 6, the value in the *Mukhtār zīj* is clearly preferable, while for no. 14, that written down in the *Īlkhānī zīj* and *ʿUmda* should be selected. For the longitude values, there are only two cases of divergence between the *Īlkhānī zīj*/ *ʿUmda* and the *Mukhtār zīj*: nos. 12 and 17. For the first, the Marāgha astronomers appear to have read the value of 190;58°, which is also registered in the Cairo manuscript of al-Maghribī's *ʿUmda*, in the archetype used by them; then, adding the increment of 2;51° to it, they obtained the value of 193;49° for the longitude of α Boo, but the *Mukhtār zīj* has 190;28° (كص ↔ كن). For the second, the value of 308;40° in the *Mukhtār zīj* is clearly a scribal mistake. The clear affinities in Ibn Yūnus's values for the ecliptical coordinates of these stars between the Cairo manuscript of al-Maghribī's *ʿUmda* and the *Īlkhānī zīj* reinforces B. van Dalen's hypothesis

---

15 See Mozaffari 2016–2017. The star table in the *Mumtaḥan zīj* and the other early Islamic star tables based on it were edited in Girke 1988, which was not published.

16 al-Maghribī, *ʿUmda*, M: f. 142v.

17 Abu ʾl-ʿUqūl, *Mukhtār zīj*, L: ff. 92v–93r. On Abu ʾl-ʿUqūl, see King 2004–2005, vol. 1, *passim*. An excerpt of the first has been published in King 2004–2005, vol. 1, p. 31. Ibn Yūnus's star table was edited in Girke 1988, but it was never published. A detailed analysis of it is in preparation by the present author.

that the Cairo manuscript of Muḥyī al-Dīn's *'Umda*, in its present form, appears to have been a side project of the work on the *Īlkhānī zīj*.[18]

10.3 Discussion of Accuracy

The 16 stars related to the observations at Marāgha should have been observed between the early 1262 CE, when the observations were begun there, and the early 1272 CE, when al-Ṭūsī dedicated the *zīj* to Abāqā, Hülegü's successor, and then departed Marāgha for Baghdad. There is no account of the observations of these 16 stars, and so nothing about their precise dates is available. The measured longitudes should have been reduced to the epoch of 1232 CE by subtracting some amount between 30/70° ($\approx$26′) and 40/70° ($\approx$34′) from them. Accordingly, if we assume that the longitudes in the 1260s were ideally error-free, this "down-dating" procedure would inevitably cause the errors in the range of $-30\cdot((1/70)-(1/71.6))°$ to $-40\cdot((1/70)-(1/71.6))°$, that is, about $-0;1°$, in the tabular longitudes (NB, the true rate of precession is equal to $1°/71.6^y$). Thus, in order to compare them with the true modern longitude values at that time, we round all the true longitudes for 1232 CE to the lower arc minute. The mean absolute errors (*MAE*) in latitude and longitude are, respectively, 12.5′ and 13.5′, and the mean errors $\mu \approx -0.4′$ and $-3.4′$, with the standard deviations $\sigma \approx 16.8′$ and 17.5′. The individual errors in longitude and latitude are shown, respectively, in Figures 10.1 and 10.2, in which the horizontal axis displays the true longitudes and the vertical axis indicates the

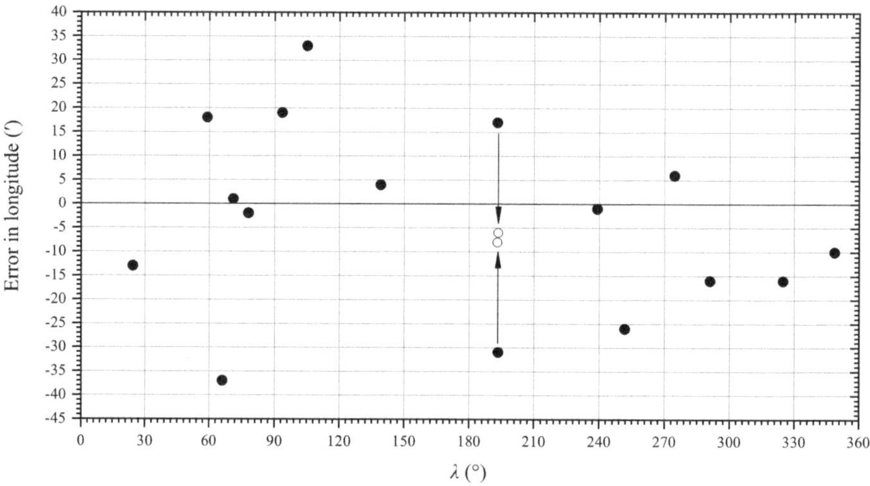

Figure 10.1 Errors in longitude in the Īlkhānid star table.

18 Private communication; it shall appear in van Dalen's *A New Survey of Islamic Astronomical Handbooks*, provisionally, entry C23.

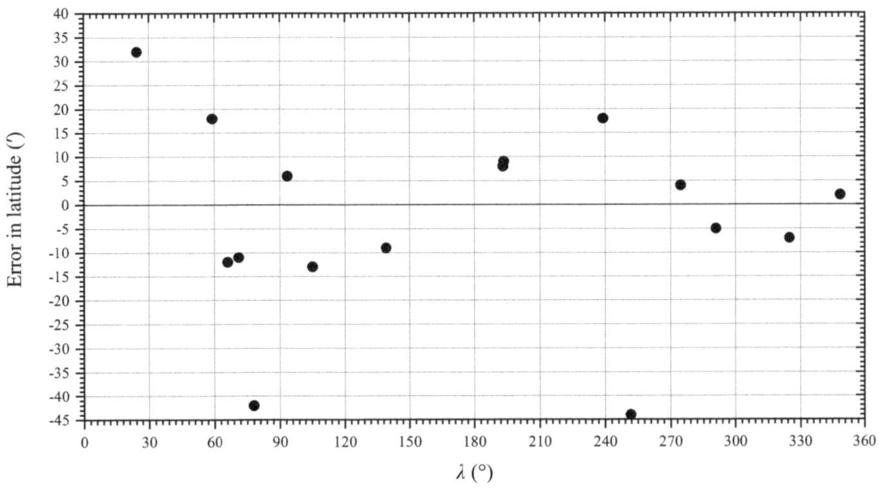

Figure 10.2 Errors in latitude in the Īlkhānid star table.

errors. The errors are purely randomly distributed over the longitudes, without showing any systematic pattern or any other specific peculiarity.

The Marāgha astronomers adopted Ibn Yunus's value for the latitude of α Ori and the Mumtaḥan values for the latitudes of β Cas, α CMa, and β Peg. The greatest difference in longitude between the Īlkhānid and Ptolemy's values are the case with β Ori (−1;5°) and α Cyg (−1;25°), whereas there is *no* difference in the case of α Vir and a negligible difference of −0;1° in the case of α Leo; also, the longitudinal differences are mostly negative, except for α Lyr and α Aql. The differences in latitude between the Īlkhānid and Ptolemy's values do not exceed ±0;10°, except for α Oph (−0;45°). If we assume that a scribal error was made in writing down the *abjad* numerals of the star's latitude in an original archetype, from which all the extant copies descended, so that the correct value of 35;5̲5° was replaced by the wrong tabular value of 35;*15*° (ﻪﻨ → ﻪﻴ), the difference in its latitude between the Īlkhānid and Ptolemy's values would amount to only −0;5°, which would then turn out to be compatible with the differences between the Īlkhānid and Ptolemy's values for the latitudes of the other stars.

It appears quite likely that the longitudes of α Vir and α Boo have been inadvertently swapped in an early key archetype, as indicated on the table. The table was clearly organized in ascending order of longitude. In reality, α Vir is, in longitude, in advance of α Boo, which is also evident in the values provided by Ptolemy, Mumtaḥan/Ibn al-Aʿlam, and Ibn Yūnus. If this is the case, the longitude errors for these two stars are almost identical (specifically, −8′ and −6′, instead of +17′ and −31′; see Figure 10.1).

For the latitudes of the 18 stars, in the case of Ptolemy, $MAE \approx 18.5′$, $\mu \approx −1.6′$, $\sigma \approx 23.2′$; in the case of the Mumtaḥan, $MAE \approx 13.2′$, $\mu \approx −0.5′$, $\sigma \approx 17.3′$; and in the case of Ibn Yūnus, $MAE \approx 14.3'/15.6'$, $\mu \approx -1.2'/-2.9'$, $\sigma \approx 18.2'/19.2'$.[19] Therefore, the Mumtaḥan and Marāgha observers achieved nearly the same degree of precision, slightly better than Ibn Yūnus's, in the determination of the stellar latitudes.

For the longitudes of the 17 stars (excluding θ Eri for having the striking outliers), in the case of the second Mumtaḥan star table, that is, Ibn al-Aʿlam's star table, $MAE \approx 37.6'$, $\mu \approx -34.1'$, $\sigma \approx 22.7'$, and in the case of Ibn Yūnus, $MAE \approx 17.2'/17.3'$, $\mu \approx -6.7'/-8.5'$, $\sigma \approx 25.5'/25.0'$ (note that his values are error-free for the four stars β Cas, α Tau, α Sco, and β Peg, and that the abnormal deviation of $-1;22°$ in the longitude of α Cyg, which exceeds the mean error by about 3σ, should be treated with caution, because of the confusion in reading this star's longitude). Therefore, it can be concluded that the Marāgha astronomers observed more exact values for the longitudes than their predecessors.[20]

The parallel observational program carried out by Muḥyī al-Dīn al-Maghribī at the Marāgha observatory has already been discussed in depth in Chapter 9. We have compared the accuracy of his observations of the eight bright stars (Table 9.3; also marked in Table 10.1 with asterisks) with the star table in the *Īlkhānī zīj* (see Section 9.6). Also, we have seen that he was not able to measure the coordinates of Vega (α Lyr) and Capella (α Aur) in the absence of a northern mural quadrant at the observatory; it is curious that the ecliptical coordinates of both the stars are found in the *Īlkhānī zīj*.

There are three crucial issues related to the Īlkhānid non-Ptolemaic star table that deserve attention, which shall be discussed in the following sections.

10.4 The Star Longitudes and the Rate of Precession

With the growth of astronomy in the early Islamic period, a *new astronomy* came into being in the Middle East by a trepidation model invented by Abū Jaʿfar al-Khāzin (900–after 970 CE), in collaboration with Ibrāhīm b. Sinān (909–946 CE), and a solar model constructed by an anonymous astronomer (preserved in Quṭb al-Dīn al-Shīrāzī's works).[21] Nevertheless, after the latter part of the tenth century, the medieval Middle Eastern astronomers unanimously rejected the idea that the displacement in the stellar longitudes consists of any motion other than a steady, continuous linear precession. The treatment of observational data and the lines of reasoning about the interpretation of them in the Eastern branch of Islamic astronomy were very often conducted toward the establishment of an astronomical system in which the parameters, including the planetary orbital elements and

---

19 In the case of Ibn Yūnus, we take 990 CE, nearly in the middle of his career, for the modern latitude values, although the star latitudes do not change so significantly in a narrow time span to noticeably affect our comparison.

20 Of course, it should be noted that a thorough analysis of Ibn Yūnus's star table should be postponed after it is fully edited, and then the observed coordinates of the remaining 41/44 stars can also be taken into consideration.

21 See Mozaffari 2016.

the rate of precession, remain constant over time, without assuming any secular or periodic variation; however, a good number of new values were observed for each and every one of them.

The epoch of Ptolemy's star catalogue in the *Almagest* is the beginning of the Antoninus era, that is, 885 Nabonassar (July 20, 137).[22] Because of an entirely faulty impression, early Islamic astronomers thought that Ptolemy borrowed his star catalogue from an alleged prototype compiled by Menelaus in Rome in 845 Nabonassar, four decades before Ptolemy. They thought that Ptolemy added a value of 0;25° to the longitude values in it (this increment is the result of using the precession rate of 1°/100^y), while all that Ptolemy mentions on Menelaus's contribution to stellar astronomy are the two observations of β Sco and α Vir. As a result, whenever they took into account Ptolemy's stellar longitudes in order to estimate the rate of precession, they decreased Ptolemy's longitudes by 0;25°, and then extended the interval of time to 40 years before the epoch of the *Almagest* star catalogue. This is taken into account, for example, by al-Battānī in his *Ṣābi'* *zīj* and by Muḥyī al-Dīn al-Maghribī in his *Talkhīṣ al-majisṭī* in the derivation of the rate of precession and is also followed by al-Ṣūfī in the *Ṣuwar al-kawākib*.[23] Concerning the medieval Middle Eastern severe doubts about the originality of Ptolemy's star catalogue in the *Almagest*, we should also mention Bīrūnī's groundless hypothesis in *al-Qānūn al-mas'ūdī* IX.3.3,[24] according to which "Ptolemy has added the difference in longitude of Regulus (α Leo) between his observation and that existing from Timocharis's time to the longitudes of all stars, which had already been registered (in an alleged catalogue of Timocharis?)."[25] However, Bīrūnī appears not to have been certain about the stellar magnitudes in the *Almagest* catalogue, because he earlier, in IX.2.1,[26] remarks that "it is equally probable that either Ptolemy registered the stellar magnitudes on the basis of what he discerned by observation, or he followed one of his predecessors, as he converted the longitudes of the stars to his time." Bīrūnī does not identify whom Ptolemy might have followed.

22 *Almagest* VII.4: Toomer 1998, p. 340.

23 al-Battānī, *Zīj al-Ṣābi'*, Chapter 51: Nallino [1899–1907] 1969, vol. 3, p. 187, al-Ṣūfī 1954, p. 25; Schjellerup 1874, pp. 42–43; Neugebauer 1975, vol. 1, p. 288. The value of 0;25° is, indeed, the difference in the longitudes of Spica (α Vir) and β Sco between the *Almagest* star catalogue and what Ptolemy reports from Menelaus (*Almagest* VII.3: Toomer 1998, pp. 336–338; Pedersen 1974, pp. 415–416, nos. 52, 53).

24 al-Bīrūnī 1954–1956, vol. 3, p. 997, F: f. 141v, B: f. 188v.

25 It is not known how Bīrūnī could think of Timocharis's alleged star catalogue as a prototype of Ptolemy's, while Ptolemy clearly states that he does not put any trust in the observations pertaining to Timocharis; for example, the statement in *Almagest* VII.3 (Toomer 1998, p. 329) that "the observations of the school of Timocharis are not trustworthy, having been made very crudely." The idea seems to have come from Abu 'l-Wafā' al-Būzjānī (940–997 CE); see Mozaffari 2023, pp. 481–484.

26 al-Bīrūnī 1954–1956, vol. 3, pp. 991–992, F: f. 140v, B: f. 187r.

Whenever the medieval Middle Eastern astronomers compared their measured stellar longitudes with those registered in Ptolemy's catalogue, always with an extension of them to Menelaus's time, they obtained the values close to 1° in 66 Persian years for the rate of precession; but when they compared their stellar longitudes either with those recorded by Ptolemy from his Greek predecessors or with those of their Islamic predecessors, they yielded the smaller values, very often about $1°/70^y$. The most important examples are briefly mentioned in the following.

The Islamic astronomers appear to have been aware of this dilemma in an early stage, as the first trace of it can be found in the Banū Mūsā's *On the solar year*. They observed Regulus in 199 Y (830–831 CE) and then compared their own (?) longitude of 133;2° (not so far from the Mumtaḥan value of 133;0° for 198 Y/829–830 CE) with the value of 119;50° determined by Hipparchus in 128–129 BCE;[27] the increment of 13;12° in longitude in the intervening 957 Egyptian/Persian years corresponds very nearly to the annual precessional motion of 0;0,49,39° or $1°/72.5^y$.[28] But when the same longitude of 133;2° is compared with Ptolemy's value of 122;30° measured on February 23, 139 CE,[29] the precession rate of 0;0,54,53° or about $1°/66^y$ results.[30] The Banū Mūsā employ these results in order to derive the length of the sidereal year by the comparison of their observations with Ptolemy's and Hipparchus's and then check the results against the corresponding value derived from Meton and Euctemon's observation. According to them, this procedure "demonstrates that the observation by Hipparchus is more correct [than Ptolemy's]."[31]

In his *Zīj al-Ṣābi'*, Chapter 51,[32] al-Battānī refers to his observations of β Sco, α Leo, and α CMa carried out at Raqqa in 1627 Nabonassar (879–880 CE). Then, assuming Menelaus as a source for Ptolemy's star catalogue, he computes the differences between his observed longitudes of these stars and Ptolemy's corresponding values minus 0;25°. The longitudinal differences are closely equal to 11;50°, which, divided by the time span of 782 years, yields the precession rate of 1° in 66 Persian/Egyptian years. Muḥyī al-Dīn compared his own derived longitudes of the eight bright stars (marked in Table 10.1) in 636–637 Y (2015–2016 Nabonassar) with Menelaus's alleged ones (i.e., *Almagest* − 0;25°) for 845 Nabonassar;

---

27 *Almagest* VII.2: Toomer 1998, p. 328, Pedersen 1974, p. 415, no. 46. Note that the earliest observation of Regulus Ibn Yūnus quotes from the Banū Mūsā comes back to 209 Y (840–841 CE), when they found its longitude to be equal to 133;49,40° (modern value at that time: **133;45°**); Ibn Yūnus also mentions (see note 32) the Banū Musā's other two values for the longitude of this star: 133;55,55° for 216 Y (847–848 CE) (**133;50°**) and 133;27° (. . .;57?) in 219 Y (850–851 CE) (**133;53°**). From these values, it is obvious that the accuracy of the Banū Mūsā's stellar observations was about ±5'.

28 Neugebauer 1962, p. 279, Goldstein 1994, p. 190.

29 *Almagest* VII.2: Toomer 1998, p. 328, Pedersen 1974, p. 420, no. 83.

30 Neugebauer 1962, p. 281–282.

31 Neugebauer 1962, p. 283. The phrase in the square brackets added for clarity is Neugebauer's.

32 Nallino [1899–1907] 1969, vol. 3, pp. 187–190.

the mean difference of 17;45° between the two sets of longitudes again results in the rate of precession of 1°/66^y in the 1,170 years between them. It is notable that in his first *zīj* written in Damascus, the *Tāj al-Azyāj* (*Crown of the zījes*), Muḥyī al-Dīn has the value 1°/72 Persian years for the rate of precession, which is the excellent medieval approximation to the true value of 1°/71.6^y.[33]

In his *Ḥākimī zīj*, Chapter 5,[34] Ibn Yūnus lists the 12 values for the longitude of Regulus measured by six individuals or groups of astronomers of the early Islamic period (e.g., the Mumtaḥan observers, the Banū Mūsā family, and the Banū Amājūr family), in the period from 826/7 CE to 975/6 CE, and nicely tries to demonstrate that all of them are in accordance with the precession rate of 1°/70^y. The last of them is Ibn al-Aʿlam's value of 135;6° for the longitude of Regulus for 365 H/344–345 Y (975–976 CE), which, compared to the longitude of 133;0° for 214 H/198 Y (829–830 CE) in the *Mumtaḥan zīj*, *strictly* results in the value of 1° in 70 Persian years for the rate of precession.[35] Ibn Yūnus's formal value for the rate of precession is 1°/70$^1/_4$ Egyptian/Persian years.[36] In his *al-Qānūn al-masʿūdī* VI.8,[37] Bīrūnī derives the value of 1°/70$^1/_3$^{y} on the basis of a comparison between Abu 'l-Wafā''s and Hipparchus's observations of Regulus made, respectively, in 974 CE and 128/129 BCE; also, he yields 16 cycles (of 360°) in 160,696,125 days (or 1° in 76.4 years) by comparing between his own and Timocharis's observations of Spica carried out, respectively, in July 2, 1009 CE, and March 9/10, 294 BCE.[38] The precession constant of 1°/70^y was dominant in the late Islamic period, as it is adopted in the official works connected to the two observatories established in Marāgha and Samarqand, that is, *Īlkhānī zīj* and Ulugh Beg's *Sulṭānī zīj* III.13, as well as in Ibn al-Shāṭir's *Jadīd zīj* and al-Kāshī's *Khāqānī zīj*.[39]

The main reason behind the derivation of a greater precession rate of 1°/66^y, whenever Ptolemy's stellar longitudes are taken into account and are extended to Menelaus's time, is that the longitudes in the *Almagest* star catalogue are often less than the true values for Ptolemy's time by about 1°.[40] For example, in the case of al-Battānī, his own and Ptolemy's values for the longitudes of the three stars and the errors ζ are as follows:

33 Dorce 2002–2003, p. 198, 2003, pp. 111, 180. The value of 1°/72^y can be found in the *Barcelona Tables* (*ca.* 1381) and also was independently measured in Italy or France in 1306 (as documented in a codex preserved in Vienna, no. 5311, f. 137r; see Goldstein 1994, pp. 193, 196–197; for the star tables in this manuscript, see Kunitzsch 1986).

34 Ibn Yūnus, *Zīj*, L: pp. 106–108; Caussin 1804, pp. 143–155.

35 NB. (135;6–133;0)/(345–198) ≡ 1°/70^y.

36 Ibn Yūnus, *Zīj*, L: p. 125.

37 al-Bīrūnī 1954–1956, vol. 2, pp. 676–678, F: ff. 100r–v, B: ff. 150r–v.

38 *Almagest* VII.3: Toomer 1998, pp. 335–336, Pedersen 1974, p. 410, no. 13. Bīrūnī finally adopted the value of 1°/69^y in order to construct his own table of the solar apogee's motion, which, also, serves for the precessional motion (al-Bīrūnī 1954–1956, vol. 2, p. 695).

39 Ulugh Beg, P1: f. 116v, P2: f. 132v. Ibn al-Shāṭir, *Jadīd zīj*, Chapter 19: O: f. 61v, L1: ff. 23r–v, L2: ff. 28v–29r, PR: ff. 30v–31r. Kāshī, *Khāqānī zīj*, IO: f. 166r.

40 The literature on the source(s) of errors in Ptolemy's star catalogue is vast; for example, for a detailed survey, see Evans 1987.

	al-Battānī	ζ_B	Ptolemy	ζ_P
β Sco	227;45° **227;35**	+0;10°	216;20° **217;17**	−0;57°
α Leo	134; 0 **134;17**	−0;17	122;30 **124;02**	−1;32
α CMa	88;50 **88;40**	+0;10	77;40 **78;29**	−0;49

It should be noted that the conversion of Ptolemy's longitudes to Menelaus's time by the use of Ptolemy's precession rate of $1°/100^y$ causes an initial error of $\zeta_0 = -8.5'$.[41] Al-Battānī's errors are all below 20′. The aggregate errors $\zeta = \zeta_B - \zeta_P + \zeta_0$, consisting of the individual errors ζ_B and ζ_P committed, respectively, by al-Battānī and Ptolemy and the initial error $\zeta_0 = -8.5'$ fall over the range of 58.5′ ± 8′, which causes an error of about +4.5″ in the annual rate of precession in the interval of time of 782 years; therefore, instead of the true value of $1°/71.6^y \approx$ 50.3″, the value of 50.3″ + 4.5″ = 54.8″ ≈ $1°/66^y$ results for the rate of precession. This situation is also correct for Muḥyī al-Dīn's analogous derivation within four centuries later; in this case, the mean aggregate error amounts to $\zeta \approx 1;30°$, which results in an error of about +4.6″ in the annual rate of precession.

The incompatibility of the rates of precession shown earlier can be seen in a slightly different way in the non-Ptolemaic star table in the *Īlkhānī zīj*. In order to bring stellar longitudes in agreement with each other for a given date, Ptolemy's longitudes needed to be adjusted by the rate of precession of $1°/66^y$, and those observed by Islamic astronomers by the rate of $1°/70^y$. The late Islamic astronomers in general and the Marāgha astronomers in particular should perhaps have seen two possible solutions for this problem: to accept either (1) that the rate of precession is not constant or (2) that there is something wrong with the longitudes in Ptolemy's star catalogue.

The variation in the rate of precession is mentioned in some cosmographic works of the Marāgha circle in connection with the variation of the obliquity of the ecliptic. However, none of them was considered seriously enough to be promoted to a practical level and to be embodied and consolidated in the medieval Middle Eastern astronomical system. This is because, in this system, the observed deviations and inconsistencies were attributed, for methodological reasons, to difficulties with observational methods and instruments (see, for example, the quotation from Muḥyī al-Dīn in the next section).[42] The other contributing factor was,

---

41 Note that in 40 years between Ptolemy and Menelaus, the precessional motion amounts to 40/71.6 ≈ 33.5′, which is, indeed, greater than the value 40/100 = 25′ derived from Ptolemy's rate of precession by 8.5′.

42 Some examples are briefly summarized in Mozaffari 2016. A further example can be added from one of Ibn al-Shāṭir's statements in the prologue to his *Jadīd zīj* (O: f. 2v): "[M]y disagreement with my predecessors on what in which there exists any difference [between my predecessors and I] is because of *observational necessities* and demonstrative precision." Emphasis is ours.

presumably, that all the Middle Eastern medieval astronomers were making use of the tropical year (i.e., the year of the seasons), the most fundamental parameter in Ptolemaic astronomy.[43] Consequently, they had to accept the constant values for the rate of precession; however, the fact that this type of solar year can by no means remain constant, because of the motion of the Sun's apogee, had been established, at least, since Bīrūnī's time, after its first appearance in the Banū Mūsā's *On the solar year*, and a concept of the "mean" solar year even emerged thereafter.[44]

However, there were also reasons that the second option was not embraced. It is generally true, as G. Grasshoff puts it,[45] that testing Ptolemy's star catalogue and estimating its systematic error require an adequate theory of precession. Nevertheless, that Ptolemy's values for the longitudes of, at least, the brighter zodiacal stars, which were utilized to derive the rate of precession, are not consistent with Timocharis', Hipparchus', and early Islamic astronomers' values could have been known to the Marāgha astronomers, or any medieval astronomer prior to them, without knowing the true rate of precession. Moreover, as any medieval astronomer could readily understand from Ptolemy's remarks in the *Almagest*, his star catalogue is intrinsically dependent upon his solar theory. Thus, whenever a discrepancy was detected in Ptolemy's solar theory, it could give rise to some doubts on the correctness of the longitudes in his star catalogue as well. It is interesting that Bīrūnī detects an error in Ptolemy's solar theory,[46] but he does not make any linkage between this discovery and any problem in the *Almagest* star catalogue in relation to the clear disagreement between the precession rate computed on the basis of it and those derived from the observations by Ptolemy's predecessors.

The main reason that such problems did not appear critical to the Islamic astronomers or why none of them was thought to be an inevitable consequence of the fact that the longitudes in Ptolemy's star catalogue are less than the correct values at his time appears, quite probably, to have been that, as mentioned earlier, the Islamic astronomers did not believe *at all* that the *Almagest* star catalogue is due to Ptolemy's observations but an update of an earlier star catalogue observed by Menelaus, as the majority of them thought, or Timocharis, as Bīrūnī did. Consequently, any possible discovery that the longitudes in it are less than the true

43 The term "tropical year" is usually used for the description of the concept of the solar year in astronomy from the ancient to the early modern time, while such solar years that are the time intervals between the two consecutive apparent passage of the Sun through an equinox or a solstice are the "season years" and have nothing to do with the tropical year according to the modern conception; see Meeus 2002, pp. 357–366.

44 Neugebauer 1962, pp. 283–285, al-Bīrūnī 1954–1956, vol. 2, pp. 668–669; also, see Hartner and Schramm 1961, p. 214f. That the lengths of the *true* solar years with respect to the ecliptic are varied is also mentioned in Naṣīr al-Dīn al-Ṭūsī's *Taḥrīr al-majisṭī* (*Exposition of the* Almagest) as a comment on *Almagest* III.1 (P1: p. 88, P2: f. 24v, P3: f. 41r; for the translation of the relevant passage, see Saliba 1987, p. 148).

45 Grasshoff 1990, pp. 17–22.

46 al-Bīrūnī 1954–1956, vol. 3, pp. 1193–1194.

values for Ptolemy's time could be attributed to the fact that Ptolemy made use of the rate of precession of $1°/100^y$ in order to update his star catalogue from a more ancient prototype. And since this value is less than the value of $1°/70^y$ – the frequently derived estimation of the medieval Middle Eastern astronomers, from about the latter part of the tenth century on – which they thought to be true, it should not come as a surprise that Ptolemy's longitudes are less than the true values in his era. After all, it was not the origin or validity of the *Almagest* star catalogue that was at issue in the medieval period; rather, the medieval astronomers needed to know the increment of precessional motion which, when added to Ptolemy's stellar longitudes, produced positions tolerably in agreement with their own observations.

Within one century after the Marāgha observatory, Ibn al-Shāṭir sets the seal on the achievement of all his Islamic predecessors by accepting the precession rate of $1°/70^y$, because "it is [the value derived] by the use of Hipparchus's [observational data]," as mentioned in the apparatus to the table of the precessional motion in his *Jadīd zīj*.[47] There is little doubt that this affirmative statement has a corresponding negative counterpart in the form that he denies the precession rate of $1°/66^y$, because it is the result obtained in association with Ptolemy's stellar longitudes. In the prologue to his work, he writes:

> Know that, in the investigation of the observations, I followed the nearest ways [i.e. simplest methods] and the longest intervals of time. In doing so, I used the ancient observations, with which all of the researchers agree, and which are taken from the observations of Hipparchus the virtuous and his predecessors, *not* the observations of Ptolemy.[48]

About one century later, in his commentary on al-Ṭūsī's *Taḥrīr al-majisṭī* (*Exposition of the Almagest*), Abd al-ʿAlī al-Bīrjandī (d. 1525/1526) casts a doubt of another kind on Ptolemy's star data in line with the general idea of medieval Middle Eastern astronomers about the *Almagest* star catalogue. The comment in question pertains to *Almagest* VII.3, devoted to Ptolemy's use of the ancient records of stellar declinations for demonstrating the axis and rate of the precessional motion. In this curious passage, we are told:

> It is probable that the longitudes and declinations of *those stars* were known from the observations of [i.e. were measured by] Hipparchus; but, Ptolemy knew [i.e., measured] only their declinations, but not their longitudes, by obtaining their meridian altitudes and taking the differences between them and the co-latitude of the place. Then, he added the difference between the two declinations [i.e., Hipparchus's and Ptolemy's

---

47 Ibn al-Shāṭir, *Jadīd zīj*, O: f. 28r, L1: f. 49v, L2: f. 62v, PR: ff. 61v, 99v.
48 Ibn al-Shāṭir, *Jadīd zīj*, O: f. 2v. Emphasis is ours.

values for the declination of a given star] to the declination of the longitude of the star as found by Hipparchus; and he derived the corresponding arc [i.e., longitude] of the resultant from the table of declination. Then, he found the second value for the longitude, which is more than the first [i.e., Hipparchus's]. The commentator [i.e., al-Bīrjandī himself] has mentioned it only as a possibility.[49]

Of course, it does not seem that al-Bīrjandī believed that this is true of Ptolemy's whole star catalogue but is only correct in the case of the six stars Ptolemy selected in order to prove his value for the rate of the precession, because the emphasized "those stars" in the first line appears to refer only to those six stars; moreover, the procedure described in this passage, as al-Bīrjandī also explains in detail, is only correct for the stars near either of the equinoxes, like the six stars taken into account by Ptolemy.

10.5 Changes in the Stellar Latitudes

No connection between the changes in stellar latitudes and the observed decrease in the obliquity of the ecliptic was established in the Marāgha astronomical tradition, because of an obvious reason: no set of latitude values given in the Īkhānid star table, either derived from the *Mumtaḥan zīj* (ascribed to Ibn al-Aʿlam) and Ibn Yūnus or observed by the Marāgha astronomers themselves, yields a pattern of differences from Ptolemy that could be attributed either to a steady decrease in the obliquity of the ecliptic (according to the Islamic observational programs carried out since the early ninth century) or to any imagined periodic variation in it. Of the nine northern stars observed by the Marāgha astronomers, five have the latitudes greater than Ptolemy's, and four less than them; and of the seven southern stars, three have the latitudes greater than Ptolemy's, and four less than them. Obviously, the differences are random;[50] in addition, as mentioned earlier, they do not exceed $\pm 0;10°$. If the deviations were owing to the variation in the tilt of the ecliptic, then entirely the opposite situations would have been taken place in the latitudes of the northern and southern stars. With a continuous decrease in the obliquity of the ecliptic, the latitudes of the northern stars in the region of longitude between 0° and 180° should always increase and, on the contrary, those of the southern stars in this region should always decrease. The diametrically opposite

---

49 al-Ṭūsī, *Taḥrīr*, P1: p. 220; al-Bīrjandī, *Sharḥ*, P: f. 228r. This statement, as rendered to a modern form, means that Hipparchus observed the declination δ_H and longitude λ_H of each star, from which the declination of the longitude $\delta(\lambda_H)$ can be computed. According to al-Bīrjandī, Ptolemy observed only the declination δ_P; then, in order to derive the longitude λ_P at his time, he took the relation $\delta(\lambda_P) - \delta(\lambda_H) = \delta_P - \delta_H$ to be correct (which Ptolemy actually did and which is nearly correct for the stars near the equinox points), and computed $\delta(\lambda_P) = \delta_P - \delta_H + \delta(\lambda_H)$, from which he derived λ_P by aid of a table of the declination of the longitudes (e.g., in *Almagest* I.15: Toomer 1998, p. 72).

50 Ibn al-Aʿlam's (i.e., Mumtaḥan) and Ibn Yūnus's latitude values show nearly the same random pattern as in the values observed by the Marāgha astronomers.

condition should be correct for the stars in the longitudinal zone between 180° and 360°. It is obvious from Table 10.1 that the five stars (nos. 4, 8, 11, 16, and 18) in the Īkhānid observations contradict this rule.

Muḥyī al-Dīn al-Maghribī precisely reached the same conclusion on the basis of a comparison of his derived latitude values with Ptolemy's. Of the five southern stars he observed in Marāgha, three (α Tau, α CMa, and α Sco) have the latitudes more than those given in the *Almagest* and two (α CMi and α Vir) have the latitudes less than them. But for the northern stars, two (α Leo and α Aql) have the latitudes greater than the *Almagest*, and the latitude of the remaining star (α Boo) is less than it (cf. Table 9.3). It is evident that the four stars α Tau, α CMa, α Vir, and α Aql contradict the conditions stated earlier.

In an interesting statement, he concludes:

> Their (i.e., the stars') latitudes as we found (i.e., measured) observationally do not differ/deviate from their latitudes as observed by the ancients, to the extent that would be significant. *They* [i.e., the differences/deviations] *are due to the observations, not owing to the* [proper/ secular] *motion.*[51]

This statement indicates a resemblance to the modern treatment of the minor deviations between the two sets of empirical data, some instances of which can, of course, be found in the works of his predecessors, notably in Bīrūnī's.[52] It does not appear to be far from the truth to assume that this constituted the main line of reasoning about the observed changes in the stellar latitudes in the Marāgha observatory, and the analogous conclusions were, also, made by the other Marāgha astronomers, as incidentally and implicitly reflected in the Īlkhānid non-Ptolemaic star table.

10.6 The Dresden Celestial Globe

The only surviving instrument connected with the Marāgha observatory is a small bronze celestial sphere made by Mohammad, a son of Mu'ayyad al-Dīn al-'Urḍī (d. 1266 CE), the instrument-maker of the first period of the Marāgha observatory.[53] Let us examine whether it is possible to establish a relation between the new ecliptical coordinates determined at the Marāgha observatory for the 16 stars and their positions on the surface of this globe.

In the analyses made from the beginning of the nineteenth century to the end of the last century, the various dates 1277, 1288, 1289, and 1304/1310 CE were estimated for its construction. The second is the most recent derivation made by G. Oestmann on the basis of the measurement of the ecliptical positions of 963

---

51 al-Maghribī, *Talkhīṣ* VII.2: ff. 114v–115r. The emphasis is ours.
52 For example, see Sheynin 1973, 1992, 1993.
53 On it, see, for example, Sezgin and Neubauer 2010, vol. 2, p. 52.

out of 997 stars marked on the globe. Oestmann gives the details of his measured longitudes and latitudes of 80 stars, for which there exists a mean longitudinal difference of 17;33° between the globe and Ptolemy's star catalogue. On a globe with a diameter of 144 mm, each degree on a great circle corresponds to an arc of length of 1.3 mm. This is so small that no precision better than 0;15°, corresponding to about 0.3 mm, can be expected in the drawing of the star marks on its surface. Oestmann claims to have been able to measure the longitudes to a precision of 0;10° and the latitudes with an accuracy of 0;15°. The differences in longitude he found between the *Almagest* star catalogue and the globe are between 15° and a bit more than 20°. The majority of the longitudinal differences fall in a zone between 17° and 18°, and thus, it seems quite reasonable to assume that the precessional increment considered in the construction of the instrument was not far from his derived mean value of 17;33°; however, it is better to take a date in the 1280s as the *lower limit* of the estimation.[54] The greatest number of the 80 stars considered by Oestmann, that is, 16 stars, are placed in the two groups with the longitudinal difference either of 18° or of 17;50° (8 stars each). Moreover, the four zodiacal stars α Tau, α Leo, α Vir, and α Sco, to which a special importance should be given, have, respectively, longitudinal differences of 18;10°, 18;00°, 17;50°, and 17;20° (with an average of 17;50°). Accordingly, it seems reasonable to take an *upper limit* somewhere in the 1300s for the epoch of the globe.[55]

In total, 13 of the 16 stars in the Īlkhānid table can be found in Oestmann's list of 80 stars engraved on the globe (the exceptions are α Lyr, α Cyg, and β Peg). The longitudinal difference between the globe and Ptolemy runs from +16;40° (α Oph and α Aql) to +19;50° (β Cas). The corresponding differences from the Īlkhānid table fall into the range between −0;10° (α Aql) and +3;25° (β Cas). Because of the scatter in the differences, the uncertainty of the epoch of the globe, and especially, its smallness, which make one unable to obtain a reliable scheme for the distribution of errors in longitude and latitude, it seems impossible to evaluate whether the longitudes of the 16 stars have anything with the new observations made at Marāgha in the 1260s. However, it does not seem too probable that the globe-maker concerned himself enough with these 16 stars to use their coordinates from the Īlkhānid table, while relying on the *Almagest* star catalogue for the other 981 stars.

---

54 See Oestmann 2002. From this value, the average longitudinal difference between the globe and the Ptolemaic star table of the *Īlkhānī zīj* would amount to 17;33° − 16;45° = 0;48°. We know that the Marāgha astronomers applied the rate of precession of 1°/66^y in the stellar longitudes associated with Ptolemy. Thus, the estimated date of the construction would be 1232 + 0;48 × 66 ≈ 1285 CE. With the precessional rate of 1°/70^y, the date would be 1288 CE, as Oestmann gives.

55 NB. 1232 + (17;50 − 16;45) × 66 ≈ 1304 CE or 1232 + (17;50 − 16;45) × 70 ≈ 1308 CE.

10.7 Conclusion

We scrutinized the comparative star table incorporated in the *Ilkhānī zīj*, including the ecliptical coordinates of 18 stars according to the *Almagest* star catalogue and the observations made by the Islamic astronomers from the early ninth century to the 1260s, including the Mumtaḥan group, Ibn al-Aʿlam, Ibn Yūnus, and the astronomers at the Marāgha observatory. Our analysis has shown that these Islamic observers attained nearly the same degree of accuracy in the measurement of the stellar latitudes, with a mean absolute error $\leq \frac{1}{4}°$, but the Marāgha astronomers, on the average, obtained more accurate values for the stellar longitudes, with a mean absolute error $< \frac{1}{4}°$ than their predecessors'. In the previous chapter, these results have been compared with the observations made by Muḥyī al-Dīn al-Maghribī simultaneously with, but independently from, his colleagues at the Marāgha observatory.[56]

Because of such methodological reasons as the awareness of the limitation of observational methods and of the observational difficulties and errors, no allowance was made in medieval Middle Eastern astronomy for any secular change, periodic variation, or proper motion in the parameters which are assumed to be constant in Ptolemaic astronomy. The Ilkhānid star table involves the two peculiar, complicated problems in this respect:

1. In order to convert the stellar longitudes to the epoch of the *Ilkhānī zīj*, the Marāgha astronomers used the precession rate of $1°/66^y$ for Ptolemaic longitude values, but $1°/70^y$ for those observed by their Islamic predecessors. It is a reflection of the fact that whenever the Muslim astronomers made their measurements of the rate of precession on the basis of Ptolemy's star catalogue, they obtained the first value, but when they employed the ancient Greek observations recorded in the *Almagest* or the longitude values observed by their medieval predecessors, they very often arrived at the latter value or other values close to it. This is because of the existence of a systematic error of about $-1°$ in Ptolemy's stellar longitudes. This problem and other issues entwined with it – most importantly, the possible change in the rate of precession over time – could not be critical to the Marāgha astronomers, because, following their medieval predecessors, they did not believe that the *Almagest* star catalogue is due to Ptolemy's observations but rather an update of a more ancient star catalogue.

2. The other problem arises from the various values observed for the stellar latitudes by the Islamic astronomers from the early ninth century to the thirteenth century. From an interesting and informative statement by Muḥyī al-Dīn al-Maghribī, we can be almost certain that the Marāgha astronomers neglected the minor differences found between various observational data,

---

56 Also, they can be compared with the accuracy of Ulugh Beg's star catalogue (see Shevchenco 1990, Krisciunas 1993, Verbunt and van Gent 2012).

which were attributed to observational difficulties and errors, instead of assuming any proper motion in latitude for the fixed stars.

Finally, we examined the Dresden celestial globe, the only instrument surviving from the Marāgha observatory, with the hope of finding a linkage between the ecliptical coordinates of the 16 stars observed by the Marāgha astronomers and the positions assigned to them on this representational device. The result is that no relation can be found between them.

References

Abu 'l-ʿUqūl, *al-Zīj al-mukhtār min al-azyāj*, MS. L: London, British Library, no. Or. 3624 (copied in 1009 H/1600–1601 CE).

Bernard, E., 1684, "The longitudes, latitudes, right ascensions, and declinations of the chiefest fixt stars according to the best observers: In a letter from Mr. Edward Bernard, to the reverend and learned Dr. Rob. Huntington provost of Trinity College near Dublin in Ireland", *Philosophical Transactions of the Royal Society* **14**, pp. 567–576.

al-Bīrjandī, Niẓām al-Dīn ʿAbd al-ʿAlī b. Muḥammad b. al-Ḥusayn, *Sharḥ* Taḥrīr al-majisṭī (*Commentary upon* al-Ṭūsī's *Exposition of the* Almagest), MS. P: Iran, Parliament Library, no. 6167.

al-Bīrūnī, Abū al-Rayḥān, 1954–1956, *al-Qānūn al-masʿūdī* (*Masʿūdic canons*), 3 Vols., Hyderabad: Osmania Bureau; MSS. F: Paris, Bibliothèque Nationale de France, no. Ar. 6840, B: Berlin, Staatsbibliothek zu Berlin, no. Or. Oct. 275 = Ahlwardt 5667.

Caussin de Perceval, J.-J.-A., 1804, "Le livre de la grande table hakémite, Observée par le Sheikh,. . ., ebn Iounis", *Notices et Extraits des Manuscrits de la Bibliothèque nationale* 7, pp. 16–240.

van Dalen, B., in preparation, *A New Survey of Islamic Astronomical Handbooks*.

Dorce, C., 2002–2003, "The *Tāj al-azyāj* of Muḥyī al-Dīn al-Maghribī (d. 1283): Methods of computation", *Suhayl* 3, pp. 193–212.

Dorce, C., 2003, "El Tāŷ al-azyāŷ de Muḥyī al-Dīn al-Maghribī", in: *Anuari de Filologia*, Vol. 25, Secció B, Número 5, Barcelona: University of Barcelona.

Evans, J., 1987, "On the origin of the Ptolemaic star catalogue", *Journal for the History of Astronomy*, 18, pp. 155–172 and 233–278.

Girke, D., 1988, *Drei Beiträge zu den frühesten islamischen Sternkatalogen. Mit besonderer Rücksicht auf Hilfsfunktionen für die Zeitrechnung bei Nacht*, Preprint Series No. 8, Frankfurt am Main: Institut für Geschichte der Naturwissenschaften.

Goldstein, B. R., 1994, "Historical perspectives on copernicus's account of precession", *Journal for the History of Astronomy* **25**, pp. 189–197.

Grasshoff, G., 1990, *The History of Ptolemy's Star Catalogue*, New York: Springer.

Hartner, W. and Schramm, M., 1961, "Al-Bīrūnī and the theory of the solar apogee: An example of originality in Arabic science", in: Crombie, A. C. (ed.), *Scientific Change. Historical Studies in the Intellectual, Social and Technical Conditions for Scientific Discovery and Technical Invention, from Antiquity to the Present*, Symposium on the History of Science, University of Oxford 9–15 July 1961, London: Heinemann, pp. 206–218.

Ibn al-Shāṭir, ʿAlāʾ al-Dīn Abu 'l-Ḥasan ʿAlī b. Ibrāhīm b. Muḥammad al-Muṭaʿʿim al-Anṣārī, *al-Zīj al-Jadīd*, MSS. K: Istanbul, Kandilli Observatory, no. 238, O: Oxford, Bodleian Library, Seld. A inf 30, D: Damascus, Asad National library, no. 3093; L1:

Leiden, Universiteitsbibliotheek, Or. 65; L2: Leiden, Universiteitsbibliotheek, Or. 530; PR: Princeton, Princeton University Library, Yahuda 145.

Ibn Yūnus, ʿAlī b. ʿAbd al-Raḥmān b. Aḥmad, *Zīj al-kabīr al-Ḥākimī*, MSS. L: Leiden, Universiteitsbibliotheek, Or. 143, O: Oxford, Bodleian Library, Huntington, no. 331.

al-Kāshī, Jamshīd Ghiyāth al-Dīn, *Khāqānī zīj*, MS. IO: London: India Office, no. 430; P: Iran: Parliament Library, no. 6198.

Kennedy, E. S., 1956, "A survey of Islamic astronomical tables", *Transactions of the American Philosophical Society*, New Series **46**, pp. 123–177.

King, D. A., 2004–2005, *In Synchrony with the Heavens: Studies in Astronomical Time-keeping and Instrumentation in Medieval Islamic Civilization*, 2 Vols., Leiden-Boston: Brill.

Knobel, E. B., 1875–1877, "The chronology of star catalogues", *Memoirs of the Royal Astronomical Society* **43**, pp. 1–23.

Krisciunas, K., 1993, "A more complete analysis of the errors in Ulugh Beg's star catalogue", *Journal for the History of Astronomy* **24**, pp. 269–280.

Kunitzsch, P., 1964, "Das Fixsternverzeichnis in der „Persischen Syntaxis" des Georgios Chrysokokkes", *Byzantinische Zeitschrift* **57**, pp. 382–409; reprinted in: Kunitzsch 1989, Trace II.

Kunitzsch, P., 1986, "The star catalogue commonly appended to the *Alfonsine Tables*", *Journal for the History of Astronomy* **17**, pp. 89–98; reprinted in: Kunitzsch 1989, Trace XXII.

Kunitzsch, P.,1989, *The Arabs and the Stars*, Northampton: Variorum.

al-Maghribī, Mūḥyī al-Dīn, *Adwār al-anwār*, MSS. M: Iran, Mashhad, Holy Shrine Library, no. 332; CB: Ireland, Dublin, Chester Beatty, no. 3665.

al-Maghribī, Mūḥyī al-Dīn, *Kitāb al-Zīj ʿUmdat al-ḥāsib wa ghunyat al-ṭālib*, MS. M: Cairo: Egyptian National Library, no. MM 188.

al-Maghribī, Mūḥyī al-Dīn, *Talkhīṣ al-majisṭī*, MS. Leiden: Universiteitsbibliotheek, Or. 110.

Meeus, J., 2002, *More Mathematical Astronomy Morsels*, Richmond: Willmann-Bell, Inc.

Mercier, R., 1994, "English orientalists and mathematical astronomy", in: Russell, G. A. (ed.), *The Arabick Interest of the Natural Philosophers in Seventeenth-Century England*, Leiden: Brill, pp. 158–214.

Mozaffari, S. M., 2016, "A forgotten solar model", *Archive for History of Exact Sciences* **70**, pp. 267–291.

Mozaffari, S. M., 2016–2017, "A revision of the Mumtaḥan and Ibn al-Aʿlam's star tables", *Suhayl* **15**, pp. 67–100.

Mozaffari, S. M., 2023, "A survey of Abu ʾl-Wafāʾ's solar and stellar observations", *Journal of Astronomical History and Heritage* **26**, pp. 469–488.

Nallino, C. A. (ed.), 1899–1907/1969, *Al-Battani sive Albatenii Opus Astronomicum*, Publicazioni del Reale osservatorio di Brera in Milano, n. XL, pte. I–III, Milan: Mediolani Insubrum. The Reprint of Nallino's edition: Minerva, Frankfurt, 1969.

Neugebauer, O., 1962, "Thabit ben Qurra 'On the Solar Year' and 'On the Motion of the Eighth Sphere'", *Proceedings of the American Philosophical Society* **106**, pp. 264–299.

Neugebauer, O., 1975, *A History of Ancient Mathematical Astronomy*, Berlin-Heidelberg-New York: Springer.

Oestmann, G., 2002, "Measuring and dating the Arabic celestial globe at Dresden", in: Dorikens, M. (ed.), *Scientific Instruments and Museums* (Proceedings of the XXth International Congress of History of Science, Liège, 20–26 July 1997, Vol. 16/De Diversis

Artibus: Collection de travaux de l'Académie Internationale d'Histoire des Sciences, Vol. 59), Turnhout: Brepols, pp. 291–298.

Pedersen, O., 1974, *A Survey of Almagest*, Odense: Odense University Press. With annotation and new commentary by A. Jones, New York: Springer, 2010.

Pingree, D., 1964, "Gregory chioniades and palaeologan astronomy", *Dumbarton Oaks Papers* **18**, pp. 133–160.

Pingree, D., 1972, "Precession and trepidation in Indian astronomy before A.D. 1200", *Journal for the History of Astronomy* **3**, pp. 27–35.

Pingree, D., 1976, "The recovery of early Greek astronomy from India", *Journal for the History of Astronomy* **7**, pp. 109–123.

Saliba, G., 1987, "The role of the *Almagest* commentaries in medieval Arabic astronomy: A preliminary survey of Ṭūsī's Redaction of Ptolemy's *Almagest*", *Archives Internationales d'Histoire des Sciences* **37**, pp. 3–20. Rep. Saliba 1994, pp. 143–160.

Saliba, G., 1994, *A History of Arabic Astronomy: Planetary Theories during the Golden Age of Islam*, New York: New York University.

Schjellerup, H. K. F. K., 1874, *Description des étoiles fixes composée au milieu du dixième siècle de notre ère, par l'astronome persan Abd-al-Rahman Al-Sufi: Traduction littérale de deux manuscrits arabes de la Bibliothèque Royale de Copenhague et de la Bibliothèque Impériale de St. Pétersbourg*, St. Petersburg: Eggers et Cie & H. Schmitzdorff, J. Issakof & Tcherkessof.

Sezgin, F. and Neubauer, E., 2010, *Science and Technology in Islam*, 5 Vols., Frankfurt: Institut für Geschichte der Arabisch-Islamischen Wissenschaften.

Shevchenco, M., 1990, "An analysis of errors in the star catalogues of Ptolemy and Ulugh Beg", *Journal for the History of Astronomy* **21**, pp. 187–201.

Sheynin, O. B., 1973, "Mathematical treatment of astronomical observations (A historical essay)", *Archive for History of Exact Sciences* **11**, pp. 97–126.

Sheynin, O. B., 1992, "Al-Bīrūnī and the mathematical treatment of observations", *Arabic Sciences and Philosophy* **2**, pp. 299–306.

Sheynin, O. B., 1993, "The treatment of observations in early astronomy", *Archive for History of Exact Sciences* **46**, pp. 153–192.

al-Ṣūfī, ʿAbd al-Raḥmān, 1954, *Ṣuwar al-kawākib al-thamāniya wa ʾl-arbaʿīn*, Hyderabad-Dn: Osmania Oriental Publications Bureau.

Toomer, G. J. (ed.), 1998, *Ptolemy's Almagest*, Princeton: Princeton University Press.

al-Ṭūsī, Naṣīr al-Dīn, *Īlkhānī zīj*, MSS. C: University of California, Caro Minasian Collection, no. 1462; T: University of Tehran, Ḥikmat Collection, no. 165 + Suppl. P: Iran, Library of Parliament, no. 6517 (Remark: The latter is not actually a separate MS, but contains 31 folios missing from MS. T. The chapters and tables in MS. T are badly out of order, presumably owing to the folios having been bound in disorder); P: Iran, Library of Parliament, no. 181; M1: Iran, Mashhad, Holy Shrine Library, no. 5332a; M2: Iran, Qum, Marʿashī Library, no. 13230; IT: Istanbul, Topkapı Sarayı Müzesi Kütüphanesi, Ahmet III, no. 3513/1 (up to f. 133v, copied for Timurid sultan Iskandar b. ʿUmar Shaykh Mīrzā I (1384–1415 CE), the ruler of central Iran from 1409 CE, who had a strong interest in knowledge and culture and was Jamshīd Ghiyāth al-Dīn al-Kāshī's (d. 22 June 1429 CE) patron, finished on 3 Ramaḍān 814/19 December 1411); L: Leiden, Universiteitsbibliotheek, Or. 75; F: Paris, Bibliothèque nationale de France, Persan 163 (copied by Aṣīl al-Dīn Ḥasan, the second son of Naṣīr al-Dīn al-Ṭūsī); B: Berlin, Staatsbibliothek Preußischer Kulturbesitz zu Berlin, Sprenger, no. 1853 (completed in 689 H/1290 CE); O: Oxford, Bodleian Library, Huntington, no. 143; Fl: Florence, Biblioteca Medicea Laurenziana, Or. 24.

al-Ṭūsī, Naṣīr al-Dīn, *Taḥrīr al-majisṭī* (*Exposition of the* Almagest), MSS. Iran, Parliament Library, P1: no. 3853, P2: no. 6357, P3: no. 6395.

Ulugh Beg, *Sulṭānī Zīj*, MS. P1: Iran, Parliament Library, no. 72; MS. P2: Iran, Parliament Library, no. 6027.

Verbunt, F. and van Gent, R. H., 2012, "The star catalogues of Ptolemaios and Ulugh Beg; Machine-readable versions and comparison with the modern HIPPARCOS Catalogue", *Astronomy & Astrophysics* **544**, p. A31.

Yaḥyā b. Abī Manṣūr, *Zīj al-mumtaḥan*, MS. E: Madrid, Library of Escorial, árabe 927, published in *The verified astronomical tables for the caliph al-Mamūn*, Sezgin, F. (ed.) with an introduction by Kennedy, E. S., Frankfurt am Main: Institut für Geschichte der Arabisch-Islamischen Wissenschaften, 1986, MS. L: Leipzig, Universitätsbibliothek, Vollers 821.

Part V

OBSERVATIONAL INSTRUMENTATION

11

GHĀZĀN KHĀN'S ASTRONOMICAL INNOVATIONS AT MARĀGHA OBSERVATORY[1]

11.1 Introduction

Everything we know about the observatory in Tabrīz founded by Ghāzān Khān (1271–1304), the seventh ruler of the Īlkhānid dynasty of Iran (r. 1295–1304), was published by Aydın Sayılı some 50 years ago.[2] Relying on primary historical sources, he presented an adequate description of its astronomical activities and of the many skills of Ghāzān Khān in the field of observational instrumentation.[3] Sayılı also presented a good overview of the Marāgha observatory from 1260 to 1283,[4] but due to a lack of reliable evidence for the period after ca. 1280, he made some statements – especially concerning Ghāzān Khān's astronomical innovations – that could not be substantiated. A recently discovered treatise has now revealed the exact type and location of Ghāzān Khān's innovations, which we enumerate in what follows. Our treatise begins where Sayılı left off and appears to give dependable information that we can use to illuminate the later period at the Marāgha observatory, during which very little is known concerning the type and extent of astronomical activity.

We begin by introducing Rashīd al-Dīn Ṭabīb's claim as to Ghāzān Khān's astronomical activities and innovations. We then examine the validity of the claim with the help of the newly discovered treatise, as well as verify the reliability of the information in the treatise as to the type, structure, and location of Ghāzān Khān's newly made instruments. For this purpose, it was necessary to examine Ghāzān Khān's innovations both in the context of the Marāgha astronomical tradition (as

1 Original publication: S. M. Mozaffari and G. Zotti, "Ghāzān Khān's astronomical innovations at Marāgha observatory," *Journal of the American Oriental Society* **132** (2012), pp. 395–425. © 2012 American Oriental Society, and republished by permission.

2 Sayılı 1960, pp. 224–232.

3 The primary sources include (1) Rashīd al-Dīn Ṭabīb, Faḍl Allāh al-Hamadhānī's *Jāmi' al-tawārīkh*, ed. 1994, vol. 2: pp. 1205f, (2) Fakhr al-Dīn Abū Sulaymān Dāwūd b. Tāj al-Dīn Abū l-Faḍl Muḥammad b. Dāwūd al-Banākitī's *Tārīkh-i Banākitī*, and (3) 'Abd Allāh Waṣṣāf al-Ḥaḍra al-Nīshābūrī's *Tārīkh-i Waṣṣāf.* For other important sources, see Mīrkhvānd's *Tārīkh-i rawẓa al-ṣafā* and Khvāndamīr's *Ḥabīb al-siyar.*

4 Sayılı 1960, pp. 187ff.

DOI: 10.4324/9781003481966-17

the dominant tradition of the time) and in the context of medieval observational instrumentation in general. We follow with a general classification of the instruments and argue their possible relation to later Western models, and we conclude by describing the instruments in the order in which they appear in the treatise.

11.2 Historical Background

According to his vizier, Rashīd al-Dīn Ṭabīb (d. 718/1318), Ghāzān Khān was a prominent artisan, an alchemist, an expert in medicine and botany (he invented a new antitoxin called *tiryāq-i ghāzānī*), and a mineralogist, as well as being interested in theology.[5] In his youth, he was taught by Mongol Buddhist monks, but he later converted to Islam.[6] Upon his victory over the Mamlūk army at the battle of Wādī l-Khaznadār, Ghāzān arrived in Marāgha, the site of the renowned observatory built in the thirteenth century under the direction of Naṣīr al-Dīn al-Ṭūsī, where he resided from 15 Ramaḍān until 24 Shawwāl 699 (June 4–July 13, 1300).[7] After he arrived,

> [o]n the next day, he went to watch the observations; he looked at all the operations (*aʿmāl*) and instruments, studied them, and asked about their procedures, which he understood in spite of their difficulty. He ordered the construction of an observatory next to his tomb in *Abwāb al-Birr* [in the district] of *al-Shām* in Tabrīz[8] for several operations. He clearly explained how to do those operations so that local wise men marveled at his intelligence, because such work (*ʿamal*) had not been done in any era. Those wise men said that constructing it [the observatory] would be extremely difficult. He guided them, whereupon they commenced building it and they finished it per his instructions. Those wise men and all the engineers agreed that nobody had done such a thing before nor had imagined doing it.[9]

Thereafter:

> On several occasions he went to Marāgha, asked for an explanation of the instruments there, examined their configuration (*kayfiyya*) carefully,

---

5 Rashīd al-Dīn, *Jāmiʿ al-tawārīkh*, vol. 2, pp. 1331–1341, 1348–49; see also Sayılı 1960, p. 227.

6 On Ghāzān, see Sykes 2003, pp. 110ff., Lambton 1988, pp. 324ff., Halm 2004, pp. 62ff.; on his conversion to Islam, see Morgan 2007, pp. 185, 196.

7 Rashīd al-Dīn Ṭabīb, *Jāmiʿ al-tawārīkh*, vol. 2, p. 1296. Khvāndamīr (*Ḥabīb al-siyar*, vol. 3, p. 154) has Ghāzān staying in Marāgha until Dhū l-ḥijja 699 (September 1300), but according to Rashīd al-Dīn Ṭabīb, Ghāzān left Marāgha for Ujān on Tuesday, 24 Shawwāl (July 13).

8 A rural area south of Tabrīz where Ghāzān founded from 696 to 702 (1297 to 1302–1303) a gigantic dodecahedral tomb around which were built 12 charitable and scholarly buildings (including the observatory). See Rashīd al-Dīn Ṭabīb, *Jāmiʿ al-tawārīkh*, vol. 2, pp. 1377–1384; al-Nīshābūrī, *Tārīkh*, pp. 229–231; Sayılı 1960, p. 226. Ghāzān himself drew the plans for this complex (Rashīd al-Dīn Ṭabīb, *Jāmiʿ al-tawārīkh*, vol. 2, p. 1376).

9 Rashīd al-Dīn Ṭabīb, *Jāmiʿ al-tawārīkh*, vol. 2, p. 1296; al-Banākitī (*Tārīkh*, p. 463) records nothing about his order to construct the Tabrīz observatory.

and studied them. He had a general idea of them. As per his nature (*ṭab*), everything having to do with the siting and the building of the [Marāgha] observatory he commanded to construct [likewise].[10]

Rashīd al-Dīn Ṭabīb also reported that he ordered the construction of a hemispherical instrument (*gunbad*) for solar observations at Tabrīz observatory and described it in technical detail for his astronomers.[11]

11.2.1 The Marāgha Observatory

The Marāgha observatory was built in 1259 by Hülegü (d. 1265), the founder of the Īlkhānid dynasty; for the circa 58 years it functioned, it represented the acme of Islamic astronomy. We have divided the astronomical activities in Marāgha into two distinct periods: the first from its beginnings to 1283, the second from *ca.* 682/1283 to 716/1317.[12] Even before the construction of the observatory, it appears that observations had been undertaken in Marāgha: in his treatise on the astrolabe (*Fī kayfiyyat al-tasṭīḥ al-basiṭ al-kurī*), Ibn al-Ṣalāḥ al-Hamadhānī (d. 1153) wrote that in Marāgha he found the magnitude of 23;35° for the obliquity of the ecliptic (*mayl al-kullī*)[13]; there also appeared three important *zīj*es (astronomical handbooks or tables) that later had bearing on work undertaken during the second period: al-Khāzinī's *Zīj al-sanjarī* (510/1116),[14] al-Fahhād's *ʿAlāʾī zīj* (*ca.* 1166),[15] and the *Zīj al-shāhī* of Ḥusām al-Dīn al-Sālār, who was killed by order of Hülegü on 8 Muḥarram 661/November 22, 1262.[16] The first two were translated into Greek, while the third was mentioned in these translated texts.[17]

During the first period of the observatory, two prominent *zīj*es were written: al-Ṭūsī's *Zīj-i īlkhānī* in Persian and Muḥyī l-Dīn al-Maghribī's *Adwār al-anwār* in Arabic (completed in 1275).[18] Both were reliably quoted in the second period, by works either connected to the observatory, such as al-Wābkanawī's *Zīj al-muḥaqqaq*

---

10 Rashīd al-Dīn Ṭabīb, *Jāmiʿ al-tawārīkh*, vol. 2, p. 1340; Sayılı 1960, p. 228.

11 Rashīd al-Dīn Ṭabīb, *Jāmiʿ al-tawārīkh*, vol. 2, p. 1340.

12 The selection of the initial date of the second period was made on the basis of the death of Muḥyī l-Dīn al-Maghribī, the last prominent scholar of the first scientific circle at Marāgha, who worked independently of Naṣīr al-Dīn al-Ṭūsī's circle, in Rabīʿ I 682/June 1283. For the final date, there is no evidence of observations being made in the observatory between 1317 and 1339, when the observatory lay in ruins (Sayılı 1960, pp. 212f.).

13 See Lorch 2000, p. 401. Ibn al-Ṣalāḥ, f. 62r: *wa-huwa ʿalā mā wajadnāhu bi-l-raṣad bi-l-Marāgha 23 juzʾ wa-35 daqīqa.*

14 See Kennedy 1956, p. 7, no. 27, King and Samsó 2001, p. 45 (the date given as 1150 is wrong). Recently, this Greek translation was edited by Leichter (2004).

15 al-Fahhād al-Dīn Abū l-Ḥasan ʿAlī b. ʿAbd-al-Karīm al-Fahhād of Shīrwān. See Kennedy 1956, p. 14, no. 84, King and Samsó 2001, p. 45, Pingree 1985, pp. 7–8, Mozaffari 2019, 2023.

16 Kennedy 1956, no. 32.

17 Pingree 1985, p. 53. Both al-Wābkanawī (*Zīj*, A: f. 3r, B: ff. 4r–5v) and al-Kamālī (*Zīj-i ashrafī*, f. 117r) referred to them.

18 King and Samsó 2001, p. 45.

al-sulṭānī[19] and Greek translations made by Gregory Chioniades,[20] or independent of it, such as al-Kamālī's *Zīj-i ashrafī*, completed by the end of the Persian year 681 (i.e., March 12, 1303) in Shīrāz.[21] The instrument-maker in the first period was Mu'ayyad al-Dīn al-'Urḍī (d. 1266) (see later).

In the second period, there were at least five outstanding scientists in the field of mathematical science connected with Ghāzān Khān's court:[22]

1. Quṭb al-Dīn al-Shīrāzī, who apparently spent his time in a solitary manner between 1283 and 1311 in Tabrīz. He revised his *Zīj-i riḍwānī* in 690/1291–1292.[23]

2. Shams al-Dīn Muḥammad al-Wābkanawī al-Bukhārī, the author of *Zīj al-muḥaqqaq al-sulṭānī*. In this work we are confronted with one of the most pre-eminent Islamic *zīj*es produced from direct observations. He sharply criticized al-Ṭūsī's *Zīj-i īlkhānī* for the reason that the calculated positions of the seven planets were never in agreement with actual observations, and because it was only a copy of earlier *zīj*es, especially in the fundamental planetary parameters. He referred to Muḥyī l-Dīn al-Maghribī's *Adwār* as being "based on the new Īlkhānid observations" (i.e., Muḥyī l-Dīn's own observations) for the sake of making a distinction between it and *Zīj-i īlkhānī*, which was assumed to be obtained through the "Īlkhānid observations" (i.e., the observational plans supervised by al-Ṭūsī and performed by his colleagues).[24]

Al-Wābkanawī introduced some new topics, such as a rule for conjunctions between Jupiter and Saturn, the report of their triple conjunction in 1305–1306 and their single one of January 1286, and the exact report of the annular eclipse of January 30, 1283, in detailed numerical values.[25] The period of his observations, as he himself says, extended over 40 years.[26]

---

19 See van Dalen 2007, Kennedy 1956, p. 8, no. 35, King and Samsó 2001, p. 46; and Chapter 3.

20 Pingree 1985, pp. 52–53.

21 Kennedy 1956, no. 4, King and Samsó 2001, p. 44.

22 For some of these facts, see Sayılı 1960, pp. 211ff. We only add some new results in what follows. For further information on the astronomers mentioned, see Rosenfeld and Ihsanoğlu 2003, with corrections given in Rosenfeld 2004; *Dictionary of Scientific Biography*; Hockey *et al.* 2007.

23 Kennedy 1956, p. 4, no. 13.

24 For example, al-Wābkanawī, *Zīj*, Book 3, Section 3, Chapter 1: A: f. 53r, B: f. 96r; Book 3, Section 9, Chapter 5: A: f. 60r, B: f. 108v; Book 3, Section 13, Chapter 6: A: f. 67r, B: 120v; and many other places. Since al-Wābkanawī rightfully contends that *Zīj-i īlkhānī* is based on the earlier astronomical tables rather than obtained from independent observations, he sometimes goes further, stating that only Muḥyī l-Dīn's *Adwār* are the "Īlkhānid Observations" (al-Wābkanawī, *Zīj*, prologue: A: f. 3r; B: 4v). This difference between the two *zīj*es became a standard and was repeated in later periods (e.g., see Sayyid Muḥammad [*fl. ca.* 1400], *Latā'if al-kalām*, pp. 322–324).

25 See Chapter 3.

26 The earliest observation mentioned in his *zīj* is that of the Moon on December 3, 1272 (IV, 14: A: ff. 89v–90r; B: f. 155r). His latest observation is that of the triple conjunctions of Jupiter and Saturn during 1305–1306 (V, 1, 4: A: 125r; B: 235r). He states that he made his monumental work by order of Ghāzān Khān and dedicated it to Sulṭān Abū Saʿīd Bahādur (the ninth Īlkhānid ruler,

3. Shams al-Dīn Muḥammad b. Mubārakshāh Mīrak al-Bukhārī al-Hirawī (d. 1340).

4. Niẓām al-Dīn al-Nīsābūrī, who is wrongly known as the author of *Zīj al-ʿalāʾī*, actually written by al-Fahhād.

5. Shams al-ʿUbaydī, a mathematician.[27]

Also during this period, Gregory Chioniades (d. *ca*. 1320) was in Tabrīz, where he translated al-Khāzinī's *Zīj al-sanjarī*, al-Fahhād's *Zīj al-ʿalāʾī*, and a text on the *ʿilm al-hayʾa*[28] into Greek.[29] According to Chioniades, he used the oral instructions of a person named Σάμψ Πουχαρής, born at Bukhārā on June 11, 1254,[30] who is most likely Shams al-Dīn Muḥammad al-Wābkanawī al-Bukhārī.[31] In his *Asʾila wa-ajwiba* (*Questions and answers*), Rashīd al-Dīn Ṭabīb answered the theological questions of a "Frankish sage" (*ḥakīm-i farang*),[32] who is probably the same Chioniades.[33]

---

27 He wrote three works: *Sharḥ-i maṭāliʿ*, *Matn-i Uqlīdus*, and *Risālat al-ḥisāb*. Ḥamd Allāh al-Mustawfī, *Tārīkh-i guzīda* (written *ca*. 730/1329–1330); Khvāndamīr, *Ḥabīb al-siyar*, vol. 3, p. 191.

28 See Paschos and Sotiroudis 1998.

29 We know for certain that he spent several years between 1295–1297 and 1310–1314 in Tabrīz. See Westerink 1980. Since Pingree (1985, p. 22) noted that he was in Constantinople in 1302, and Ghāzān Khān received envoys from Emperor Andronicus II in September 1302, Chioniades was probably among them.

30 Pingree 1985, pp. 16–17.

31 According to Pingree (1985, P. 15), this name should be read as Shams al-Bukhārī, but he distinguished him from both al-Wābkanawī (each now has a separate entry in the recently published *Biographical Encyclopedia of Astronomers*) and Shams al-Dīn Mīrak al-Bukhārī, who died in 1340. On the other hand, Kennedy and Kunitzsch believed that some tables existing in Greek manuscripts and attributed to Shams al-Bukhārī (see Pingree 1985, pp. 23–29) are fragments of al-Wābkanawī's *Zīj* (Kennedy 1956, p. 8, no. 35, Kunitzsch 1964). A comparison of Chioniades's Greek *Revised Canons* and al-Wābkanawī's *Zīj* shows that several fragments of these Greek translations are direct and faithful translations of the statements of al-Wābkanawī in his own *Zīj*. For an example, see Pingree 1985, pp. 307, 311, and al-Wābkanawī's *Zīj*, B: f. 130r. According to Chioniades, Shams wrote a treatise about the astrolabe dedicated to Emperor Andronicus II, although no manuscript of this treatise in Arabic or Persian has survived. It now appears likely, however, that this treatise is the one attributed to al-Wābkanawī (MS Turkey, Topkapı Saray, no. 3327/4). An exact comparison between the Persian (al-Wābkanawī's *Zīj* and his treatise on the astrolabe) and the Greek manuscripts could clarify this identification.

32 Rashīd al-Dīn Ṭabīb, *Asʾila wa-ajwiba*, vol. 1, pp. 28–50, vol. 2, pp. 52–94.

33 This assumption – along with the dedication of a treatise to Andronicus II (see earlier, no. 31) – clearly shows a variety of cultural relations existing between these dynasties, more prevalent than the scientific relations between Īlkhānid and European rulers during the first period of Marāgha activities. For a study of international scientific relations of the first period, see Comes 2004.

11.3 The Treatise

The anonymous Persian treatise studied in this chapter is titled *Risālat al-ghāzāniyya fī l-ālāt al-raṣadiyya* (*Ghazan's treatise on observational instruments*) and provides entirely new information about Ghāzān's approach to observational instruments and his innovations in this field. We find a full description of 12 instruments allegedly invented by him which were employed at the observatory at Marāgha instead of at his "new" observatory in Tabrīz (our author clearly says this, and he also has included tables of oblique ascensions for the latitude of Marāgha).

Three copies of this treatise are preserved in libraries in Tehran[34] – in Sipahsālār Library (no. 555D, ff. 15v–49v, henceforth *S*), in the library of the Parliament (no. 791, pp. 29–97; henceforth *P*), and – an incomplete one – in Malik National Library (no. 3536, pp. 41–56; henceforth *M*).[35] The manuscript collections of which S and P are a part are similar in many aspects; they appear to have been copied by a single scribe, with the title *ra'īs al-kuttāb*, on Thursday, 23 Jumādā II 1294 (July 5, 1877).[36]

After praising God and introductory words on the necessity of astronomical instrumentation, the treatise begins with the description of the five instruments mentioned in Ptolemy's *Almagest*: *ḥalqa nuḥāsiyya* (two meridian rings, I, 12); *lubna* (quadrant, I, 12); another *ḥalqa nuḥāsiyya* (equinoctial ring, III, 1); *dhāt al-ḥilaq* (armillary sphere, V, 1); and *dhāt al-shu'batayn* (parallactic instrument, V, 12).

The description of no. 3 makes clear that it has been wrongly given the name of no. 1 in the list (in all MS copies); it should be correctly named *ḥalqat al-i'tidāl*. Although both instruments were mostly made of copper (*nuḥās*), only the instrument known as "two meridian rings" was customarily called *ḥalqa nuḥāsiyya*.[37]

34 Two other copies – unseen – are in the Sharqi Asafiya Library, Hyderabad, and in the Egyptian National Library, Cairo; see Storey 1958, vol. 2, part 1, p. 64, no. 96, and King 1986, p. 166. Our thanks to the *JAOS* reviewer who brought these to our attention.

35 In M, the prologue of the treatise is wrongly attributed to Ghiyāth al-Dīn Jamshīd al-Kāshī (d. 1436), probably due to its similarity with al-Kāshī's *Sharḥ-i ālāt-i raṣad* (*Description of observational instruments*), completed in Dhū l-Qaʿda 818/January 1416. Al-Kāshī's two treatises are found in S and M before this treatise: *Risāla fī istikhrāj jayb daraja wāḥida* (*Treatise on calculating the sine of one degree*; S: ff. 1v–8v) and *Sharḥ-i ālāt-i raṣad* (*Description of the observational instruments*; S: ff. 9v–14v; M: pp. 31–39); see Kennedy 1961. For al-Kāshī's *Zīj*, see Kennedy 1956, no. 12.

36 Both S and P include four treatises: the aforementioned two works by al-Kāshī, the treatise presented here, and an anonymous treatise on mechanics. In both S and P, the handwriting, the figures, the scribal errors, and the repetitions are identical. The folios have been paginated in an appreciably similar fashion. The only difference is that the last two tables in S (nos. 6 and 7 in Section 11.3.1, later) are filled in, while the others, as well as all the tables in P, are blank. Although the date of copy appears as Thursday, 23 Jumādā II 194 (S: f. 49r; P: p. 96), on the opening page of P (p. 1), the scribe has mentioned explicitly his name and the date of copy as 1294 and set his own seal below it. Since the two manuscripts are identical, when we refer herein to S, P can be read as well.

37 In his treatise, al-Kāshī designated *ḥalqat al-i'tidāl* as *ḥalqa iskandariyya* ('Alexandrian circle'), which is connected with Ptolemy's famous observation (*Almagest* III.1) which our author relates: "as happened to the author of the *Almagest* [i.e., Ptolemy] in Alexandria's 'Gymnasium' or 'Palaestra' (παλαίστρά, *al-riwāq al-mal'ab al-iskandariyya*), the [Sun's] light appeared in the [equinoctial]

The author of the treatise reviews the classical instruments and rejects the adequacy of each. In the case of the equinoctial ring, he repeats (S: ff. 17r–v, M: p. 44) the same difficulty that Ptolemy mentioned in the *Almagest* (III.1), namely, that the very weight of the ring causes it to veer from its true position of angle $90-\varphi$ in respect to the horizon.[38]

In the case of the armillary sphere, used for measuring the longitude of any star, he notes that we must first have the coordinate of a reference star. In the *Almagest* (V, 1), Ptolemy described a method by which one considers the position of the Sun determined beforehand (from the tables based on the solar theory) as the primary reference and goes on to measure the unknown coordinate of a given star with the Moon's position as an intermediate reference.[39] Thus, says our author, the measurements will be approximate, not certain, or not completely derived from observation. The Sun's true position can be obtained by the armillary sphere, but Ptolemy made no mention of this. To do this, the instrument should be set in its correct position (regarding the geographical latitude, the zenith [vertical orientation], and the placement of the instrument in the meridian plane). The ecliptic ring of the instrument should be placed in such a position that its upper, sunward limb obscures its lower limb, so that the ecliptic ring's inner surface will be in shadow. In this case, the instrumental ecliptic will be in the plane of the true ecliptic. Now the Sun's position can be obtained by placing the outer latitudinal ring in line with the Sun.

This method has some practical difficulties – as is evident, in this method only an object on the ecliptic can be used as the primary reference. But only the Sun is both on the ecliptic and so luminous that it can place the inner surface of the instrumental ecliptic in shadow, thus determining whether or not the ecliptic ring is exactly in its true position. (Accordingly, if we do not use the Sun, we need a source giving us the beforehand determined position of such an object.) That the reference star must be an object on the ecliptic (which should be the Sun) causes some limitations in utilizing the instrument. This is where the treatise's author criticizes the instrument (S: ff. 18r–v, M: p. 45). In his judgment, that procedure will make the measurements approximate (*bī taqrīb*), not certain (*bī taḥqīq*).[40]

---

circle twice in one equinox" (S: ff. 17r–v; M: p. 44). (It should be noted that "Palaestra" was not a stoa, *riwāq*, as our author claims.) Some lines earlier, our author speaks of *al-riwāq al-murabba'*, which is identical to "square, τετραγών, stoa" in *Almagest*; see Toomer 1984, p. 133, n. 7, and p. 134). As stated in *Almagest* III, 1, Ptolemy had "one" equinoctial circle in the square stoa and some in the Palaestra. The terminology of our treatise is adopted from the Arabic translation of *Almagest* (Arabic *Almagest*, ff. 28r–v). It is interesting that the passage in question from the *Ghāzānid treatise* is marginally quoted in an extant manuscript of al-Marrākushī's *Jāmi' al-mabādī wa 'l-ghāyāt fī 'ilm al-mīqāt* (Istanbul, Süleymaniye Library, Nuruosmaniye Collection, no. 2902, f. 153r).

38 This problem was also noted by the thirteenth-century astronomer Mu'ayyad al-Din al-'Urḍī (d. 1266), who introduced a solution for preventing this insufficiency by placing the ring into a larger ring erected in the plane of the meridian for supporting the ring. See Seemann 1928, pp. 57ff., instrument IV.

39 Toomer 1984, p. 219.

40 See also Wlodarczyk 1987, pp. 177, 182. It is worth mentioning that in his *Talkhīṣ al-majistī*, Muḥyī l-Dīn al-Maghribī gave a rather complicated method for determining the ecliptical coordinates

With regard to the parallactic instrument, the author states:

> With this instrument, the ultimate altitudes of the stars on the meridian circle that are not in excess of 30° will [only] be known approximately; because this instrument's third rule, by which the chord of the [zenith] angle is known, truly does not show the chord of [this large] angle.
>
> (S: f. 18v, M: p. 46; Appendix, §1)[41]

With Ptolemy's parallactic instument with a chord rule of only 60$^\mathrm{p}$, altitudes below 30° could not be measured, only estimated.[42]

About Ghāzān Khān's new approach, the treatise's author says:

> Over the years I have been praying for the imperial government of the king of the world, the great Īlkhān, the King of Kings on Earth, the patron Ghāzān Khān – may God perpetuate his kingdom and spread his shadow over all the inhabitants of the world. I sometimes thought about and searched for observational instruments by which observations can be produced precisely and certainly without suffering and trouble, until in the government of the world's king [i.e., Ghāzān Khān], twelve kinds of observational instruments appeared that had not appeared with any of his antecedents and their descendants, and had not been possible [to conceive] for them. By them [i.e., these instruments], all observations can be exactly and certainly known with the least cost and effort, because these instruments consist entirely of rules and straight lines. Although they are long, constructing them fully straight and dividing them into minutes and seconds is possible and by them those [observational] matters can be found exactly, whereas finding them by the five classical instruments is not possible, as we will describe.
>
> (S: ff. 18v–19r, M: pp. 47–48; Appendix, §2)

The panegyrical introduction in the preceding quote assures us that the description that follows is of the 12 instruments invented by Ghāzān, and also that they are in their original forms, because it implies that the treatise was written during Ghazan's lifetime (or between 1300 and 1304).

---

of fixed stars and planets, instead of using the armillary sphere. He measured the time interval between the meridian transit of the Sun (or another reference star with known longitude) and that of the celestial object whose coordinates were desired, and the meridian altitude of the object. Then, first, the longitude of the culminating ecliptic degree (midheaven, *medium coeli*) was determined by time and reference to the star's longitude, and second, the ecliptical coordinates of the object were determined by altitude and midheaven. For an example of Mars, see Mozaffari 2018–2019, pp. 182–188. On Muḥyī al-Dīn, see Chapter 9.

41 For the original Persian text of this citation and others, see the Appendix.

42 Seemann 1928, p. 107. Our author's critique is similar to al-'Urḍī's opinion of Ptolemy's parallactic rule given at the end of his own treatise. For the configuration of Ptolemy's parallactic instrument, see Toomer 1984, pp. 244–247.

The author then introduces the two parts (*qism*) of his treatise (S: f. 19r; M: p. 48): "To mention observational instruments and to know their applications" and "To determine the stars' positions in ecliptical longitude and latitude." The second part, however, is not found in the extant copies S, P, and M.

On the superiority of construction over the older instruments, the new ones being based on long straight beams and avoiding circular structures, he notes:

> This instrument [no. 2, below] is preferable and superior to all [observational] instruments for four reasons. First, each [older] instrument, which is well known and in common use, is dedicated to an important [application] – as we said earlier – while with this [new] instrument the determination of all quantities that can be found by those [previous] instruments is possible. Second, the expenditure, cost, effort, and occupation of [constructing] this instrument are less than for the preceding instruments, as a whole. Third, what was observed was not revealed with certainty and exactitude by the [previous] instruments, because all those instruments are made up of arcs and circles, so that if they are small, it is not possible to divide them into minutes and seconds, and the results are approximate. If they are large, it is not possible to make them completely circular, as they ought to be, and then their defect (*fasād*) and disorder (*khalal*) are more than their benefit, whereas these [new] instruments are made up of straight rules (*misṭara-hāy-i mustaqīm*) and straight lines (*khuṭūṭ-i mustaqīm*), [so that] however long, to construct them being straight, without disorder and trouble, is possible. And, fourth, in those [old] instruments, the arcs are determined [directly], whereas in these [new] instruments, the parts [functions of arcs; i.e., sin, tan, etc.] whose corresponding arc (*ḥiṣṣa-yi qaws*) is smaller are determined; therefore the arcs are often determined [more] exactly and with certainty.
>
> (S: ff. 23r–v, M: pp. 54–56; Appendix, §3)

A comparison with instruments described before this era confirms our author's claims of new instruments. In addition to portable instruments (e.g., astrolabes), of which a variety of models are known and plenty of examples are extant, observational instruments before the founding of the Marāgha observatory appear to have been large models of instruments known from classical times, modified to a certain degree, and named after patrons. For example, in the first observational program established by the ʿAbbāsid caliph al-Maʾmūn (r. 812–833), under the directorship of Yaḥyā b. Abī Manṣūr (*fl. ca.* 820), an armillary sphere with divisions for each 10 arc minutes and a circle "whose nature is unclear"[43] had been used (see later), while in the second observational program, a mural quadrant (*lubna*) and a gnomon (*shākhiṣ*) were used. From the Buwayhid period (932–1062), we find a large version of Ptolemy's "two circles" named *ḥalqat al-ʿaḍudiyya* (after ʿAḍud al-Dawla,

---

43 Charette 2006, p. 125.

d. 983), which al-Ṣūfī mentioned in his *Ṣuwar al-kawākib al-thābita*.[44] There was also a sextant built by Abū Maḥmūd al-Khujandī (d. 1000) for his patron, Fakhr al-Dawla (r. 976–997), called *suds al-Fakhrī*, which was an instrument similar to the *lubna* erected in the meridian plane but consisting instead of one-sixth of a circle.[45] From the Ghaznawid period (975–1187), al-Bīrūnī describes a *ḥalqat al-yamīnī* (after Yamīn al-Dawla Maḥmūd of Ghaznā, d. 1030), which seems to have been a solstice ring established along the meridian. And al-Khāzinī left a few treatises from the period of the Saljūq sultan Malikshāh I (r. 1073–1092) in which he describes the classical instruments.[46]

More importantly, in his *Risāla fī kayfiyyat al-irṣād*, the Damascene astronomer Muʾayyad al-Dīn al-ʿUrḍī described the instruments built in Marāgha for al-Ṭūsī:[47] (1) a great mural quadrant; (2) an armillary sphere (with some improvements over that of Ptolemy); (3) solstitial and (4) equinoctial armillae; (5) Hipparchan dioptra (with improvements to observe eclipsed diameters of the Sun or the Moon); (6) *dhāt al-rubʿayn*, a double quadrant from copper inside a circular wall, capable of measuring, at the same time, the azimuth and altitudes of two objects; (7) an improved version of Ptolemy's parallactic instrument; (8) *dhāt al-jayb wa l-samt*, an instrument to determine sine (of zenith distance) and azimuth using a wooden bar rotating on an iron axis inside another circular wall, on which one end the alidade can slide, the other end sliding up a vertical central pillar; (9) *dhāt al-jayb wa l-sahm*, a similar instrument to determine sine and versine; and (10) *ālat al-kāmila*, "perfect instrument," consisting of a rotating parallactic rule inside a graduated ring.

The four new instruments al-ʿUrḍī introduces (6, 8, 9, and 10) appear to be a combination of quadrants or rulers (in different ways in order to employ various trigonometric functions) for determining the star's altitude and, by mounting the complex on an elevated circle, for measuring azimuth. However, azimuth instruments (in the sense of the instruments specifically used in the simultaneous measurements of altitude and azimuth) were apparently not so novel. The author of our treatise ascribes an *ālat-i samtiyya* ("azimuth instrument") to Abū l-ʿAbbās al-Lawkarī,[48] and we know that Ibn Sīnā built a model of it in Iṣfahān between

44 ʿAbd al-Raḥmān al-Ṣūfī, *Ṣuwar al-kawākib*, p. 302.
45 For the applications of the meridian instruments, see nos. 1 and 7 in the list of Ghazan's instruments, later.
46 For al-Khāzinī and his works, see Lorch 1995, Articles XI, XIV, Sayılı 1956, Mozaffari 2022.
47 This treatise has been thoroughly described by Seemann (1928).
48 Abū l-ʿAbbās al-Lawkarī (d. 464/1071–1072), a well-known figure of Islamic philosophy, was a contemporary of the Persian astronomer and poet ʿUmar Khayyām (d. 515/1121). He is one of a famous chain of Islamic peripatetic philosophers: Ibn Sīnā – Bahmanyār – Abū l-ʿAbbās al-Lawkarī – ... – Naṣīr al-Dīn al-Ṭūsī. Older sources provide little information about him. See al-Bayhaqī, *Tatimma*, pp. 110–111; al-Shahrazūrī, *Nuzha*, vol. 2, p. 54. There is no evidence of his appearance in the scientific circle around Malikshāh I headed by Khayyām, to which all historical matters about the scientific/astronomical activities of that period are connected. Thus, we know nothing about al-Lawkarī's astronomical activities. He is not mentioned by Ibn al-Athīr as one

1024 and 1037.[49] Muḥyī l-Dīn al-Maghribī traced the use of an azimuth instrument in the Islamic period even further back, ascribing it to Yaḥyā b. Abī Manṣūr[50] and suggesting that the azimuth ring was used in order to determine the ecliptical coordinates, which, however, is far beyond the usual applications of the known versions of an azimuth instrument (e.g., those of Ibn Sīnā and al-ʿUrḍī). While the nature of the instrument is as yet unclear, it appears to have been an instrument having the same applications as the armillary sphere.[51]

Therefore, we can consider at most five observational instruments (apparently models of azimuth instruments) to have been innovative in the Islamic period up to Ghāzān Khān's time, over a period of five centuries. In this light, our author's claim of a new approach taken by Ghāzān to observational instruments, and of his new models and their differences from the older ones, is at least not inconsistent with (if not justified by) the historical sources.[52] Ghāzān's new instruments appeared within one decade.

Besides the principal difference in the basic approach to observational instrumentation, another explicit distinction between al-ʿUrḍī's instruments and those described in our treatise is that the former are mostly made of teak from India,[53] with only circular parts cast from copper, whereas metals play a more important role in the latter.[54]

---

of Khayyām's collaborators in the task of reforming the Iranian solar calendar which Malikshāh ordered to commence in 467/1074 and which was inaugurated at the spring equinox of 1079. See Ibn al Athīr, *al-Kāmil fī l-Tārīkh*, p. 98 and Kennedy 1968, pp. 671–672. We do know, however, that he wrote an encyclopedia titled *Bayān al-ḥaqq fī ḍimān al-ṣidq*, which included an epitome of Ptolemy's *Almagest* and to which Quṭb al-Dīn al-Shīrāzī referred in the prologue on the astronomical part of his own encyclopedia dated 24 Rabīʿ I 674 (September 17, 1275): *Durrat al-tāj*, vol. 2, p. 1.

49 Wiedemann and Juynboll 1926, pp. 105ff.

50 "Yaḥyā b. Abī Manṣūr observed the Sun around the time of the autumnal equinox on Sunday, 25 Murdhādh 198 of the Yazdgirdī era (September 19, 829); he then found it by the azimuth circle, *dāʾirat al-samtiyya*, to be on Virgo 29;43." Al-Maghribī, *Talkhīṣ*, f. 58r.

51 Al-Maghribī calculates the instant of autumnal equinox from Yaḥyā's observational data to be 54 minutes after the Sun transits Baghdad's meridian on the given day. At this time, the Sun was positioned on Libra 0;17 (or true longitude = 180;17°), namely, an error of around 17 arc minutes in Yaḥyā's measurement of the solar longitude, or an error of around 6;54^h in his determination of the time of autumnal equinox. Based upon al-Maghribī's further explanations, one can also determine an error of around 10 arc minutes in the solar longitude – or a corresponding error of around 4;14^h – in Yaḥyā's data for vernal equinox of the year 198 of the Yazdgirdī era (March 16 or 17, 830). This gives the impression that the accuracy of Yaḥyā's azimuth circle was that of a typical armillary sphere, like the very model used by him.

52 It is curious that our treatise does not refer to al-ʿUrḍī and his instrument descriptions, but since both appear to be of the same tradition and to belong to the same observatory, we will note parallels between the two where appropriate.

53 Seemann 1928, p. 29.

54 Iron and pure tin (*arzīz*) were brought from Minor Asia (al-Nīshābūrī, *Tārīkh*, p. 229). We also find in a seventeenth-century treatise on the construction of the astrolabe that brass was brought from Hashtarkhān (today's Astrakhan) in the north of Azerbaijan. Muḥammad Ḥusayn b. Muḥammad Bāqir al-Yazdī, *Mīzān al-ṣanāʾiʿ*, f. 13v (written in 1072/1661–1662).

In all three manuscripts – S, P, and M – the spaces for the names of the new instruments are left blank. We find the names of instruments 1 and 2 in the description of the latter, where they are called, respectively, the "triangle instrument" (*ālat-i muthallath*) and the "perfect instrument" (*ālat-i kāmila*).[55] Instrument 12 is introduced with a rather long description. The treatise's author adds that the best observational instrument to date was the "azimuth instrument" (*ālat-i samtiyya*) invented by Abū l-ʿAbbās al-Lawkarī, which, however, also suffered from the aforementioned issues.

11.3.1 The Tables in the Treatise

The following tables are only present in manuscript S (and P):

1. *Ẓill-i mustawī* (cotangent, or umbra recta; S: f. 44v)
2. *Mayl-i awwal* and *mayl-i thānī* (first and second declinations; S: f. 45r)
3. *Ẓill-i maʿkūs* (tangent, or umbra versa; S: f. 45v)
4. *Jayb* (sine; S: f. 46r)
5. *Sahm-i qaws-i niṣf-i dawr* (versine for the arcs 0,. . ., 180°; S: f. 46v)
6. *Maṭāliʿ-i mustaqīm wa mayl* (right and oblique ascension of the parts of the ecliptic for the latitude of Marāgha; S: f. 47r)
7. *Hādhā al-jadwal al-samt alladhī li-hādhihi l-ʿurud̲ li-kull khums rubʿ rubʿ dāʾirat al-ufuq*, an unusual table titled "The table of azimuths of these [geographical] latitudes for each quadrant of the horizon circle per five degrees," whose purpose is obscure.

The author writes that he copied these tables from *Zīj-i īlkhānī*, in which we were unable to find table no. 7, however. In the manuscript, tables 6 and 7 are the only complete tables; 1–5 were left empty. From the text it can be deduced that the treatise had also contained other tables for the equation of daylight and the number of daylight hours for the equator and for Marāgha, which are also not available in these copies.

11.3.2 Authorship

Two centuries after the fall of the Buwayhids, Persian scholars again began writing in their native language, such that the very large majority of mathematical and astronomical treatises written after the mid-thirteenth century are in Persian. The language of these treatises is not as pure, however, as that of their tenth-century

---

55 Note that al-ʿUrd̲ī also called one of his instruments, a rotating parallactic rule usable for azimuth and altitude measurements, "the perfect instrument" (Seemann 1928, pp. 96–104). He built this instrument for Malik Manṣūr, the ruler of Ḥimṣ (the antique Emesa, now Homs), in 650/1252–1253, in the presence of Manṣūr's vizier, Najm al-Dīn al-Lubūdī. Clearly, these "perfect instruments" are not the same.

counterparts; Arabic and Mongolian are used liberally (although the latter is less frequent in astronomical treatises, except in descriptions of the Chinese Uighur calendar). This has a corrupting influence on phraseology, and since mathematical and astronomical treatises, whether in Arabic or Persian, also usually adopted a simple unitary style, the combination of both makes the definitive distinction of a literary style very difficult.

It is clear that our author is a professional astronomer with ample knowledge of observational astronomy and instrumentation and capable of sound comparative discussion and critical conclusions about them. We can therefore propose al-Wābkanawī, with high probability, and al-Nīsābūrī, with less, as our most probable candidates for authorship. In some aspects – particularly the type and order of explanation, the frequency of technical terminology, the sentence structure, and the remarkably sparse descriptive material – our treatise is so similar to al-Wābkanawī's *zīj* that the authorial voice seems to be identical; in equal measure, it is less like al-Nīsābūrī's treatises, in which, for example, there is an excess of explanation. In our treatise, we also find the same description of instrument 12, accompanied by some additional notes, as in al-Wābkanawī's *zīj* (IV, 15, 8),[56] where he describes the instrument as one of the marvels of observational work. In addition, the style of both treatises seems to be identical. All this, along with the fact that al-Wābkanawī was Ghāzān's astronomer-royal and had his royal order, *yarlīgh*, to complete a *zīj* in that period, leaves little room for doubt in the identification of our author as al-Wābkanawī.

11.4 The Instruments

We can classify the 12 new instruments into six types, as shown in Table 11.1. In types A, D, and E, we see that the latter instrument of each type is an improved and more sophisticated version of the first one of the respective type. Nos. 9 and 4 in type F can be compared in a similar way. In other words, the treatise follows an approximately evolutionary model for describing the instruments. The

Table 11.1 The types of the 12 instruments

Type	Involved functions	Nos.
A (triangle-type instruments)	Special function described in the treatise	1, 2
B	Chord arm on circle	3
C (wire instruments)	Sine, chord, and tangent	5, 6
D (square-type instruments)	Tangent	7, 8
E	Sine and versine	10, 11
F (improved Ptolemy's parallactic instrument)	Chord	9, 4
G	Pinhole device	12

56 A: f. 92r–v, B: ff. 159r–v.

instruments of type C form a related couple, where one instrument is better for higher elevations than the other.

Some of the instruments allegedly designed by Ghāzān Khān show a great similarity with instruments constructed in Europe in the following centuries. The most remarkable similarity is between nos. 7 and 8 and Georg von Peuerbach's geometrical square (described in *Canones Gnomonis*, MS Vienna no. 5292, ff. 86v–93r, printed in Nuremberg, 1516). Instrument no. 12 is a pinhole image device, and the factors taken into consideration and applied in its construction are similar to those described at least two decades later by Levi ben Gerson (1288–1344). As we will see later, instrument no. 12 is the link between the dioptra and the instruments that were later used as a camera obscura. Although there is no concrete evidence of how such knowledge was transferred, there were widespread politico-cultural relations between Iran and Europe at that time, which might account for easy access.[57]

Five circular traces have been discovered to the south, southeast, and north of the central building of the observatory in Marāgha. Al-ʿUrḍī described only three big circular instruments at Marāgha. More importantly, with an average rainfall of 300–1,000 mm during the spring and long periods of freezing weather (over 100 days), the wooden instruments noted three decades earlier by al-ʿUrḍī, and especially their fine gradations required for observations, had probably been destroyed by Ghāzān Khān's time. Al-ʿUrḍī died eight years before al-Ṭūsī; thus, he clearly could not reconstruct his own instruments. Therefore, it now seems quite likely that these five locations had been given over to the "new" instruments.

Except for three instruments – nos. 1, 2, and 12 – the instruments are described in the treatise without being named, so we will refer to them by their sequence number. In what follows we provide a short description of each, following the treatise, and a virtual reconstruction, remarking on difficulties with the translation and practical considerations. Instrument no. 12 will be presented in more detail with a full translation.

The treatise gives numerous dimensions of the instruments and their parts. The *dhirāʿ* used in the reconstruction is the royal cubit (*dhirāʿ al-malik* or *al-dhirāʿ al-hāshimī*) of 665 mm, as used in al-ʿUrḍī's treatise and shown in Table 11.2.

Table 11.2 Units of measurement used in the treatise

Symbol	Unit	Arabic	Persian	Relation		Length (mm)
cb	cubit	*dhirāʿ*	*gaz*		24 fg	665
sh	handspan	*shibr*	*wajab*	1/3 cb	8 fg	221.67
hd	handbreadth	*qabḍa*	—	1/6 cb	4 fg	110.83
fg	finger	*aṣbaʿ*	*angusht*	1/24 cb	1 fg	27.71

57 On a possible connection of Peuerbach and Marāgha astronomy, albeit not concerning instruments, see Dobrzycki and Kremer 1996.

Frequently, however, not all dimensions are given, and for the reconstruction, we had to estimate useful dimensions ourselves. In addition, some numbers are clearly copying errors and do not make sense. We note such errors in our remarks.

Several instruments are based on the concept of rigid right isosceles triangles, or moving triangle legs with a chord rule for measuring the chord of the angle, $\mathrm{crd}\,\alpha = 2\sin(\alpha/2)$. As was common, lengths and angles are defined so that the base length of the triangle legs is divided into 60 parts which are each divided into 60 "minutes," and the hypotenuse/chord of the triangle shows a gradation in the same units, so we will write $\mathrm{Crd}\,\alpha = 60\mathrm{crd}\,\alpha$ and, similarly, $\mathrm{Sin}\,\alpha = 60\sin\alpha$, etc. The hypotenuse of the right triangle will then be of length $\mathrm{Crd}90° = \sqrt{2}.60^{\mathrm{p}} = 84^{\mathrm{p}}51'10''$. Most instrument descriptions with chord rules mention chord rules with regular gradations up to 85^{p}, obviously for practical purposes.

11.4.1 Instrument 1

The applications of this instrument are altitude and zenith distance of a (culminating) star, determination of the obliquity of the ecliptic, finding local latitude. The instrument consists of a right isosceles triangle with horizontal basis, fixed in the meridian. A centrally mounted alidade carries two sights. Its fiducial edge is formed by omitting one half of its outermost one-third of length. The triangle legs are graduated regularly in units of one-sixtieth the triangle's height from the top and both lower ends toward the center points of the legs, where the scales meet at $42;25^{\mathrm{p}}$. The scales are split into three parallel bands showing parts, minutes, and seconds.[58] The description omits absolute sizes for the instrument but adds a detailed description to compute the altitude or zenith angles, respectively, from the values read on the scale. Effectively, it replaces a mural quadrant with its circular scale (Figure 11.1).

11.4.2 Instrument 2

The second instrument adds the capability of measuring altitude in any azimuth to instrument 1. On a base cross of 15 cb teakwood or copper bars, aligned with the meridian and supported with cross beams, a vertical triangle of copper bars with base length 7.5 cb is erected. The triangle is mounted on a central iron shaft 3 fg in diameter (Figure 11.2). Alternatively, the triangle can be put on a 2 fg iron shaft mounted on a wooden cylinder of 0.5 cb diameter. The copper alidade's fiducial edge is again formed by removing half of it outward of two-thirds of its length from its axis. The triangle can be moved in azimuth, and the altitude measured on the scale as on instrument 1.

---

58 We must assume simply "smaller parts," as this gradation seems pointless.

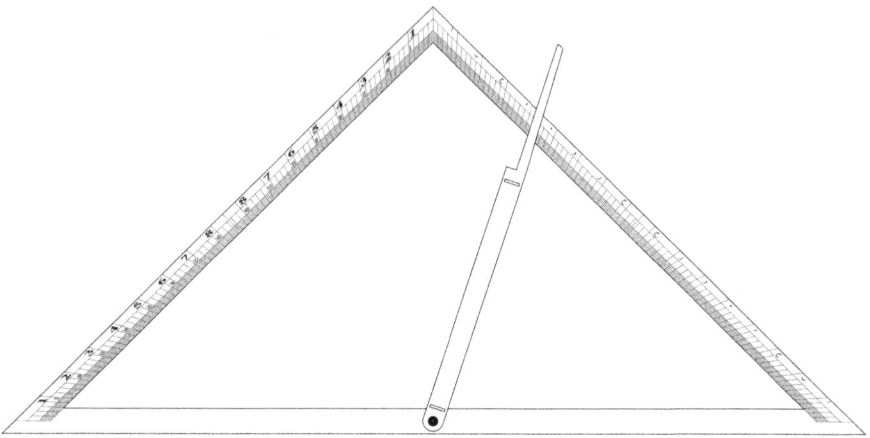

Figure 11.1 Instrument 1: a triangle placed in the meridian.

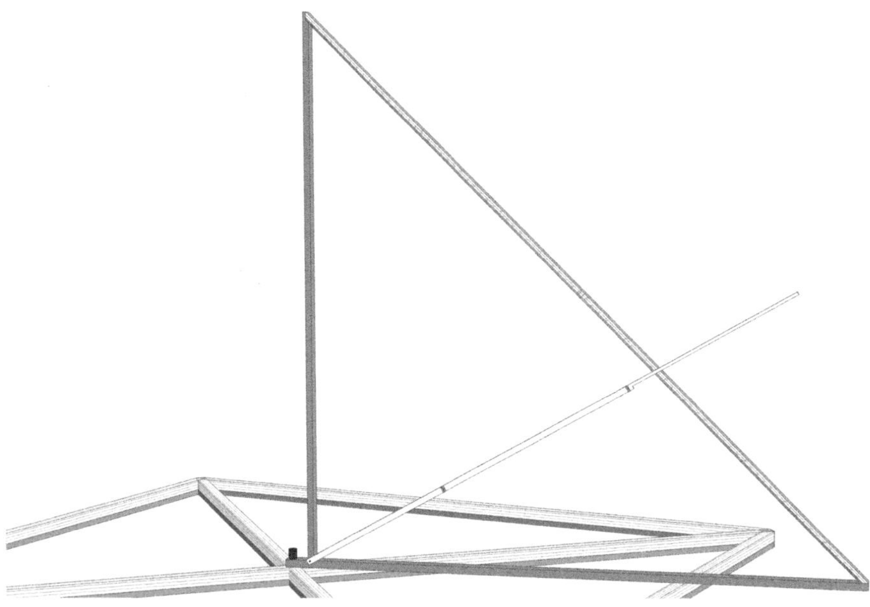

Figure 11.2 Instrument 2: a rotating triangle.

Our author gives definite sizes and dimensions for many of the rules, and directions as to which material to use for construction. If made from copper, the 4 fg ×
2.5 fg × 8 cb base bar would weigh about 364 kg. The perpendicular rule (no thickness given explicitly) of (assumed) 2 fg × 2 fg × 7.5 cb adds 137 kg, and the

312

chord rule of 2 fg × 2 fg × 7.5 $\sqrt{2}$ cb another 193 kg. An alidade of only (assumed) 1 fg × 1 fg × 7.5 cb would weigh another impressive 28 kg but should be even heavier if the dimensions of the rules are larger. In total, this gives a moving mass of more than 700 kg. It seems clear that this instrument could only have been used to measure the altitude of a celestial object in a preselected azimuth, held by several strong assistants. For high altitudes, a ladder was obviously required to read the scale value.

11.4.3 Instrument 3

Contradicting the straight-rule concept, this instrument uses a graduated ring for reading azimuths.[59] It is 3 fg wide and 2 fg thick and of "large" diameter, to be installed 0.5 cb [*sic*] above ground. A central wooden cylinder topped by an iron plate supports an iron shaft on which rotates a wooden azimuth bar, which protrudes over the ring and carries a nail indicating the azimuth on the ring's graduated scale. An alidade is mounted in an excavated hole close to the axis and carries a chord rule jointed to its other end. The graduated chord rule slides through a dovetail slit on the outer end of the azimuth rule (Figure 11.3).

The length of the downward-pointing chord rule dictates that the ring must have a certain height. Therefore, we must assume that *nīm* (0.5) for the height of the ring is a copyist's error for *dū wa nīm* (2.5). As shown in the reconstruction, the moderate size with a ring of 3 cb diameter in 2.5 cb corrected height provides a comfortable instrument for a single user, aided by an assistant standing outside the ring and reading the values from the scales. Made from copper, however, the alidade mass for 3 fg × 3 fg × (2 cb − 1 hd) would be more than 75 kg, requiring again at least another strong assistant, but for teakwood, we can estimate about one-tenth of this mass.

11.4.4 Instrument 4

This is a rotating parallactic rule: a base cross of 4 fg square rules is aligned in the meridian and east–west lines. Two equal brass or copper rules of 2 fg square and length are hinged on their upper ends. One is attached vertically with several rings on a central iron shaft of 3 fg diameter. On the lower end of the vertical rule, the chord rule is attached. The other rule is the alidade, with sights and a slit on its lower end for the chord rule. On the east and west ends of the base cross, azimuth rules are attached. After measuring altitude, the chord rule is laid horizontally on the ground. On its 60^p mark, a nail protrudes downward. One azimuth rule is rotated toward this nail, and from the length read on the nail, the azimuth can be derived (Figure 11.4).

---

59 Such circular walls with graduated rings had already been used by al-ʿUrḍī in three instruments. Maybe this instrument served as replacement in an existing ring.

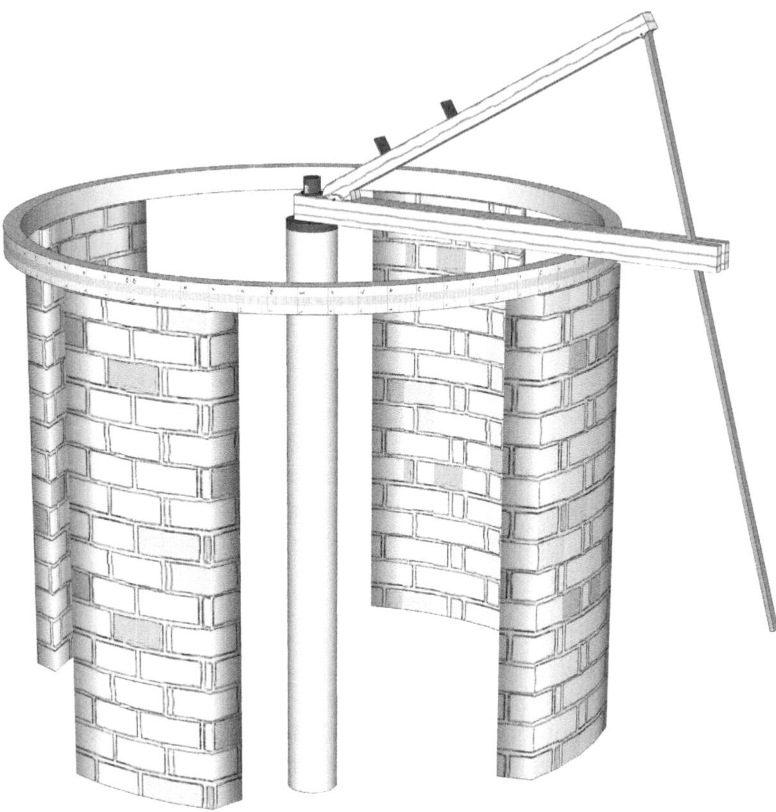

Figure 11.3 Instrument 3: alidade with chord on copper or brass ring.

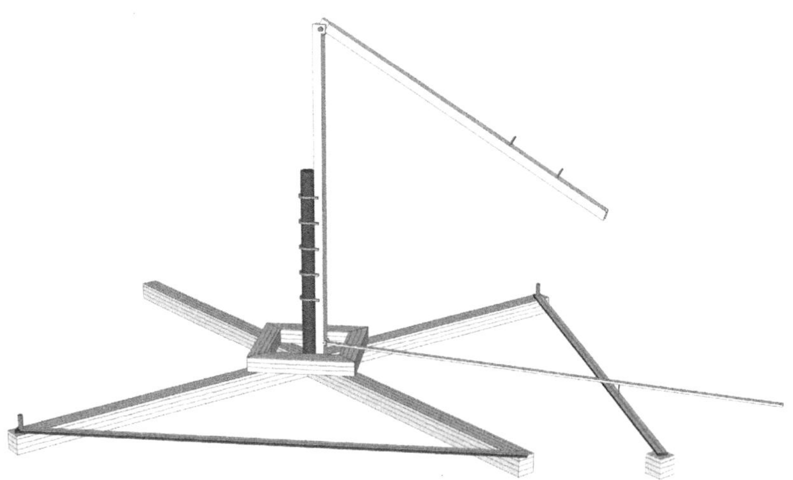

Figure 11.4 Instrument 4: a rotating parallactic rule.

Although rule lengths are not given, we estimate 3 cb length for the vertical and alidade rules in order for the instrument to be functional. It is interesting to note that al-ʿUrḍī's "perfect instrument" likewise used a parallactic rule, but inside a circle, so the azimuth was read by putting the chord rule onto the azimuth ring.[60]

11.4.5 Instrument 5

Although not explicitly mentioned, instruments 5 and 6 form a pair, usable for different altitude ranges. Contrary to what was recommended by al-ʿUrḍī,[61] here and with instrument 6, a catgut rope or copper wire (*khayṭ*) is used for measuring parts.

Instrument 5 includes a central high round or rectangular pillar (height P) of sun-dried bricks, topped by a conical copper or iron roof of 1 sh height. On its peak, a *kawkabah*-shaped shaft[62] supports a ring that holds the *first khayṭ*. An Indian circle encloses the pillar.[63] On the base of the pillar placed underground, there is a hidden thickness (*thikhan-i nahānī*) of firm wood or iron.[64] Around the base of the pillar, there is a circular excavation of 0.5 fg (depth and width) in which an iron circle can freely move around the pillar (Figure 11.5).[65]

A second *khayṭ*, called shadow *khayṭ* (*or* tangent *khayṭ, khayṭ-i ẓill*), is connected with the top ring[66] and leads through the base ring, and a third *khayṭ* is placed on the meridian line, close to the base of the pillar, and secured with a nail and a ring, representing the meridian line. An alidade of 0.25 cb length is mounted on the first *khayṭ* and has a ring on its outer end, through which a conical plummet on a fourth *khayṭ* can be dropped to the ground. A long graduated rule of wood is used for measurements, and steps of stones for lower altitudes if needed.

To measure altitude, the first *khayṭ* is stretched, and the plummet dropped when the star is seen through the alidade. The distance from the pillar's foot is the cotangent of the altitude and can be measured along the shadow *khayṭ*.[67] To measure the

---

60 Seemann 1928, p. 96.
61 Seemann 1928, p. 107.
62 The Arab term *kawkab* means "star," but here, apparently, the Persian *kawkabah* is meant, which is a staff with an incurved head, used to prevent the top ring from falling off.
63 An *Indian circle* is a circle drawn on the ground, of an arbitrary radius, at the center of which a gnomon is installed. The Sun's shadow crosses the circle at two points in the morning and afternoon. The line connecting those points represents a true east–west line (at least at the solstices, when the Sun's declination does not change during the few hours between the events).
64 Its purpose is not given, but it is likely to increase the stability of the instrument.
65 It is not completely clear whether the circle is attached to the pillar as shown or it is attached in some undocumented way to the ground (maybe to the "hidden thickness" just mentioned), where it could also encircle a non-circular pillar. The instrument's purpose would not change if this circle was fixed to the ground, but its size must always match the base diameter of the conical roof.
66 This connection seems unnecessary, but it may permit the verification of the vertical alignment of the first and second *khayṭ*.
67 Actually, it is cot h $(P-l_4)$ for altitude h and length l_4 of the plummet *khayṭ*.

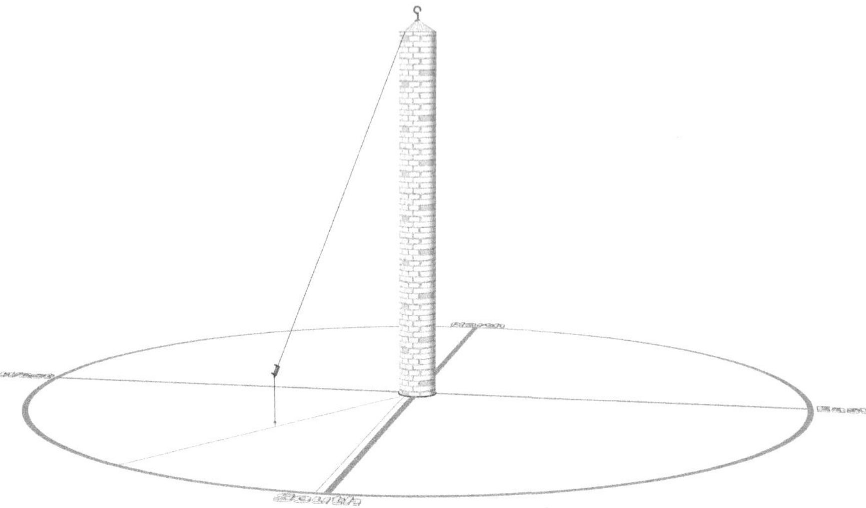

Figure 11.5 Instrument 5: pillar with ropes.

co-azimuth,[68] we stretch the meridian *khayṭ* parallel to the meridian and measure its perpendicular distance from the plummet mark.[69]

The description of this instrument is very confusing and leaves many details out; constructing an instrument that fits the description and still provides correct and usable measurement results was not easy. We propose the following, which only contradicts the description in one detail: the attachment point of the meridian *khayṭ*.

Our pillar shown here is 10 cb high, with a diameter of 1 cb, both dimensions not explicitly given. The key purpose of this instrument as we interpret it – although not described in the treatise as such and even contradicting the original sketch – appears to be that the fiducial triangle does not use the central axis of the pillar but a vertical line that is formed by the point where the first *khayṭ* bends around the corner of the flat conic roof on top of the pillar and the small ring (*zirih*) on the base ring, where the shadow *khayṭ*, and quite likely also the meridian *khayṭ*, are attached. The only dimensional instruction found in the treatise is a roof height of 1 sh, which would form the low angle shown with our (assumed) pillar diameter of 1 cb. This flat cone appears to be a required component, its angle defining the minimum altitude of observation, and its diameter must be identical to the base ring's diameter to form the vertical fiducial line.

---

68 We find azimuth *a* from measured length $l = l_2 \cdot \sin a$. This is described as a co-azimuth in the treatise, which counts azimuths from the east–west line.

69 In our reconstruction, we have understood this instruction to create a true meridian line for the current azimuth setting, namely, attaching the meridian *khayṭ* from the same small ring (*zirih*) on the base ring where the shadow *khayṭ* is attached outward, parallel to the meridian line drawn on the ground in the instrument's center line. The literal text, which has the meridian *khayṭ* directly attached to the pillar, suffers from obvious problems.

The meridian *khayṭ* is described in the text as being attached to a fixed point on the pillar's base on the meridian line. In this event, it would be pointless to have another movable *khayṭ*, because the meridian line permanently drawn on the ground would fulfill the same purpose. Also, the measurements and calculation of the azimuth would have to take the radius of the pillar (or, for a rectangular pillar, the radius of the base ring) into account. To be functional as described, it would seem that this *khayṭ* was likewise attached to the base ring at the same place as the shadow *khayṭ*, so that a "local coordinate system" was cleverly carried around the pillar.

11.4.6 Instrument 6

A small central shaft inside an Indian circle with meridian and east–west lines harbors a *khayṭ* with an alidade on its end. From this, a plummet on a second *khayṭ* can be dropped to the ground, and its impact point measured by a long ruler (Figure 11.6). For high altitudes, an observing platform can be used.

It seems that this instrument complements no. 5 for objects in low altitudes. Clearly, however, all instruments involving rope lengths suffer from their flexibility and slack.

11.4.7 Instrument 7

A square is made up of four graduated rules and fixed in the meridian plane. In its top corners, an alidade can be mounted, depending on northern or southern targets. The scales provide the tangent or cotangent of the meridian altitude. In effect, it also replaces a mural quadrant, but with a graduation different from no. 1 (Figures 11.7(a) and 11.7(b)).

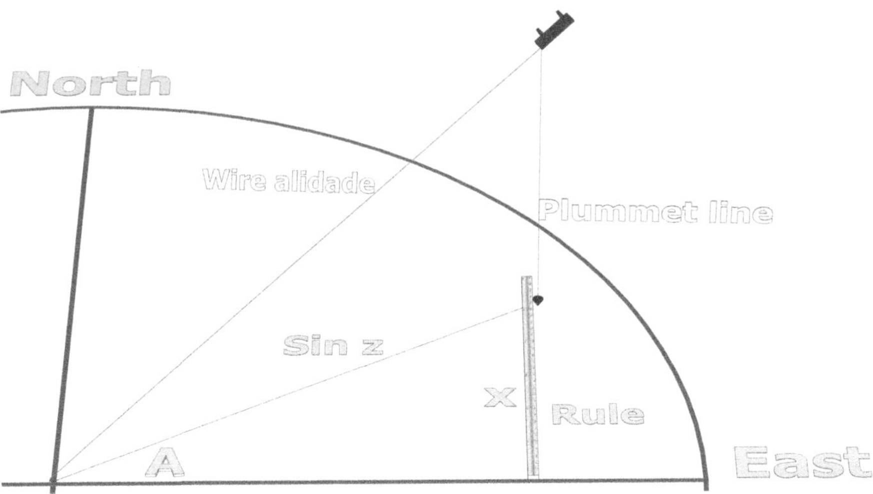

Figure 11.6 Instrument 6: ropes.

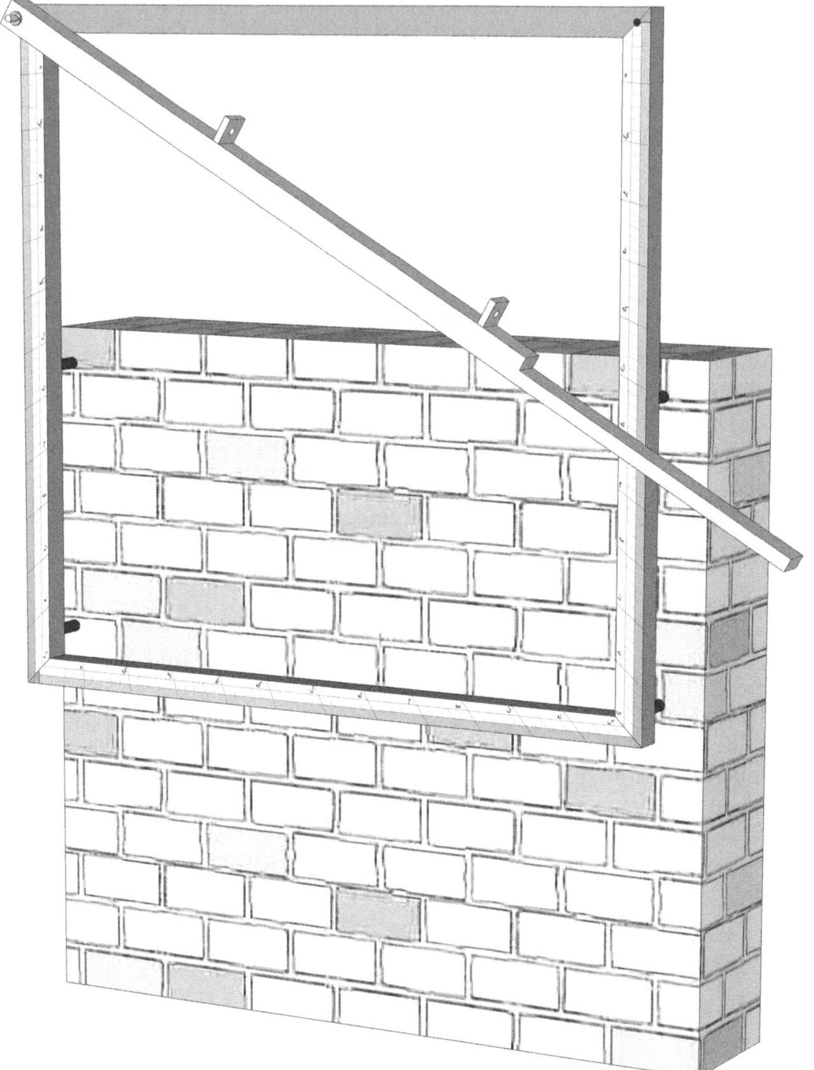

Figure 11.7(a) Instrument 7: meridian square.

This instrument is practically identical to the *quadratum geometricum*, or geometrical square, of Georg von Peuerbach (1423–1461) (Figure 11.7(c)) and an instrument used by Tycho Brahe (1546–1601) (Figure 11.7(d)). We do not know whether Peuerbach knew about this instrument or invented it independently. It might also have not been completely new to both authors: Dieter Lelgemann describes a reconstruction of a *skiotherikós gnomon* (shadow frame) as precursor of the geometric square, for use as a (terrestrial) trigonometric survey

318

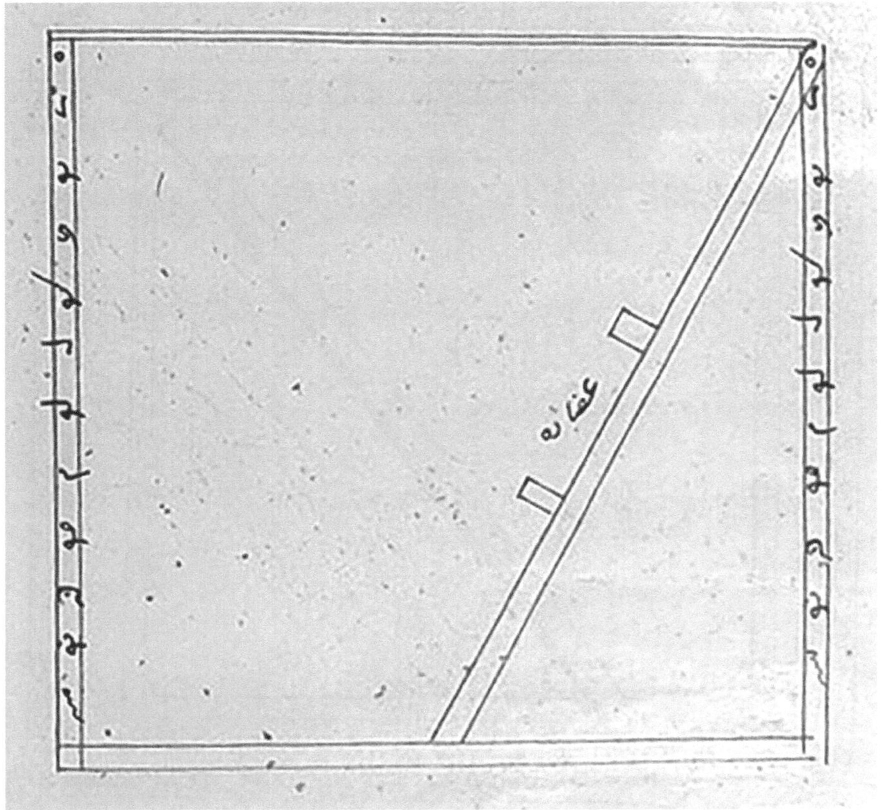

Figure 11.7(b) A meridian square, early fourteenth century (S: f. 36v).

Source: Scan of a part of a folio from a manuscript kept in a library in Iran. The metadata has been given in the chapter. No copyright.

instrument.[70] The eleventh-century polymath al-Bīrūnī used a similar instrument for terrestrial surveys.[71]

11.4.8 Instrument 8

This instrument extends no. 7 by mounting it on an iron shaft of 3 fg diameter and surrounding it with a "large" cardinally aligned copper square with gradations up to 60ᵖ (and parallel smaller gradations) from each side's center point to its corners. The inner square, on which the alidade is mounted in the lower corner, rotates on top of the outer to read tangent of azimuth (Figure 11.8). Using viable dimensions,

70 Lelgemann *et al.* 2005.
71 al-Bīrūnī, *Nihāyāt al-amākin*, pp. 210–212.

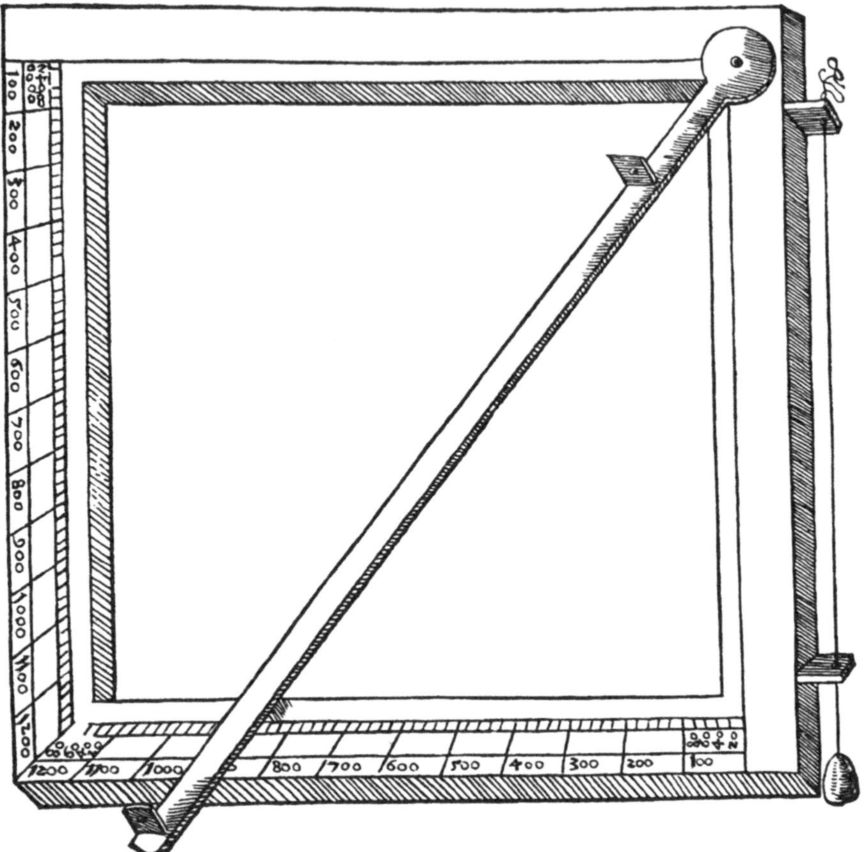

Purbach's quadratum geometricum, um 1450,

nach Regiomontan's Scripta.

Figure 11.7(c) Peuerbach's *quadratum geometricum*, mid-fifteenth century.

Source: From J. Schöner, *Scripta clarissimi mathematici . . .*, Nuremberg 1544, f. 62v. Scan of a part of a page from Peurbach's work (fifteenth century). No copyright.

this smaller instrument allows better handling than instrument no. 2. We estimate the upper frame to have 2 cb × 2 fg × 2 fg bars, giving still about 160 kg of moving mass for the frame with alidade.

Our author sets great store by this instrument:

This instrument provides the altitude and the azimuth of a given star with complete accuracy and certainty. By means of this instrument, all observations

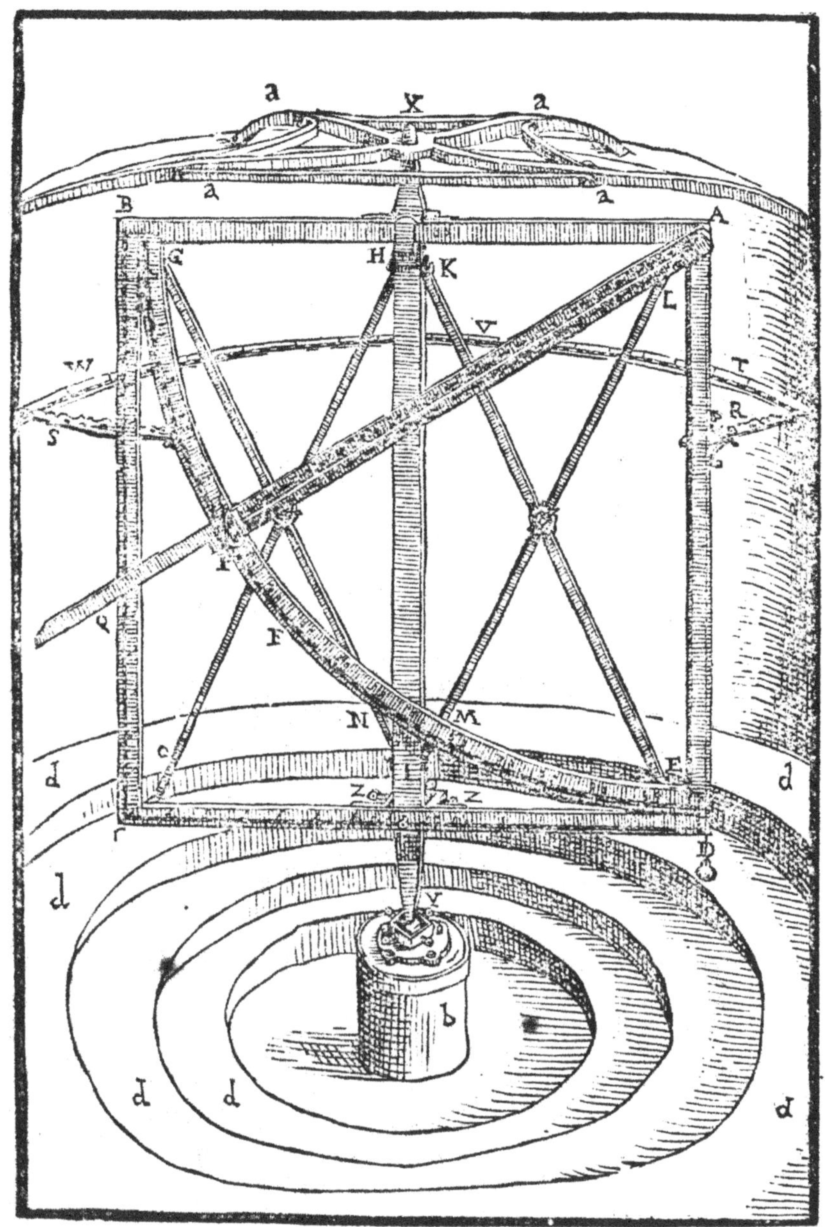

Figure 11.7(d) A movable model by Tycho Brahe.

Source: Tychonis Brahe astronomiae instauratae mechanica, Nuremberg 1602, between ff. B3 and B4. Scan of a part of a page from Tycho Brahe's work (sixteenth century). No copyright.

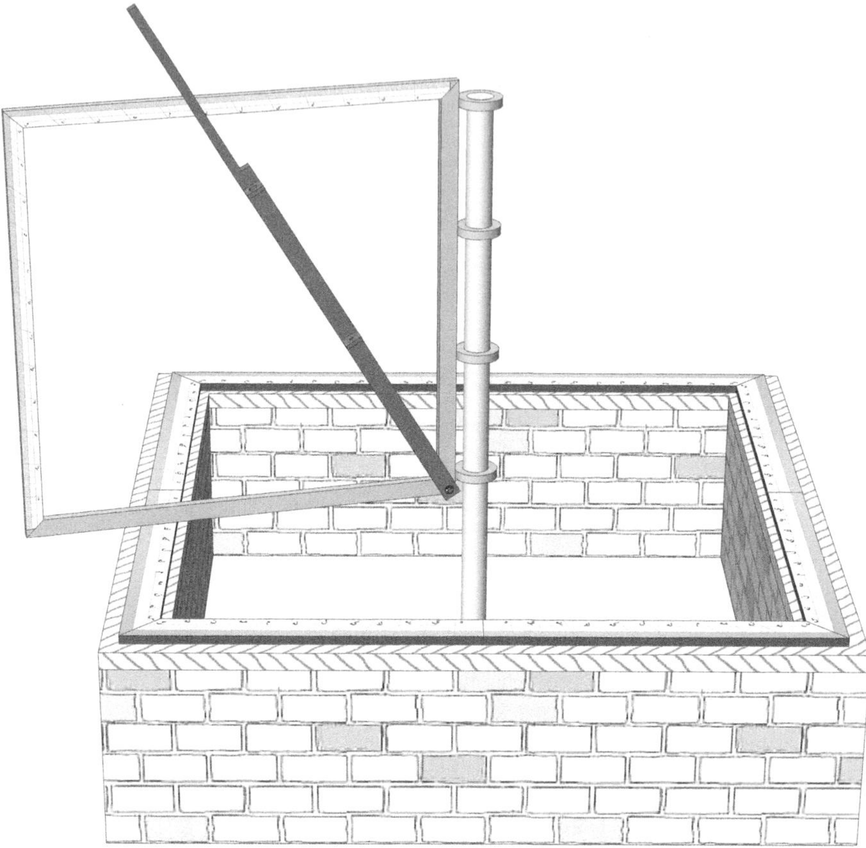

Figure 11.8 Instrument 8: rotating square.

that were known throught the observational instruments, as well as several other matters, are known and witnessed. This instrument is preferable to other observational instruments for the reasons mentioned earlier.

(S: f. 70v; Appendix, §4)

11.4.9 Instrument 9

A double pillar of teakwood is set up in the meridian with a gap of 4 fg. A slightly shorter alidade of < 4 fg × 2 fg with two sights is joined to the pillar by a nail to leave 0.5 cb[72] space to the ground. A graduated chord rule of 2 fg × 1 fg thickness and 85/60 the alidade's length is mounted by a nail in a split on the alidade's lower end, where also the scale begins. The chord rule's free end is fed through the dou-

---

72 Again, this is probably a scribal error or omission for 1.5 (?); see later.

ble pillar, where the alidade ends when hanging vertical. A rope of catgut (*khayṭī az rūdkish*) runs from the alidade's end over a pulley wheel (*bakra*) mounted on another pillar of equal height, standing in the meridian outside the alidade's swing area (Figure 11.9).

This instrument is used for determining the maximum altitude of a given star. We pull the rope (lifting the alidade) until a given star is seen through both sights of the alidade. The distance between the halving split end of the alidade and the other end of the chord rule[73] is the chord of the zenith distance (co-altitude) of that star.

This is obviously another variant of Ptolemy's parallax instrument (*Almagest* V.12). The interesting differences are the full-length chord rule and the pulley (both already described by al-ʿUrḍī,[74] who mounted the latter onto an arm extending from a wall) and the mounting of the chord rule on the alidade. The latter, however, makes it necessary to raise the point where the free end of the chord intersects the double pillar: for a chord rule of length $\sqrt{2}$ of the base length, the chord will point one-quarter of the base length below this contact point at zenith distance $z \approx 41.41°$ and may collide with the ground if it is not high enough.[75] We chose 5 cb as base length, this size being identical to the similar instrument described by al-ʿUrḍī, which, however, has the chord attached to the base of the double pillar. This length provides us with 55 mm per part on the chord rule, or just short of 1 mm per minute. The text says literally, "the lower end of the alidade is 0.5 cb above ground," which would allow for no more than 2 cb base length. Assuming another scribal error for 1.5 cb true elevation solves the problem; otherwise, a trench of almost 1 cb depth for the chord rule just south of the double pillar must be assumed, of which there is, however, no mention in the text. Thus, the double pillar may have been as tall as 7 cb, with a base length of 5 cb, and a chord contact point at 1.5 cb height.

11.4.10 Instrument 10

This instrument consists of two copper rules of equal length, where the thicker and wider vertical sine rule has a hole and on which is attached the smaller horizontal versine rule (or *sahm*, lit. "arrow"). A rope of catgut (*zih*) connecting the top of the sine rule to the end of the *sahm* prevents it from being detached. An alidade of size equal to the *sahm* is connected to its outer end. The sine rule is graduated

---

73 This description is inexact: we should draw a mark on the two parallel rules where the lower end of the alidade swings through. At the moment of observation, we move the chord rule right below this mark, and then the distance between the head of the alidade and this mark will be the chord of the zenith distance of the star.

74 Seemann 1928, p. 86.

75 In an instrument of base length 1, the chord rule of length $\sqrt{2}$ extends below the contact point by $y = \sin\frac{z}{2}.\left(\sqrt{2} - \operatorname{crd} z\right)$. The maximum extent is where $\frac{d}{dz}\sin\frac{z}{2}.\left(\sqrt{2} - 2\sin\frac{z}{2}\right) = 0$, thus $z = 2\arcsin\frac{\sqrt{2}}{4} = 41.41°$, where this amount is $y = 1/4$.

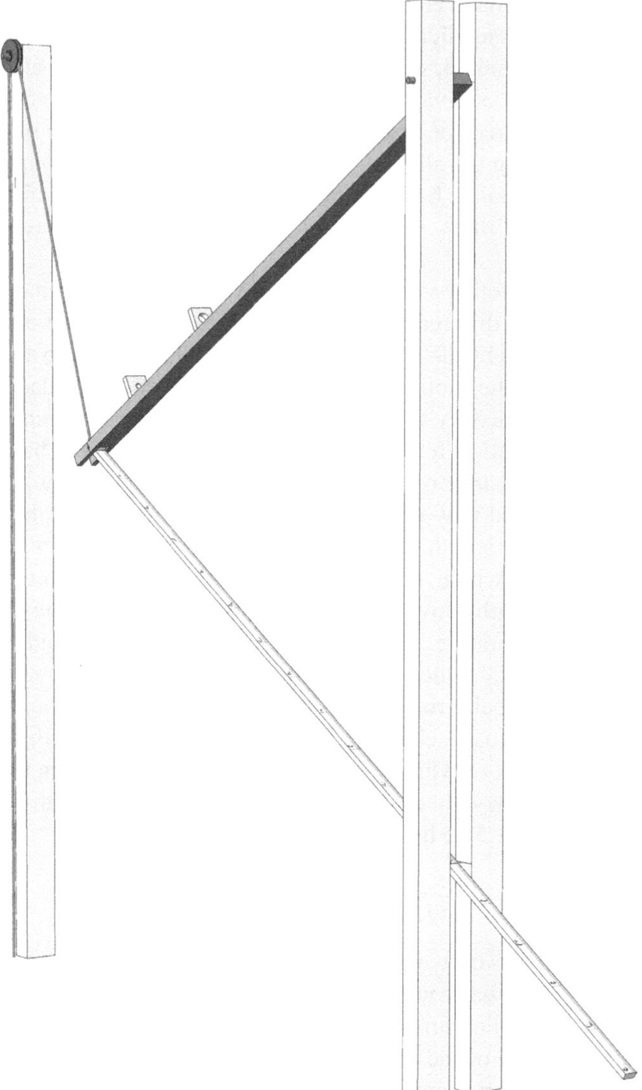

Figure 11.9 Instrument 9: improved parallactic rule.

from the hole to the top, the *sahm* from the end opposing the alidade joint. This instrument is installed in the meridian plane so that the sine rule is perpendicular on the horizon and the versine rule is parallel to it (Figure 11.10).

We move the alidade until a given star is seen through both of its sights, then the sine rule is moved to touch the alidade on the point that gives the sine of the star's altitude. The length of the *sahm* protruding from the hole shows the versine of altitude.

324

Figure 11.10 Instrument 10: sine/versine in the meridian.

Dimensions of the rules are not presented in the text. From the description of instrument no. 11, we estimate for this instrument also a length for the rule of 3 cb, so that the "minute" divisions are about 0.55 mm apart. Based on this length and dimensions for the rules found earlier in this text, an alidade and versine rules of 1 fg × 1 fg × 3 cb is about 14 kg each; the sine rule, with estimated 1.5 fg × 1.5 fg × 3 cb, would be about 31 kg.

The author does not state how the instrument is to be installed, just directing it to be placed in the meridian. It would be reasonable to assume that the instrument was installed higher than the horizon, for example, on a wall. About 1.5 cb seems to be a practical height. Given the sliding sine rule and the purpose of the catgut

325

to prevent the sine rule from falling off the versine rule, the only fixed point can be the end of the versine rule, where also the alidade is attached. Such a solution would suffer problems from flexure, however, so a support for the other end would be required.

In all likelihood, this description was only included as a prelude to the next instrument, and instrument no. 10 was never built. The same principle was described and used by al-ʿUrḍī in his "instrument with sine and versine."[76]

11.4.11 *Instrument 11*

This instrument again is built on a base cross of teakwood or copper rules aligned on the meridian and east–west lines and further strengthened by supportive rules of 1 cb length. In its intersection, an iron shaft of 2 fg diameter is mounted firmly on the ground. The versine rule, an extremely round copper rule of 2.5 fg diameter, is mounted with one end on the central shaft. Its outer end should coincide with the ends of the base cross rules. A rectangular sine rule of equal length, 2 fg × 0.5 fg, is mounted on a short pipe (*anbūba*) of 3 fg length which slides along the versine rule. Another copper rule of equal length is mounted as an alidade on the outer end of the versine rule, bearing two sights.

The versine and sine rules are again graduated in three bands, one of 60 parts, the others into finer fractions. To measure altitude, the sine rule is held in a vertical position and slid along the versine rule until it touches the targeting alidade (Figure 11.11(a)). To gain azimuth, we keep the versine rule in place and rotate the sine rule into the horizontal plane. Then the sine rule is moved until its graduated edge aligns with the outer end of the east–west rule (Figure 11.11(b)).

At this point,

$$\overline{VW} = \mathrm{Sin}A \qquad \text{sine of azimuth of observed object, } \textit{jayb-i samt-i irtifāʿ}$$

$$\overline{OV} = \mathrm{Sin}(90° - A) \qquad \text{co-sine of azimuth of observed object, } \textit{jayb-i samt-i tamām-i irtifāʿ}$$

The original, very confusing figure in the treatise shows catgut between the sine and versine rules, which is, however, not mentioned in the text and appears unnecessary. Objects close to the horizon (h < 15°) and zenith cannot be measured due to collisions between the pipe and the ends of the versine rule.

The original drawing shows three concentric circles labeled as having semi-diameters of 1.6 cb, 2 cb, and 3 cb, and with the largest being of equal radius to the rule lengths, from which we assume this base length of 3 cb for the versine, sine, and alidade rules. A copper alidade of 1 fg × 1 fg × 3 cb weighs about 14 kg, which seems reasonable.

---

76 Seemann 1928, p. 92.

The purpose of these circles is not documented. The circle of 1.6 cb radius would indicate an altitude angle of 62° 11′, or a culminating declination of $\delta = 9°$ 31′, while the circle with 2 cb radius indicates $h = 70° 31′$, or a culminating declination of $\delta = 17° 52′$, equivalent to ecliptic longitudes of ♈24° or ♉20°, respectively; the purpose of these data is, however, unclear.

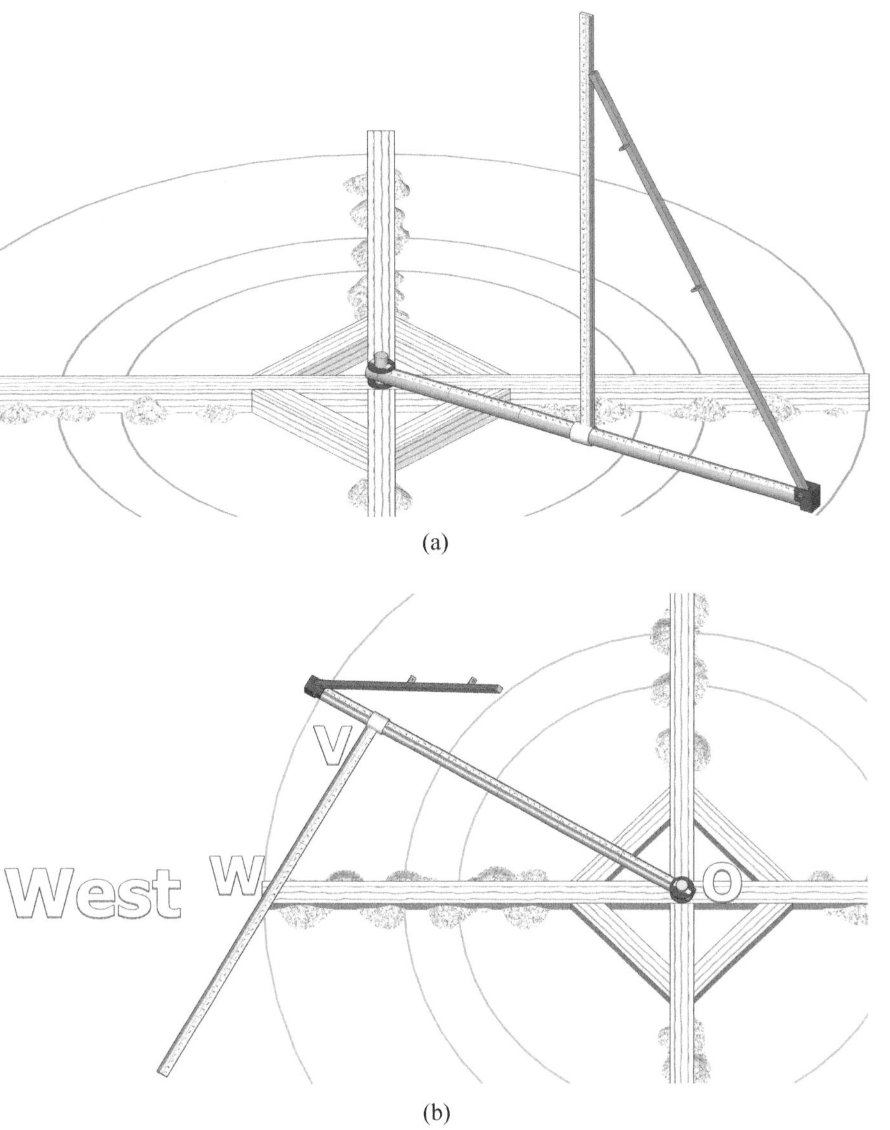

(a)

(b)

Figure 11.11 Instrument 11: rotating sine/versine, measuring altitude (a), measuring azimuth (b).

11.4.12 Instrument 12

The description of this instrument in our manuscript appears as an appendix to the treatise, after the date of copy (S: ff. 49r–v). However, a figure apparently intended to illustrate the description has been drawn five folios earlier (S: f. 44r) among the sketches of instrument no. 11, although the figure actually appears to depict al-'Urḍī's dioptra for eclipse observation.[77] Nevertheless, the number of instruments will only match 12, as given by the author, if we include this instrument. As an example of an instrument description with instructions for its use, we provide a close translation in this case.

The rule with which by the image of the ray (*'aks-i shu'ā'*) the radial magnitude of the eclipsed Sun is found.[78] From a straight and rigid wood, we make a rule like the astrolabe's alidade. Two pinnulae (*lubna*) are on both of its two ends. The width of one of them is 4 fg, and the other is wider than the first by 2 fg [total: 6 fg]. On the center of the greater pinnula, there is a perfectly round hole (*thuqba*). Around the center of the smaller sight, which is aligned with the center of the greater sight, we draw a circle whose radius is equal to the apparent radius of the Sun. To draw this circle, two days before the eclipse we place this instrument facing the Sun so that the size of [the circle of] the sunlight coming through the hole of the great pinnula and shining on the lesser pinnula is known. Then the circle around the center of the lesser sight is drawn with the dimensions of the illuminated circle. Then we divide this circle's diameter into twelve equal parts, by which the digits of the diameter [of the eclipsed Sun] (*aṣābi'-i quṭr*) are revealed. Then the circle's circumference is divided into twelve parts, by which the digits of the area [of the eclipsed Sun] (*aṣābi'-i jirm*) are revealed. We draw semi-diameter lines from the center of the circle onto the parts of the circle's circumference parts.[79] We draw circles on the digits of the diameter, by which the digits of the area [of eclipsed Sun] are made clear. If we need high accuracy,

---

77 Seemann 1928, p. 61.

78 Instead of a specific term such as *dioptra*, this long qualitative description may be due to the fact that the term διόπτρά had apparently not been translated or had not entered into Arabic. For Ptolemy's statement "We too constructed the kind of *dioptra* which Hipparchus described, which uses a four-cubit rod" (*Almagest*, V.14; see Toomer 1984, pp. 251–252; Heiberg, *Syntaxis*, p. 417), the Arabic translation by Ḥunayn–Thābit reads: "We, too, constructed the *miqyās* that Hipparchus made, with a four-cubit rod/rule, *misṭara*" (Arabic *Almagest*, f. 74r). In Arabic and Persian, *miqyās* ("scale," "measuring tool") is a general term not specifying specialized usage, and in Ghāzān's treatise, it would have been applied for different things. (In instrument no. 4, *miqyās* is used for both the central iron shaft and the chord rule's nail pointer.) Therefore, it appears that the Islamic astronomers, following the linguistic style of the Arabic *Almagest*, appealed to the qualitative practical descriptions such as "the rule by which" for naming the dioptra-shaped instruments.

79 The text does not say whether this partition should be regular.

we will divide the digits of the diameter and those of the circumference into minutes.

On the day of the eclipse, we place the sights facing the Sun and wait for the appearance of a slight shadow, like a fly's wing. This is the beginning of the eclipse. By means of an astrolabe or clepsydra (*shīsha-i sā'at*, lit. "hourglass"),[80] we determine the time of the start and end of the eclipse. We wait while the shadow increases in size, until it no longer grows and starts to decrease in size. By the darkness of the circle of the diameter of the Sun, the digits of the diameter and of the area are revealed. By the times given by the clepsydra or determined by the altitude of the Sun, the begin and end times of the eclipse are found. When darkness vanishes from the circle of light, this is the time of complete luminosity [end of the eclipse].

(Appendix, §5)

This instrument is apparently intended to replace the dioptra of antiquity. An early dioptra seems to have already been described by Archimedes (third century BC) in *The Sand Reckoner*.[81] Ptolemy used a dioptra originally described by Hipparchus with four cubits of length.[82] This dioptra has a fixed (lower) pinnula, on which there is a hole for sighting, and a movable one (outer pinnula), which is placed in front of the Sun. The solar/lunar angular diameter is calculated based on the movable pinnula's width and the distance between the two pinnulae.

The classical dioptra was employed to determine the apparent angular diameter of the Sun and the Moon. Like other medieval scholars, our author noted that Ptolemy said nothing about its construction, but his predecessors did. For instance, in his commentary on Book V of Ptolemy's *Almagest*, Pappus of Alexandria gave a description of this instrument. Proclus described it slightly differently.[83] Moreover,

---

80 Wābkanawī uses the term *pangān* in the otherwise similar paragraph in his own *zīj*, which is originally a simple Iranian clepsydra (Arabicized as *bankām*) in the shape of a floating bowl (*ṭās*), having a hole in its apex and two graduated scales (usually drawn with the aid of an astrolabe) for both equal and unequal hours on its peripheral surface. The bowl was placed in a vessel of water, and the level of water flowing into it determined the time. The first description of it in the Islamic period seems to appear in *al-'Amal bi-l-asṭurlāb* of al-Ṣūfī (d. 376/986), pp. 299–302, Chapters 354–57. On the one hand, this instrument was apparently very commonly used for timekeeping when one did not need to know the positions of the celestial objects, a process called "science of *bankāmāt*." On the other hand, it seems that it was modified in Marāgha in such a manner that it was able to determine even the minutes of an hour, since Muḥyī l-Dīn al-Maghribī used it for his observations (see Saliba 1986, and Chapter 9). The origin of this instrument dates back to both Babylonian and Indian texts of the first millennium BC (Pingree 1973, pp. 3–4). Archaeological excavations have unearthed early models in India, apparently belonging to the same period (Rao 2005, pp. 505–506).

81 Heath 1897, pp. 221–232; see also Shapiro 1975.

82 4 cb = 185.28 cm in his case: 1 Greek fg = 19.3 mm; thus, 1 cb = 46.32 cm. *Almagest* V.14: Toomer 1984, p. 56.

83 Goldstein 1987, pp. 174–175.

Heron of Alexandria promoted the dioptra and constructed two types (vertical and horizontal).[84] None added details concerning the use of this instrument for determining the eclipsed diameter or area of the Sun or the Moon by drawing a circle on the lower pinnula (as our treatise did) or using a circular plate on the lower pinnula (as al-ʿUrḍī did). This number is usually gained by calculations (*Almagest*, VI, 7). During antiquity and the early Islamic period, astronomers estimated it without an instrument of any kind and used their own estimate to check the results of their calculations.[85]

In his treatise, al-ʿUrḍī presented an improvement on the classical dioptra for determining the eclipsed diameter of the Sun or the Moon.[86] Similar to the ancient dioptra, he uses both a movable pinnula and a fixed one, but there is a conical hole on each of them. The angular diameter of the Sun and the Moon is calculated based on the width of the hole on the outer pinnula and the distance between the pinnulae. For al-ʿUrḍī, however, the most important application of the instrument is the measurement of the eclipsed diameter of the Sun and the Moon. For this, he uses two circular brass plates (*mir ʾāt*, "mirrors"), one for each type of eclipse. Before the eclipse, the upper pinnula is shifted until the luminary of interest exactly fills its visible diameter. A scale allows one to read the value of its visible diameter. During the eclipse, the respective brass aperture is brought in front of the entrance of the upper pinnula to cover the bright part of the luminary.[87] In any event, the instrument requires looking directly through the pinnulae, which is known to be dangerous in the event of solar observations. Also, use of a device with two conical holes, a movable pinnula with graduated scale, and additional apertures appears to be overly complicated.

Our author presents a new instrument that basically fulfills the same purpose, that is, measuring solar eclipses, but is significantly easier to produce and does not harm the eyes. The upper pinnula is described as 2 fg larger than the lower, most likely so as to provide good shading for the lower projection screen.

Our author attributed the instrument under discussion to al-Ṭūsī (d. 1274), but we have not found any mention of this in his works. Only in his *Exposition of Almagest* does al-Ṭūsī note concerning the dioptra described in the *Almagest* that "it is possible that errors will occur [in the calculation of the apparent diameter] if the length of the rule is much longer than the width of the sight."[88]

Al-Wābkanawī described this same instrument in his *zīj*, calling it "one of the marvels of the observational works" (*min jumla gharā ʾib-i a ʿmāl-i raṣadī*).[89] The details he gives are the same as in our treatise. Since he worked in the same period

<hr>

84 Lewis 2001, pp. 41–42, 51f.
85 See Stephenson and Said 1991, Said and Stephenson 1991.
86 Seemann 1928, pp. 61–71.
87 Seemann (1928, pp. 66f.) notes that this description is not completely clear, since a mechanism for measuring the amount of the aperture's shift is not described explicitly.
88 al-Ṭūsī, *Taḥrīr al-majistī*, f. 37v.
89 al-Wābkanawī, *Zīj*, book IV, sec. 15, ch. 8, A: ff. 159r–v.

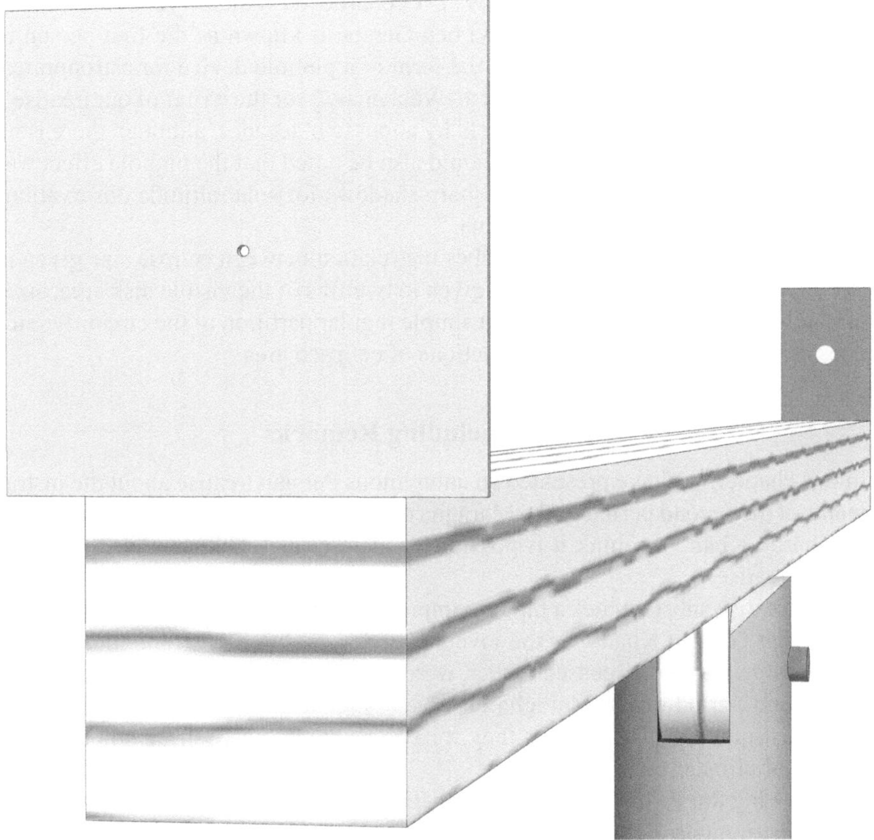

Figure 11.12 Instrument 12: Pinhole device.

as the most important royal astronomer in Ghāzān's court, we can infer that the astronomers of that era were aware of this instrument and that it was a new device. In his *zīj*, al-Wābkanawī usually called his own innovations a "marvel."

The prevalent information regarding the basic principle of the construction and early use of the pinhole device is derived from *Kitāb al-Manāẓir* of Ibn al-Haytham (Alhazen, d. *ca.* 1038).[90] His work was translated into Latin in 1272 by Vitellio, but the application of pinhole images in astronomical observations, especially for the eclipsed diameter/surface of the luminaries, was known in the West from at least *ca.* 1187 by Roger of Hereford. He was followed by such figures as William of Saint-Cloud (d. *ca.* 1292), Levi ben Gerson (Gersonides,

90 Sabra 1989, pp. 90–91.

331

d. 1344),[91] Henry of Hesse (d. 1397), Leonardo da Vinci, Tycho Brahe, and Johannes Kepler. Nevertheless, Levi ben Gerson is known as the first person to construct a dedicated instrument in the form of a pinhole device for astronomical purposes. However, it is evident that al-Wābkanawī – or the writer of our treatise – preceded Levi ben Gerson in this field by about two decades, although there is no evidence about any connection. It should also be noted that the pinhole effect was used in the same period to achieve sharp shadows for solar altitude observations at the Gaocheng observatory in China.[92]

In *Almagest* VI.7, Ptolemy describes the relation between eclipse size given in twelfths of solar diameters and size given in twelfths of the visible disk area, sizes our author also refers to. However, a simple regular partition of the circumference does not seem to help determine fractions of eclipsed area.

11.5 Concluding Remarks

In this chapter, we have presented an anonymous Persian treatise about the instruments of the second period of the Marāgha observatory, built under the supervision of Ghāzān Khān. We think it is possible to identify al-Wābkanawī as the author of this treatise.

The treatise substantiates a hitherto unproven historical claim by Rashīd al-Dīn Ṭabīb that Ghāzān Khān was the inventor of several "new" astronomical instruments, which, as we discussed earlier, were installed during what we have termed the second period of the Marāgha observatory. Ghāzān utilized a new approach in his instruments, insisting that they were to be made mostly from straight rules instead of circular components.

Following the treatise as closely as possible, we created virtual reconstructions of the instruments, but we found several inconsistencies that must be the result of copying errors in the three copies of the treatise available to us. With minor corrections, the instruments could be shown to work. However, if made from copper, the weight of some instruments might have rendered them hard to use if the large size recommended to achieve the required accuracy was taken into account. One instrument, no. 12, seems to be the first pinhole device specifically described for solar eclipse observations. On the other hand, the author breaks with the recommendations of his precursor, al-ʿUrḍī, not to use ropes for measuring lengths.

In addition, there appears to be continuity or succession for a few instruments after the second period of the Marāgha observatory, notably the geometrical square, which appeared in Europe in the mid-fifteenth century. However, despite known contacts between Ghāzān Khān's court and European rulers, we cannot, at this time, provide evidence that this manuscript might have been brought to Europe.

91 See Goldstein 1985, p. 146, Mancha 1992, p. 293.
92 Aslaksen 2001, pp. 23–25.

References

Aslaksen, H., 2001, *Calendars, Interpolation, Gnomons and Armillary Spheres in the Work of Guo Shoujing (1231–1314)*, Department of Mathematics, National University of Singapore, retrieved from www.math.nus.edu.sg/aslaksen/projects/nst-urops.pdf (retrieved October 24, 2010).

al-Banākitī, Fakhr al-Dīn Abū Sulaymān Dāwūd, 1969/2000, *Tārīkh-i Banākitī*, Shuʿār, Jaʿfar (ed.), Tehran: Society for the Appreciation of Cultural Works and Dignitaries.

al-Bayhaqī, Ẓahīr al-Dīn, 1994, *Tatimmat al-Ṣiwān al-ḥikma*, Rafīq al-ʿAjam (ed.), Beirut: Dār al-Fikr.

*EI*₂: Bearman, P. J., *et al.*, 1960–2004, *Encyclopaedia of Islam*, 2nd edn., Leiden: Brill.

al-Bīrūnī, Abū al-Rayḥān, 1962, *Kitāb Taḥdīd nihāyāt al-amākin li-taṣḥīḥ al-masāfāt al-masākin*, Muḥammad al-Ṭanjī (En. tr.), Ankara: Doğuş.

Brahe, Tycho, 1602, *Tychonis Brahe Astronomiae instauratae mechanica*, Noribergae: Apud L. Hulsivm.

Charette, F., 2006, "The locales of Islamic astronomical instrumentation", *Journal for the History of Science* **44**, pp. 123–138.

Comes, M., 2004, "The possible scientific exchange between the courts of Hūlāgū and Alfonso X", in: Pourjavady, N., and Vesel, Ž. (eds.), *Sciences, techniques et instruments dans le monde iranien*, Tehran: Institut Français de Recherche en Iran, pp. 29–50.

van Dalen, B., 2007, "Wābkanawī", in: *The Biographical Encyclopedia of Astronomers*, Hockey, T., *et al.* (eds.), London: Springer, 2007, pp. 1187–1188.

Dobrzycki, J. and Kremer, L. R., 1996, "Peurbach and Marāgha astronomy? The ephemerides of Johannes Angelus and their implications", *Journal for the History of Astronomy* **27**, pp. 187–237.

Goldstein, B. R., 1985, *The Astronomy of Levi Ben Gerson (1288–1344)*, New York: Springer Verlag.

Goldstein, B. R., 1987, "Remarks on Gemma Frisius's *De Radio Astronomico et Geometrico*", in: Berggren, J. L., and Goldstein, B. R. (eds.), *From Ancient Omens to Statistical Mechanics: Essays on the Exact Sciences Presented to Asger Aaboe*, Copenhagen: Copenhagen University Library, pp. 167–179.

Halm, H., 2004, *Shiʿism*, Watson, J. and Hilm, M. (En. trs.), Edinburgh: Edinburgh University Press.

Heath, T. L. (ed.)., 1897, *The Works of Archimedes*, Cambridge: Cambridge University Press.

Heiberg, J. L. (ed.), 1898, *Syntaxis Mathematica, Claudii Ptolemaei Opera Quae Exstant Omnia*, Vol. 1., Leipzig: Teubner.

Hockey, T., *et al.* (eds.), 2007, *Biographical Encyclopedia of Astronomers*, London: Springer.

Ibn al-Athīr, ʿIzz al-Dīn, 1966, *al-Kāmil fī l-tārīkh*, Beirut: Dār Ṣādir.

Ibn al-Ṣalāḥ, Aḥmad b. Muḥammad, *Fī kayfiyyat al-tasṭīḥ al-basīṭ al-kurī*, MS Tehran, Majlis Library, no. 6412.

al-Kamālī, Muḥammad b. Abī ʿAbd Allāh Sanjar, *Zīj-i ashrafī*, MS Paris, Bibliothèque nationale, Suppl. Pers., no. 1488/1.

al-Kāshī, Ghiyāth al-Dīn Jamshīd, *Sharḥ-i ālāt-i raṣad*, MS Tehran, Univ. of Tehran, Collection of Ḥikmat, no. 159/4, ff. 115v–116v; MS Tehran, National Library of Malik, no. 3536, pp. 31–39; MS Tehran, Library of Sipahsālār, no. 555D, ff. 9v–14v; MS Leiden, Or. 945, ff. 12r–13r (published in Kennedy 1961).

Kennedy, E. S., 1956, *A Survey of Islamic Astronomical Tables*, Philadelphia: American Society Publishing.

Kennedy, E. S., 1961, "Al-Kāshī's treatise on astronomical observational instruments", *Journal for Near Eastern Studies* **20**.2, pp. 98–108 (repr. in idem, 1983, *Studies in the Islamic Exact Sciences*, Beirut, pp. 394–404).

Kennedy, E. S., 1968, "The Exact science in Iran under the Saljuqs and Mongols", in: *The Cambridge History of Iran*, Boyle, J. A. (ed.), Cambridge: Cambridge University Press, Vol. 5, pp. 659–680.

Khvāndamīr, 1954, *Ḥabīb al-sīyar*, Tehran: Khayyam.

King, D. A., 1986, *A Survey of the Scientific Manuscripts in the Egyptian National Library*, Winona Lake, IN: Eisenbrauns.

King, D. A. and Samsó, J., 2001, "Astronomical handbooks and tables from the Islamic world", *Suhayl* **2**, pp. 9–106.

Kunitzsch, P., 1964, "Das Fixsternverzeichnis in der 'Persischen Syntaxis' des Georgios Chrysokokkes", *Byzantinische Zeitschrift* **57**, pp. 382–409 (repr. in idem, *The Arabs and the Stars*, Northampton: Variorum, 1989).

Lampton, A. K. S., 1988, *Continuity and Change in Medieval Persia*, New York: Bibliotheca Persica,.

Leichter, J. G., 2004, *The Zīj as-Sanjarī of Gregory Chioniades: Text, Translation and Greek to Arabic Glossary*. Ph.D. dissertation. Providence, Brown University.

Lelgemann, D., *et al.*, 2005, "Zum antiken astro-geodätischen Messinstrument Skiotherikos Gnomon", *Zeitschrift für Vermessungswesen* **130**.4, pp. 238–247.

Lewis, M. J. T., 2001, *Surveying Instruments of Greece and Rome*, Cambridge: Cambridge University Press.

Lorch, R., 1995, *Arabic Mathematical Sciences*, Aldershot: Variorum.

Lorch, R., 2000, "Ibn al-Ṣalāḥ's treatise on projection: A preliminary survey", in: Folkerts, M., and Lorch, R. (eds.), *Sic Itur ad Astra: Studien zur Geschichte der Mathematik und Naturwissenschaften*, Wiesbaden: Harrassowitz, pp. 401–408.

al-Maghribī, Muhyī al-Dīn, *Talkhīs al-Majistī*, MS Leiden, Or. 110.

Mancha, J. L., 1992, "Astronomical use of pinhole images in William of Saint-Cloud's *Almanach Planetarum* (1292)", *Archive for History of the Exact Sciences* **43**, pp. 275–298.

al-Marrākushī, Abū ʿAlī al-Ḥasan b. ʿAlī b. ʿUmar, *Jāmiʿ al-mabādī wa ʾl-ghāyāt fī ʿilm al-mīqāt* (*Comprehensive collection of principles and objectives in the science of timekeeping*), MS Istanbul, Süleymaniye Library, Nuruosmaniye Collection, no. 2902.

Mīrkhvānd, M., 2002, *Tārīkh-i rawẓat al-ṣafā*, Kiyānfar, J. (ed.), Tehran: Aṣāṭīr.

Morgan, D., 2007, *The Mongols*, London: Blackwell.

Mozaffari, S. M., 2018–2019, "Muhyī al-Dīn al-Maghribī's measurements of Mars at the Maragha observatory", *Suhayl* **16**, pp. 149–249.

Mozaffari, S. M., 2019, "Ibn al-Fahhād and the Great Conjunction of 1166 AD", *Archive for History of Exact Sciences* **73**, pp. 517–549.

Mozaffari, S. M., 2022, "A mechanical concentric solar model in Khāzinī's *Muʿtabar zīj*", *Archive for History of Exact Sciences* **76**, pp. 513–529.

Mozaffari, S. M., 2023, "Sources of the planetary theories in Fahhād's *ʿAlāʾī zīj*: Solving a medieval case of intellectual fraud", *Suhayl* **20**, pp. 143–221.

al-Mustawfī, Ḥamd Allāh, 1960, *Tārīkh-i guzīda*, Nawāʾī, ʿAbd al-Ḥusayn (ed.), Tehran: Amīr Kabīr.

al-Nīshābūrī, ʿAbd Allāh Waṣṣāf al-Ḥaḍra, 1967, *Tārīkh-i Waṣṣāf*, Āyatī, A. (ed.), Tehran: Iranian Culture Foundation.

Paschos, E. A., and Sotiroudis, P., 1998, *The Schemata of the Stars: Byzantine Astronomy from A.D. 1300*, Singapore and River Edge, NJ: World Scientific.

Pingree, D., 1973, "The Mesopotamian origin of early Indian mathematical astronomy", *Journal for the History of Astronomy* **4**, pp. 1–12.

Pingree, D. (ed.), 1985, *Astronomical Works of Gregory Chioniades*. Vol. 1, *Zīj al-ʿalā ʾī*, Amsterdam: Gieben.

Rao, N. K., 2005, "Aspects of prehistoric astronomy in India", *Bulletin of the Astronomical Society of India* **30**.4, pp. 499–511.

Rashīd al-Dīn Ṭabīb, Faḍl Allāh al-Hamadhānī, 1993, *As ʾila wa-ajwiba*, Shaʿbānī, R. (ed.), Lahore: The Centre for Persian Research.

Rashīd al-Dīn Ṭabīb, Faḍl Allāh al-Hamadhānī, 1999/1994, *Jāmiʿ al-tawārīkh*, Rawshan, M. and Mūsawī, M. (eds.), 4 Vols., Tehran: Alborz. Thackston, W. M. (En. Tr.), Cambridge, MA: Harvard University.

Rosenfeld, B. A., 2004, "A supplement to *Mathematicians, astronomers, and other scholars of Islamic civilization and their works*", *Suhayl* **4**, pp. 87–158.

Rosenfeld, B. A. and Ihsanoğlu, E., 2003, *Mathematicians, Astronomers, and Other Scholars of Islamic Civilization and Their Works*, Istanbul: IRCICA.

Sabra, A. I., 1989, *The Optics of Ibn al-Haytham*, London: Warburg Institute and University of London.

Said, S. S. and Stephenson, F. R., 1991, "Accuracy of eclipse observations recorded in medieval Arabic chronicles", *Journal for the History of Astronomy* **22**, pp. 297–310.

Saliba, G., 1986, "The determination of new planetary parameters at the Maragha observatory", *Centaurus* **29**, pp. 249–271; repr. in idem, *A History of Arabic Astronomy: Planetary Theories during the Golden Age of Islam*, New York: New York University Press, 1994.

Sayılı, A., 1956, "Al-Khāzinī's treatise on astronomical instruments", *Ankara Üniversitesi Dil ve Tarih Coğrafya Fakültesi Dergisi* **14**.1–2, pp. 15–19.

Sayılı, A., 1960, *The Observatory in Islam*, Ankara: Türk Tarih Kurumu Basimevi.

Sayyid Muḥammad, the Astrologer, *Latāʾif al-kalām fī aḥkām al-aʿwām*, MS Tehran, Majlis Library, no. 6347.

Schöner, J., 1544, *Scripta clarissimi mathematici M. Ioannis regiomontani, . . .*, Norimberga (Nuremberg): Ioannes Montanus and Ulricus Neuber.

Seemann, H. J., 1929/1928, "Die Instrumente der Sternwarte zu Marāgha nach den Mitteilungen von al-ʿUrḍī", in: *Sitzungsberichte der Physikalisch-medizinischen Sozietät zu Erlangen*, Vol. 60, Erlangen: Kommissionsverlag von Max Mencke, pp. 15–126.

al-Shahrazūrī, Shams al-Dīn Muḥammad, 1976, *Nuzhat al-arwāḥ wa-rawḍat al-afrāḥ fī tārikh al-ḥukamā ʾ wa-l-falāsifa*, Hyderabad: Osmania Oriental Publications.

Shapiro, A. E., 1975, "Archimedes's measurement of the Sun's apparent diameter", *Journal for the History of Astronomy* **6**, pp. 75–83.

al-Shīrāzī, Quṭb al-Dīn, 1938–1944, *Durrat al-tāj li-ghurrat al-dubāj*, 5 Vols. Tehran: The Iran Ministry of Culture.

Stephenson, F. R. and Said, S. S., 1991, "Precision of medieval Islamic eclipse measurements", *Journal for the History of Astronomy* **22**, pp. 195–207.

Storey, C. A., 1958, *Persian Literature*, London: Luzac.

al-Ṣūfī, A. R., 1954, *Ṣuwar al-kawākib al-thābita*, Hyderabad: Dāʾirat al-Maʿārif al-ʿUthmāniyya.

al-Ṣūfī, A. R., 1995, *al-ʿAmal bi-l-asṭurlāb*, Morocco: ISESCO.

Sykes, P. M., 2003, *A History of Persia*, London: Routledge.

Toomer, G. J., 1984, *Ptolemy's Almagest*, Princeton: Princeton University Press (Arabic *Almagest*, tr. Ḥunayn b. Isḥāq and Thābit b. Qurra, MS Tehran, Sipahsālār Library, no. 594, copied in 480/1087–1088).

al-Ṭūsī, Naṣīr al-Dīn, *Taḥrīr al-Majisṭī*, MS. Iran, Mashhad, no. 453 (copied 1092/1681–1682).

Wābkanawī, Shams al-Dīn Muḥammad, *Zīj al-muḥaqqaq al-sulṭānī ʿalā uṣūl al-raṣad al-Īlkhānī* (*The testified zīj for the sultan on the basis of the parameter values of the Īlkhanid observations*), MSS. A: Turkey, Aya Sophia Library, No. 2694, MS. B: Iran, Yazd, Library of ʿUlūmī, no. 546 (its microfilm is available in the central library of the University of Tehran, no. 2546), P: Iran, Library of Parliament, no. 6435.

Westerink, L., 1980, "La Profession de foi de Gregoire Chioniades", *Revue des Études Byzantines* **38**, pp. 233–245.

Wiedemann, E. and Juynboll, T. W., 1926, "Avicennas Schrift über ein von ihm ersonnenes Beobachtungsinstrument", *Acta Orientalia* **11**, pp. 81–167.

Wlodarczyk, J., 1987, "Observing with the Armillary sphere", *Journal for the History of Astronomy* **18**, pp. 73–195.

al-Yazdī, Muḥammad Ḥusayn b. Muḥammad Bāghir, *Mīzān al-Sanāyiʿ*, MS Tehran University Central Library, no. 2084.

Appendix

§1. M: p. 46, ll. 5–8; P: p. 35, ll. 3–6; S: f. 18v, ll. 3–6

از این آلت غایتِ ارتفاعِ کواکب معلوم می‌گردد از دایرهٔ نصف‌النّهار که بیشتر از سـی درجـه باشـد، و آن نیـز بـه تقریب؛ بدان سبب که مسطرهٔ ثالثهٔ این آلت، که مقدار وتر زاویه از [ا]و معلوم می‌شود، مؤثّر حقیقت زاویه نیست.

§2. M: pp. 47, l. 9–48, l. 4; P: pp. 35, l. 17–36, l. 13; S: ff. 18v, l. 17–19r, l. 12

و بندهٔ کمینهٔ کمالْ سالهاست که به دعاگویی دولت همایون پادشاه عالم، ایلخان |P:36| اعظم[1] شهنشـاه روی زمین |S:19r| سلطان غازان خان – خلّد الله ملکه و دوام علی العالمین ظلّه – مشغول است، مدّتی[2] در این فکر بود و طالب آنکه آلات رصدی دست دهد که بی رنجی و مشقّتی[3] بـه تحقیـق و تـدقیق مطالـب ارصـاد از [ا]و حاصل شود تا به دولت پادشاه عالم – أدام الله ظلّه – دوازده نوع آلت رصد[4] روی نمود کـه پیـش از ایـن هـیچ کس از متقدّمان و متأخّران [را] مثل چنین آلات دست نداد و میسّر نگشت که تمامت مطالب ارصاد از ایـن آلات معلوم می‌شود به اندک مؤونت[5] و سعی به تحقیق و تدقیق تمام، زیرا که ایـن آلات تمامـت مسـاطر و خطـوط مستقیم‌اند و هر چند که دراز باشد، ممکن است که استقامت او را به جای[6] آورند |M:48| و بـه اجـزاء و کسـورْ قسمت و[7] اعتبار کنند تا مطالب[8] [به تحقیق] و تدقیق معلوم می‌شود و چندین مهمّات دیگر از این آلات معلوم می‌شود[9] که ممکن نیست که به آلات پنجگانه قدما معلوم گردد – چنانچه شرح داده آید [...]

(1) M : + سه / (2) SP : مدّت / (3) M : مشقیتی / (4) SP : رصدی / (5) SP : مؤنت / M : مونت

(6) M : جا / (7) SP : قسمت و = او / (8) SP : + او / (9) SP : – «و چندین مهمّات دیگر از این آلات معلوم می‌شود.»

§3. M: pp. 54, l. 18–56, l. 3; P: pp. 44, l. 5–46, l. 2; S: ff. 23r, l. 4–25r, l. 2

و این آلت تفضیل و ترجیح دارد بر جمیع آلات به چهار وجه:

وجه اوّل آنکه، آلاتی که مشهور است و مستعمل و اعتقاد متقدّمان و متأخّران بر آن است[1] هر یک مخصوص‌اند به مهمّی[2] – چنانچه پیش از این |M:55| گفته آمد. و در این آلت جمیع مطالب آن آلات ممکن نگردد – چنانچه بعد از این شرح داده شود.

وجه دوم[3] آنکه، مؤونت و اخراجات[4] و سعی و شغل در این آلت کمتر از جمیع آلات قدماست؛

وجه سوم[5] آنکه، مطالب از آن آلات به تحقیق و تدقیق معلوم نمی‌گردد. بدان سبب که جمیع آن آلاتْ قسی و حلقه‌اند؛ اگر کوچک می‌سازند، قسمت او به دقایق و ثوانیْ اعتبار نمی‌توان کرد و مطالب به تقریب[6] حاصل می‌شود؛ اگر بزرگ می‌سازند، ممکن نیست که استدارهٔ او را کما ینبغی به جای توان آورد. پس فساد و خلل او زیاده از فایدهٔ[7] او می‌گردد. و در این آلت مسطره‌های مستقیم است و خطوط مستقیم. و هرچند که دراز باشد ممکن است استقامت |P:45| او[8] به جای آوردن بی‌خللی و رنجی؛

وجه چهارم |S:23v| آنکه، در آن آلاتْ قسی معلوم می‌شود و در این |M:56/P:46/S:24r|[9] اجزائی که معلوم می‌گردد که حصّهٔ قوس او کمتر از او باشد بیشتر اوقات. پس، به تحقیق و تدقیقْ قوس معلوم می‌شود.

(1) MPS : (2) SP : این نسبت / M : بهمین / SP : بهمی (3) SP : دویم (4) M : اجزاجات (5) SP : سیوم
(6) M : تقریبی (7) SP : قاعده (8) SP : – او (9) M : + صورت

§4. P: pp. 73, l. 3–74, l. 2; S: ff. 37v, l. 3–38r, l. 2

[...] ارتفاع کوکب با سمت ارتفاع از این آلت معلوم می‌شود به غایت استقصا[1] و تحقیق. و جمیع[2] مطالب که از آلات ارصاد معلوم گردد، از این آلت دانسته |P:74| می‌شود |S:38r| و محقّق با چندین مطالب دیگر. و این آلتْ ترجیح و تفضیل دارد بر آلات ارصاد قدما از چهار وجه مذکور.

(1) SP : استقضا (2) SP : جمع

§5. P: pp. 96, l. 9–97; S: f. 49v, l. 9–49v

مسطره‌ای که به عکس شعاع کمیّت مقدار منکسف معلوم گردد:

طریق صنعتش

که از چوبی راست که تغییر و اعوجاج‌پذیر نباشد، مسطره‌ای سازند مثل عضادهٔ اسطرلاب که بر دو طرف آن دو لبنه باشد. عرض یکی قریب چهار انگشت و دیگر از [ا]و اوسع به قدر دو انگشت. و در میان لبنهٔ بزرگ ثقبه‌ای دقیق باشد مستدیر. و در لبنهٔ کوچکتر [بر] مسامت مرکز آن ثقبه، نقطه‌ای تعیین باید کرد که بر آن نقطهْ دایره رسم نمایند به بُعدِ نصف قطر صفحهٔ آفتاب؛ طریقش آنکه: لبنهٔ بزرگتر را دو روز پیش از کسوف محاذی آفتاب سازند و نظر کنند که شعاع آفتاب از لبنهٔ بزرگتر چقدر از لبنهٔ خوردتر منوّر |P:97| می‌گرداند و لامحاله قدرِ منوّر دایره خواهد |S:49v| بود. پس، [بر] نقطهٔ مسامت مشارالیها به بعد نصف قطر دایرهٔ مضیّهٔ دایره‌ای رسم کنند.

337

پس، قطر آن دایره را به دوازده قسم متساوی تقسیم نمایند که از اقسام اصابع قطـر باشـد و همچنـان محیط آن دایره را به دوازده قسم منقسم سازند که از آن اصابع جرم معلوم شود. و از اقسام محیطْ انصـاف اقطـار را به مرکز دایره اخراج نمایند. و بر اقسام [اصابع] قطرْ دوایر رسم کنند تا اصابع جـرم ظـاهر شـود. و اگـر زیـاده تدقیق خواهند، هر یک از اصابع قطر و اصابع جرم را به دقایق قسمت کنند.

[طریق کاربردش]

پس، در روز کسوف، آن دو لبنه را به دست گیرند[1] مسامت آفتاب و مترصّد باشند تا ظلّ ضعیف مثل پر مگـس بر کنارهٔ شعاع پیدا شود. و آن زمان ابتدای کسوف باشد. به اسطرلاب در آن حال از شعاع آفتاب بدانند تا سـاعات بدو کسوف معلوم شود و اگر در اوّل طلوع آفتابْ شیشهٔ ساعت وضع[2] نموده باشد، ساعت بدو کسوف به وجهـی دیگر معلوم شود[3] و مصحَّح گردد. و مترصّد باشد تا سیاهی به غایت رسد که زیاده نشود و میل به نقصان کنـد. از مقدار سیاهی دایرهٔ نور آفتابْ کمیّتِ اصابع قطر و جرم معلوم شود و از سـاعات و ارتفـاعْ کمیّتِ سـاعات بـدو کسوف و غایتش ظاهر شود. و چون سیاهی از دایرهٔ نور زایل شود، وقت تمام انجلاء باشد. والله أعلم.

(1) SP : بدستور (2) SP : وضوع (3) SP : + «و اگر ... معلوم شود.»

INDEX